NANOTECHNOLOGY SCIENCE AND TECHNOLOGY

QUANTUM WIRES

AN OVERVIEW

NANOTECHNOLOGY SCIENCE AND TECHNOLOGY

Additional books and e-books in this series can be found on Nova's website under the Series tab.

NANOTECHNOLOGY SCIENCE AND TECHNOLOGY

QUANTUM WIRES

AN OVERVIEW

KAMAKHYA PRASAD GHATAK
EDITOR

nova
science publishers
New York

NOTICE TO THE READER

Library of Congress Cataloging-in-Publication Data

Names: Ghatak, Kamakhya Prasad, editor.
Title: Quantum wires : an overview / Kamakhya Prasad Ghatak, editor.
Description: Hauppauge, New York : Nova Science Publishers, 2020. | Series: Nanotechnology science and technology | Includes bibliographical references and index. |
Identifiers: LCCN 2020011885 (print) | LCCN 2020011886 (ebook) | ISBN 9781536176766 (hardcover) | ISBN 9781536176926 (adobe pdf)
Subjects: LCSH: Nanowires.
Classification: LCC TK7874.85 .Q36 2020 (print) | LCC TK7874.85 (ebook) | DDC 621.3815--dc23
LC record available at https://lccn.loc.gov/2020011885
LC ebook record available at https://lccn.loc.gov/2020011886

Published by Nova Science Publishers, Inc. † New York

I stand in front of my wife See,
And clearly she wants nothing from me,
But when she speaks through her eyes,
The very "I" becomes the sea.

CONTENTS

PREFACE

It is well-known that in quantum wires (QWs), the restriction of the motion of the carriers along two directions may be viewed as carrier confinement by two infinitely deep 1D rectangular potential wells, along any two orthogonal directions leading to quantization of the wave vectors along the said directions, allowing 1D carrier transport. With the help of modern experimental techniques, such one dimensional quantized structures have been experimentally realized and enjoy an enormous range of important applications in the realm of nano-science in the quantum regime. They have generated much interest in the analysis of nano-structured devices for investigating their electronic, optical and allied properties in general. Examples of such new applications are based on the different transport properties of ballistic charge carriers which include quantum resistors, resonant tunneling diodes, quantum switches, quantum sensors, quantum logic gates, quantum transistors and sub tuners hetero-junction FETs, high-speed digital networks, high-frequency microwave circuits, optical modulators, optical switching systems, and other devices..

The importance of Fowler Nordheim Field Emission (FNFE) which is basically a quantum phenomena is already well-known, since its inception in 1928 and, in chapter 1, H. Seikh et al. have investigated the FNFE in QWs of nonlinear optical, on the basis of a newly formulated electron dispersion law by talking all types of anisotropies within the frame work of

k. p. formalism. All the results of III-V QWs form a special case of the generalized analysis. They have studied the FNFE in QWs of II-VI semiconductors on the basis of Hopfield model and Bi on the basis of the models of McClure and Choi, Hybrid, Cohen and Lax respectively. The FNFE from QWs of IV-VI, stressed materials, Te, GaP, $PtSb_2$, Bi_2Te_3, Ge, GaSb and II-V semiconductors have further been investigated on the basis of the appropriate band models. It has been found talking QWs of $CdGeAs_2$, InSb, $Hg_{1-x}Cd_xTe$, InAs, GaAs, $In_{1-x}Ga_xAs_yP_{1-y}$ lattice matched to InP, CdS, Bi, stressed InSb, Ge, GaSb, GaP, Bi_2Te_3, PbTe and Te as examples that the field emitted current depends strongly on the thickness of the quantum-confined materials in contrast with the corresponding bulk specimens. The current changes with increasing film thickness in an oscillatory way with different numerical magnitudes. It appears from the diagrams of chapter 1that the FNFE exhibits spikes for particular values of film thickness which, in turn, depends on the particular band structure of the specific semiconductor. Moreover, the FNFE from QWs of different compounds can be smaller than of bulk specimens of the same materials, which is also a direct signature of quantum confinement. This oscillatory dependence will be less and less prominent with increasing film thickness. For bulk specimens of the same material, the FNFE will be found to increase continuously with increasing electron degeneracy in a non-oscillatory manner. The appearance of the discrete jumps in the respective figures are due to the redistribution of the electrons among the quantized energy levels when the size quantum number corresponding to the highest occupied level changes from one fixed value to the others. All the results for all the models gets simplified to the expression of the field emitted current from QWs having parabolic energy bands under certain limiting condition which exhibit the mathematical compatibility of their generalized formalism.

It is well-known that the concept of the Effective Mass (EM) in quantum materials is one of the basic pillars in the realm of modern low dimensional quantum science and technology. Among the various definitions of the effective mass (e.g., effective acceleration mass, density-of-state effective mass, concentration effective mass, conductivity effective mass, Faraday rotation effective mass, etc), it is the effective momentum mass that should be regarded as the basic quantity, since it is this mass which appears in the description of transport phenomena and all other properties of the carriers in materials having arbitrary band structures together with the fact that only the basic definition of mass generates the Einstein's $E = mc^2$ (where *E, m* and *c* are the energy, mass and velocity of light respectively) equation for photons and under the condition of stationary frame of reference. In chapter 2, B. Chatterjee et al. have discussed the EM in heavily doped (HD) QWs of non-linear optical, III-V, II-VI and stressed Kane type materials by formulating the electron statistics including all types of anisotropies of the energy band constants within the frame work of Kane's theory. It appears that the EM in HDNWs of $Cd_3As_2$and $CdGeAs_2$ at the lowest subband in both the cases are strong cross-sectional functions of the dimensions which converge to their corresponding bulk values at larger dimensions. The effect of crystal field splitting in case of dispersion relation of Cd_3As_2 lets the asymptotic fall to be closer to the bulk value. It appears from their investigations that the energy spectrum in HD nonlinear optical materials is complex. The complex nature of the electron dispersion law in HD materials occurs from the existence of the essential poles in the corresponding electron energy spectrum in the absence of band tails. It may be noted that the complex band structures have already been studied for bulk materials and super lattices without heavy doping and bears no relationship with the complex electron dispersion law as formulated in chapter 2. The physical picture behind the formulation of the complex energy spectrum in HDS is the interaction of the impurity atoms in the tails with the splitting constants of the valance bands. More is the interaction; more is the prominence of the complex part than the other case. It appears from their study that the EM depends on the size quantum number and the scattering potential in many cases. The EM exists in the forbidden zone, which is impossible without the effect of band tailing. For semiconductors, in the absence of band tails the EM in the band gap is infinity. Under certain limiting conditions all the results for all the models get simplified into the well-known expressions of the parabolic energy bands.

The concept of Dispersion Relations (DR) of the carriers in quantized structures occupies a singular position in Nano Technology and whose important has already been established since the inception of the theory of band structure of Solid State Physics. The concept of DRs is of fundamental importance for not only the characterization of semiconductor nano-structures but also the study of the carrier transport in semiconductors and their quantized counter parts through the proper formulation of the Boltzmann Transport equation which, in turn, needs the corresponding carrier energy spectra. In chapter 3, M. Debbarma et al. have formulated the DR's of QWs of

n-GaP, IV-VI, Ge, Zero Gap, GaSb, $PtSb_2$, GaAs, n-InSb, $Pb_{1-x}Ga_xTe$, Te, Graphite, Sb, Bi_2Te_3, p-InSb, II-V, CuCl

and stressed p-type semiconductors having diamond structure valance bands respectively. The influence of quantum confinement is immediately apparent from the 1D DRs since the DR depends strongly on the thickness of the quantum-confined materials in contrast with the corresponding bulk specimens. Finally, it may be noted that the basic aim of this chapter is not solely to demonstrate the influence of quantum confinement on the DR from QWs of the said materials but also to formulate the appropriate DOS in the most generalized form, since the transport and other phenomena in quantized structures having different band structures and the derivation of the expressions of many important electronic properties are based on the DR in such materials.

It is well known that Keldysh first suggested the fundamental concept of a superlattice (SL), although it was successfully experimental realized by Esaki and Tsu. The importances of SLs in the field of nanoelectronics have already been described in the literature. The most extensively studied III-V SL is the one consisting of alternate layers of GaAs and $Ga_{1-x}Al_xAs$ owing to the relative ease of fabrication. The GaAs layers forms quantum wells and $Ga_{1-x}Al_xAs$ form potential barriers. The III-V SL's are attractive for the realization of high speed electronic and optoelectronic devices. In addition to SLs with usual structure, SLs with more complex structures such as II-VI, IV-VI and HgTe/CdTe SL's have also been proposed. The IV-VI SLs exhibit quite different properties as compared to the III-V SL due to the peculiar band structure of the constituent materials]. The epitaxial growth of II-VI SL is a relatively recent development and the primary motivation for studying the mentioned SLs made of materials with the large band gap is in their potential for optoelectronic operation in the blue. HgTe/CdTe SL's have raised a great deal of attention since 1979, when as a promising new materials for long wavelength infrared detectors and other electro-optical applications. Interest in Hg-based SL's has been further increased as new properties with potential device applications were revealed. These features arise from the unique zero band gap material HgTe and the direct band gap semiconductor CdTe. In addition to it, for effective mass SLs, the electronic subbands appear continually in real space.

The diffusivity-mobility ratio of the carriers in semiconductors (DMR) occupies a central position in the whole field of semiconductor science and technology since the diffusion constant (a quantity very useful for device analysis where exact experimental determination is rather difficult) can be obtained from this ratio by knowing the experimental values of the mobility. The classical value of the DMR is equal to $(k_B T/|e|)$, (k_B, T, and $|e|$ are Boltzmann's constant, temperature and the magnitude of the carrier charge respectively). This relation in this form was first introduced to study of the diffusion of gas particles and is known as the Einstein relation. Therefore, it appears that the DMR increases linearly with increasing T and is independent of electron concentration. This relation holds for both types of charge carrier only under non-degenerate carrier concentration although its validity has been suggested erroneously for degenerate materials. Landsberg first pointed out that the DMR for semiconductors having degenerate electron concentration are essentially determined by their energy band structures. This relation is useful for semiconductor homo-structures, semiconductor-semiconductor hetero-structures metals-semiconductor hetero-structures and insulator-semiconductor hetero-structures. In chapter 4, Biswas et al. have studied the DMR in QW

$GaAs / Ga_{1-x}Al_xAs, CdS / CdTe,\ PbTe / PbSnTe\ and\ HgTe / CdTe$ superlattices with graded interfaces by formulating the appropriate electron statistics per unit length in each cases. They also investigate the DMR in QW Effective Mass Superlattices of the said materials by deriving appropriate 1D Dispersion Relation. It appears that the DMR increases for both types of SLs as considered in this chapter with decreasing film thickness and increasing electron concentration per unit length respectively, although the numerical values are totally band structure dependent.

With the advent of Quantum Hall Effect there has been considerable interest in studying the Magneto Thermoelectric power (MTP) in low dimensional materials whose wave vector space are being quantized forming 1D, 2D and 0D systems. In chapter 5 Mitra et al. have made an attempt is made to study the MTP in QWs of optoelectronic materials in the presence of microwave radiation by formulating the new dispersion relations for such technologically importance compounds. They have investigated MTP taking QWs of $InSb,$ $GaAs, Hg_{1-x}Cd_xTe\ and\ In_{1-x}Ga_xAs_yP_{1-x}$ as functions of film thickness, electron concentration, light intensity and wavelength respectively. It appears that MTP decreases with increasing electron concentration in oscillatory manners. For large values of electron concentration per unit length, the MTP exhibits very quick oscillatory decrement whereas for small values of the carrier degeneracy, the MTP shows the converging tendencies. It may be noted that the MTP increases with increasing light intensity and due to both quantum limits, the variation is again non-oscillatory. Because of the quantum limits, the MTP changes slowly with wavelength in a non-oscillatory manner although the wide difference in the MTP versus wave length plot for the perturbed two and three band models of Kane together with parabolic energy bands exhibit the influence of energy band models.

With the advent of modern nano devices, there has been considerable interest in studying the electric field induced processes in semiconductors having different band structures. It appears from the literature that the studies have been made on the assumption that the carrier dispersion laws are invariant quantities in the presence of intense electric field, which is not fundamentally true. The Density-of-States function (DOS) affects nearly all the physical properties of different semiconductors and their nano structures. In chapter 6, Mondal et al. studied the DOS in the presence of intense electric field in Heavily Doped (HD) Quantum Wires (QWs) of III-V, ternary and quaternary semiconductors. They have also derived the expressions of the DOS in QWHD super-lattices of Kane type semiconductors with graded interfaces in the presence of intense electric field together with QWHD super-lattices with graded interfaces under strong electric field.

In chapter 7, the editor as the principal author has studied the quantum capacitance in quantum wire field effect transistors (QWFET's) made of nonlinear optical, III-V and II-VI semiconductors on the basis of various band models. The quantum capacitance in QWFET's of I V-VI, stressed materials, Te, GaP, $PtSb_2$, Bi_2Te_3, Ge, GaSb and II-V semiconductors have been investigated on the basis of the appropriate band models. It has been found talking QWFET's of $CdGeAs_2$, InSb, $Hg_{1-x}Cd_xTe$, InAs, GaAs and $In_{1-x}Ga_xAs_yP_{1-y}$ lattice matched to InP as examples that the quantum capacitance depends strongly on the thickness of the quantum-confined transistors in contrast with the corresponding bulk specimens. The quantum capacitance changes with increasing film thickness in quantized steps and saw-tooth manners with different numerical magnitudes. It appears from the aforementioned figures that the quantum capacitance exhibits spikes for particular values of film thickness which, in turn, depends on the particular band structure of the specific semiconductor. The quantum capacitance for the QWFET's of all the materials as considered here increases with increasing electron statistics in step like fashion and decreases in an oscillatory way with the increasing gate voltage. The appearance of the discrete jumps in the respective figures are due to the redistribution of the electrons among the quantized energy levels when the size quantum number corresponding to the highest occupied level changes from one fixed value to the others. All the results for all the models gets simplified to the expression of the Quantum capacitance of QWFET having parabolic energy bands under certain limiting condition which exhibit the mathematical compatibility of our simplified theory.

In chapter 8 Devi et al. have investigated the Einstein's Photoemission (EP) from Heavily Doped (HD) quantum wires (QWs) of non-linear optical, III-V, II-VI, Gallium Phosphide, Germanium, Platinum

Antimonide, stressed, IV-VI, Lead Germanium Telluride, Tellurium, II-V, Zinc and Cadmium diphosphides and Bismuth Telluride respectively by formulating the respective expressions of EP in each case. The influence of quantum confinement is immediately apparent since the EP depends strongly on the thickness of the quantum-confined materials in contrast with the corresponding bulk specimens. The EP decreases with increasing film thickness in an oscillatory way with different numerical magnitudes for HD QWs. It appears from the aforementioned figures that the EP exhibits spikes for particular values of film thickness which, in turn, depends on the particular band structure of the specific material. Moreover, the EP from HD QWs of different compounds can become several orders of magnitude larger than of bulk specimens of the same materials, which is also a direct signature of quantum confinement. This oscillatory dependence will be less and less prominent with increasing film thickness. It can be concluded that the EP increases with increasing degeneracy and also exhibits spikes. For bulk specimens of the same material, the EP will be found to increase continuously with increasing electron degeneracy in a non-oscillatory manner. The EP increases with increasing photon energy in a step like manner. The appearance of the discrete jumps is due to the redistribution of the electrons among the quantized energy levels when the size quantum number corresponding to the highest occupied level changes from one fixed value to the others. With varying electron degeneracy, a change is reflected in the EP through the redistribution of the electrons among the size-quantized levels. It may be noted that at the transition zone from one sub band to another, the height of the peaks between any two sub-bands decreases with the increasing in the degree of quantum confinement and is clearly shown in all the curves. It should be noted that although, the EP varies in various manners with all the variables the rates of variations are totally band-structure dependent.

In chapter 9 Das et al. have studied the Landau Dia and Pauli's Para magnetic susceptibilities in quantum wires of IV—VI materials under a parallel magnetic field on the basis of a newly derived one-dimensional electron dispersion law. It is found, taking quantum wires of PbS, PbSe and PbTe as examples, that both the susceptibilities increase with decreasing electron concentration and decreasing film thickness respectively. The paramagnetic-to-diamagnetic susceptibility ratio for electrons in IV-VI quantum wires deviates from the well-known 1/3 rule together with the fact that there is a critical zone within which quenching of the diamagnetic susceptibility occurs.

In the last chapter 10, Chakrabarti et al. have studied the carrier contribution to the second and third order elastic constants (ΔC_{44} *and* ΔC_{456}) in heavily doped quantum wires of optoelectronic materials in the presence of light waves taking n-InAs, InSb, $n-Hg_{1-x}Cd_xTe$, $In_{1-x}Ga_xAs_yP_{1-y}$ and lattice matched to InP as examples. We have also suggested an experimental method of determining ΔC_{44} and ΔC_{456} for materials having arbitrary dispersion laws. The influence of light is immediately apparent from the plots since ΔC_{44} and ΔC_{456} depend strongly on the light intensity (I) and wane length (λ) which is in direct contrast as compared with I=0. The variations of ΔC_{44} and ΔC_{456} reflect the direct signature of the light wave on the electronic, optic and the other band structure dependent properties of HD semiconducting materials in the presence of light waves and the photon assisted transport for the corresponding semiconductor devices. Although both ΔC_{44} and ΔC_{456} tends to increase with intensity and the wave length but the rate of increase is totally band structure dependent. We wish to note that in view of large changes of the elastic constants with n_0 detailed experimental work on second and third-order elastic constants as functions of n_0 would be interesting for the present case. It may be suggested that the experiments on the velocity of sound involving the shear mode as function of n_0 may exhibit the carrier contribution to the elastic constants for materials having arbitrary carrier energy spectra. This fact again suggests another experimental determination of $\Delta C44$ and ΔC_{456} beside the suggested methods of determining them as said above.

The success of an edited book largely depends on the contribution of the contributors of the different chapters. The job of the present editor is to present it in a systematic and nice way in which he can insert his own line of research in this pin-pointed topic. Errorless first edition of any book from every point of view enjoys permanently the domain of "Impossibility Theorems" and the same is very true for this edited book also. This edited book is written for graduate and post graduate students, researchers, engineers and professionals in the fields of Mechanical Engineering, Electrical and Electronic Engineering, Semiconductors and related areas, Nano-electronics, condensed matter physics, solid state sciences, materials science, nano-science and technology and nano-structured materials in general. Last but not the least we do hope that our humble effort will fire the desire of anyone engaged in the said fields, either in academics or in industries to delve deeper into this beautiful Dirac Delta Functional topic of modern research.

ACKNOWLEDGMENTS

The editor is grateful in the real sense of the term to Professors W. L. Freeman, D. Bimberg, P. Hartnagel, A. K. Roy, A. N. Chakrabarti and S. Chakrabarti for reviewing the chapters in a critical way full with in-depth analysis for this edited book on the pin pointed research topic on Quantum Wires. He is grateful to Mrs. Kara Feagaiga of the Department of acquisitions of NOVA Science publishers Inc. for her continuous help without which the completion of this edited book would be a mere dream. The editor expresses his gratitude to Mrs. N.S.G. Columbus, the President of NOVA Science publishers Inc. for accepting this edited book and also to Mrs. Aleksandra Bator, of Department of Acquisitions. Last but not the least I wish to thank Mrs. T. Dutta and Mr. M. Ghosh for placing their combined effort towards the development of this book in the DO-LOOP of a computer.

Dr. Engg. Kamakhya Prasad Ghatak
December 2019

In: Quantum Wires: An Overview
Editor: Kamakhya Prasad Ghatak
ISBN: 978-1-53617-676-6

Chapter 1

QUANTUM WIRES AND FOWLER-NORDHEIM FIELD EMISSION

H. Seikh[1], N. Alharthi[2], P. K. Bose[3], M. Mitra[4], R. Paul[5] and K. P. Ghatak[6,*]

[1]Centre of Excellence for Research in Engineering Materials, Deanship of Scientific Research, King Saud University, Riyadh, Saudi Arabia

[2]Department of Mechanical Engineering, College of Engineering, King Saud University, Riyadh, Saudi Arabia

[3]Department of Mechanical Engineering, Swami Vivekananda Institute of Science and Technology, Sonarpur, Dakshin Gobindapur, Kolkata, West Bengal, India

[4]Department of Electronics and Telecommunication Engineering, IIEST, Shibpur, Botanic Garden, Howrah, India

[5]Department of Computer Science and Engineering, University of Engineering and Management, Newtown, Kolkata, India

[6]Department of Basic Science Institute of Engineering and Management, Kolkata, West Bengal, India

---

[*] Corresponding Author's Email: kamakhyaghatakcu@gmail.com.

ABSTRACT

The importance of Fowler Nordheim Field Emission (FNFE) which is basically a quantum Phenomena is already well-known, since its inception in 1928. In this chapter an attempt is made to study FNFE in quantum wires of nonlinear optical, on the basis of a newly formulated electron dispersion law by talking all types of anisotropies within the frame work of k. p. formalism. All the results of III-V quantum wires form a special case of our generalized analysis. We have studied the FNFE in quantum wires of II-VI semiconductors on the basis of Hopfield model and Bi on the basis of the models of McClure and Choi, Hybrid, Cohen and Lax respectively. The FNFE from quantum wires of IV-VI, stressed materials, Te, GaP, $PtSb_2$, Bi_2Te_3, Ge, GaSb and II-V semiconductors have been investigated on the basis of the appropriate band models. It has been found talking quantum wires of $CdGeAs_2$, InSb, $Hg_{1-x}Cd_xTe$, InAs, GaAs, $In_{1-x}Ga_xAs_yP_{1-y}$ lattice matched to InP, CdS, Bi, stressed InSb, Ge, GaSb, GaP, Bi_2Te_3, PbTe and Te as examples that the field emitted current depends strongly on the thickness of the quantum-confined materials in contrast with the corresponding bulk specimens. The current changes with increasing film thickness in an oscillatory way with different numerical magnitudes. It appears from the aforementioned figures that the FNFE exhibits spikes for particular values of film thickness which, in turn, depends on the particular band structure of the specific semiconductor. Moreover, the FNFE from QWs of different compounds can be smaller than of bulk specimens of the same materials, which is also a direct signature of quantum confinement. This oscillatory dependence will be less and less prominent with increasing film thickness. For bulk specimens of the same material, the FNFE will be found to increase continuously with increasing electron degeneracy in a non-oscillatory manner. The appearance of the discrete jumps in the respective figures are due to the redistribution of the electrons among the quantized energy levels when the size quantum number corresponding to the highest occupied level changes from one fixed value to the others. All the results for all the models gets simplified to the expression of the field emitted current from quantum wires having parabolic energy bands under certain limiting condition which exhibit the mathematical capability of our generalized formalism.

INTRODUCTION

Quantum Wires and Fowler-Nordheim Field Emission (FNFE) is a well-known quantum-mechanical phenomenon that involves tunneling of electrons through a surface barrier due to the application of an intense external electric field. Normally, at field strengths of the order of 10^8 V/m (below the electrical breakdown), the potential barriers at the surfaces of metals and semiconductors usually become very thin and result in field emission of electrons due to the tunnel effect [1]. This has been well-investigated with reference to three dimensional electron gases in metals and semiconductors and the FNFE from quantum confined structures has also been studied in this context [2-8]. Some of significant features of the FNFE which have emerged from these investigations are:

1) The FNFE increases with increasing electron concentration in bulk materials and are significantly influenced by the carrier energy spectra of different electronic materials;
2) The FNFE increases with increasing electric field;
3) The FNFE oscillates with film thickness for quantum confined systems;
4) The FNFE oscillates with inverse quantizing magnetic field in the presence of magnetic quantization due to the Shubnikov de Haas effect;
5) For various types of superlattices of different materials, the FNFE shows composite oscillations with different system variables.

In recent years, with the advent of fine lithographical methods [9] molecular beam epitaxy [10], organometallic vapor-phase epitaxy [11] and other experimental techniques, the restriction of the motion of the carriers of bulk materials in one (quantum wells in ultrathin films, NIPI structures, inversion and accumulation layers), two (quantum wires) and three (quantum dots, magnetosize quantized systems, magneto accumulation layers, magneto inversion layers quantum dot superlattices, magneto quantum well superlattices and magneto NIPI structures) dimensions have in the last few years, attracted much attention not only for their potential in uncovering new phenomena in nanoscience but also for their interesting quantum device applications [12-14]. In ultrathin films, the restriction of the motion of the carriers in the direction normal to the film (say, the z direction) may be viewed as carrier confinement in an infinitely deep 1D rectangular potential well, leading to quantization [known as quantum size effect (QSE)] of the wave vector of the carrier along the direction of the potential well, allowing 2D carrier transport parallel to the surface of the film representing new physical features not exhibited in bulk semiconductors [15]. The low-dimensional heterostructures based on various materials are widely investigated because of the enhancement of carrier mobility [16] .These properties make such structures suitable for applications in quantum well lasers [17], heterojunction FETs [18], high-speed digital networks [19], high-frequency microwave circuits [20], optical modulators [21], optical switching systems [22], and other devices. The constant energy 3D wave-vector space of bulk semiconductors becomes 2D wave-vector surface in ultrathin films or quantum wells due to dimensional quantization. Thus, the concept of reduction of symmetry of the wave-vector space and its consequence can unlock the physics of low dimensional structures.

It is well-known that in quantum wires (QWs), the restriction of the motion of the carriers along two directions may be viewed as carrier confinement by two infinitely deep 1D rectangular potential wells, along any two orthogonal directions leading to quantization of the wave vectors along the said directions, allowing 1D carrier transport [23]. With the help of modern fabrication techniques, such one dimensional quantized structures have been experimentally realized and enjoy an enormous range of important applications in the realm of nanoscience in quantum regime. They have generated much interest in the analysis of nanostructured devices for investigating their electronic, optical and allied properties [24-30]. Examples of such new applications are based on the different transport properties of ballistic charge carriers which include quantum resistors [31, 32, 33], resonant tunneling diodes and band filters [34, 35], quantum switches [36], quantum sensors [37-38], quantum logic gates [39-40], quantum transistors and sub tuners [41-42], heterojunction FETs [43], high-speed digital networks [44], high-frequency microwave circuits [45], optical modulators [46], optical switching systems [47], and other nanoscale devices.

In this chapter, we study the FNFE from QWs of non-parabolic semiconductors having different band structures. At first we shall investigate the FNFE from QWs of non-linear optical compounds which are being used in non-linear optics and light emitting diodes [48]. The quasi-cubic model can be used to investigate the symmetric properties of both the bands at the zone center of wave vector space of the same compound .Including the anisotropic crystal potential in the Hamiltonian, and special features of the non-linear optical compounds, Kildal [49] formulated the electron dispersion law under the assumptions of isotropic momentum matrix and the isotropic spin orbit splitting constant, respectively, although the anisotropies in the two aforementioned band constants are the significant physical features of the said materials [50]. In Section 1.2.1, the FNFE from QWs of non-linear optical semiconductors has been investigated by considering the combined influence of the anisotropies of the said energy band constants together with the inclusion of the crystal field splitting respectively within the frame-work of **k.p** formalism. The III-V compounds find applications in infrared detectors [51], quantum dot light emitting diodes [52], quantum cascade lasers [53], quantum well wires [54], optoelectronic sensors [55], high electron mobility transistors [56] etc. The electron energy spectrum of III-V semiconductors can be described by the three and two-band models of Kane [57], together with the models of Stillman et al. [58], Newson and Kurobe [59] and Palik et al. [60] respectively. In this context it may be noted that the ternary and quaternary compounds enjoy the singular position in the entire spectrum of optoelectronic materials. The ternary alloy $Hg_{1-x}Cd_xTe$ is a classic narrow gap compound. The band gap of this ternary alloy can be varied to cover the spectral range from 0.8 to over 30μm [61] by adjusting the alloy composition. $Hg_{1-x}Cd_xTe$ finds extensive applications in infrared detector materials and photovoltaic detector arrays in the 8-12μm wave bands [62]. The above uses have generated the Hg_{1-}

$_xCd_xTe$ technology for the experimental realization of high mobility single crystal with specially prepared surfaces. The same compound has emerged to be the optimum choice for illuminating the narrow sub-band physics because the relevant material constants can easily be experimentally measured [63]. Besides, the quaternary alloy $In_{1-x}Ga_xAs_yP_{1-y}$ lattice matched to InP, also finds wide use in the fabrication of avalanche photo-detectors [64], hetero-junction lasers [65], light emitting diodes [66] and avalanche photodiodes [67], field effect transistors, detectors, switches, modulators, solar cells, filters and new types of integrated optical devices are made from the quaternary systems [68]. It may be noted that all types of band models as discussed for III-V semiconductors are also applicable for ternary and quaternary compounds. In Section 1.2.2, the FNFE from QWs of III-V, ternary and quaternary semiconductors has been studied in accordance with the said band models and the simplified results for wide gap materials having parabolic energy bands under certain limiting conditions have further been demonstrated as a special case and thus confirming the compatibility test.

The II-VI semiconductors are being used in nanoribbons, blue green diode lasers, photosensitive thin films, infrared detectors, ultra high-speed bipolar transistors, fiber optic communications, microwave devices, solar cells, semiconductor gamma-ray detector arrays, semiconductor detector gamma camera and allow for a greater density of data storage on optically addressed compact discs [69-71]. The carrier energy spectra in II-VI compounds are defined by the Hopfield model [72] where the splitting of the two-spin states by the spin-orbit coupling and the crystalline field has been taken into account. The Section 1.2.3 contains the investigation of the FNFE from QWs of II-VI compounds.

In recent years, Bismuth (Bi) nano-lines have been fabricated and and Bi also finds use in array of antennas which leads to the interaction of electromagnetic waves with such Bi-nano-wires [73]. Several dispersion relations of the carriers have been proposed for Bi. Shoenberg [74] experimentally verified that the de Haas-Van Alphen and cyclotron resonance experiments supported the ellipsoidal parabolic model of Bi, although, the magnetic field dependence of many physical properties of Bi supports the two-band model [75].The experimental investigations on the magneto-optical and the ultrasonic quantum oscillations support the Lax ellipsoidal non-parabolic model [75]. Kao [76], Dinger and Lawson [77] and Koch and Jensen [78] demonstrated that the Cohen model [79] is in conformity with the experimental results in a better way. Besides, the hybrid model of bismuth, as developed by Takoka et al. also finds use in the literature [80]. McClure and Choi [81] derived a new model of Bi and they showed that it can explain the data for a large number of magneto-oscillatory and resonance experiments. In section 1.2.4, the FNFE from QWs of Bi has been formulated in accordance with the aforementioned energy band models for the purpose of relative assessment. Besides, under certain limiting conditions all the results for all the models of 1D systems are reduced to the well-known result of the FNFE from QWs of wide gap materials. This above statement exhibits the compatibility test of our theoretical analysis.

Lead chalcogenides (PbTe, PbSe, and PbS) are IV-VI non-parabolic semiconductors whose studies over several decades have been motivated by their importance in infrared IR detectors, lasers, light-emitting devices, photovoltaics and high temperature thermoelectrics [82]. PbTe, in particular, is the end compound of several ternary and quaternary high performance high temperature thermoelectric materials [83]. It has been used not only as bulk but also as films [84], quantum wells [85], superlattices [86], nanowires [87], and colloidal and embedded nanocrystals [88] and PbTe films doped with various impurities have also been investigated [89]. These studies revealed some of the interesting features that had been seen in bulk PbTe, such as Fermi level pinning and, in the case of superconductivity [90]. In Section 1.2.5, the FNFE from QWs of IV-VI semiconductors has been studied taking PbTe, PbSe, and PbS as examples.

The stressed semiconductors are being investigated for strained silicon transistors, quantum cascade lasers, semiconductor strain gages, thermal detectors and strained-layer structures [91]. The FNFE from QWs of stressed compounds (taking stressed n-InSb as an example) has been investigated in Section 1.2.6. The vacuum deposited Tellurium (Te) has been used as the semiconductor layer in thin-film transistors (TFT) [92] which is being used in CO_2 laser detectors [93], electronic imaging, strain sensitive devices [94] and multichannel Bragg cell [95]. The Section 1.2.7 contains the investigation of FNFE from QWs of Tellurium.

The n-Gallium Phosphide (n-GaP) is being used in quantum dot light emitting diode [96], high efficiency yellow solid state lamps, light sources, high peak current pulse for high gain tubes. The green and yellow light emitting diodes made of nitrogen-doped n-GaP possess a longer device life at high drive currents [97]. In Section 1.2.8, the FNFE from QWs of n-GaP has been studied. The Platinum Antimonide ($PtSb_2$) finds application in device miniaturization, colloidal nanoparticle synthesis, sensors and detector materials and thermo-photovoltaic devices [98]. The Section 1.2.9 explores the FNFE from QWs of $PtSb_2$.. Bismuth telluride (Bi_2Te_3) was first identified as a material for thermoelectric refrigeration in 1954 [99] and its physical properties were later improved by the addition of bismuth selenide and antimony telluride to form solid solutions [100]. The alloys of Bi_2Te_3 are useful compounds for the thermoelectric industry and have been investigated in the literature [100]. In Section 1.2.10, the FNFE from QWs of Bi_2Te_3 has been considered.

The usefulness of elemental semiconductor Germanium is already well known since the inception of transistor technology and, it is also being used in memory circuits, single photon detectors, single photon avalanche diode, ultrafast optical switch, THz lasers and THz spectrometers [101]. In Section 1.2.11, the FNFE has been studied from QWs of Ge. Gallium Antimonide (GaSb) finds applications in the fiber optic transmission window, heterojunctions and quantum wells. A complementary hetero-junction field effect transistor in which the channels for the p-FET device and the n-FET device forming the complementary FET are formed from GaSb. The band gap energy of GaSb makes it suitable for low power operation [102]. In Section 1.2.12, the FNFE from QWs of GaSb has been studied. The II-V semiconductors are being used in photovoltaic cells constructed of single crystal semiconductor materials in contact with electrolyte solutions. Cadmium selenide shows an open-circuit voltage of 0.8V and power conservation coefficients near 6 percent for 720-nm light

[103]. They are also used in ultrasonic amplification [104]. The development of an evaporated thin film transistor using cadmium selenide as the semiconductor has also been reported [105-106]. In Section 1.2.13, we shall study the FNFE from QWs of II-V semiconductors. The section 1.3 contains the result and discussions pertaining to this chapter.

THEORETICAL BACKGROUND

The FNFE from Quantum Wires of Non-Linear Optical Semiconductors

The form of **k.p** matrix for nonlinear optical compounds can be expressed extending Bodnar [50] as

$$H = \begin{bmatrix} H_1 & H_2 \\ H_2^+ & H_1 \end{bmatrix} \tag{1}$$

where,

$$H_1 \equiv \begin{bmatrix} E_{g_0} & 0 & P_{\|}k_z & 0 \\ 0 & (-2\Delta_{\|}/3) & (\sqrt{2}\Delta_{\perp}/3) & 0 \\ P_{\|}k_z & (\sqrt{2}\Delta_{\perp}/3) & -(\delta+\frac{1}{3}\Delta_{\|}) & 0 \\ 0 & 0 & 0 & 0 \end{bmatrix}, \qquad H_2 \equiv \begin{bmatrix} 0 & -f_{,+} & 0 & f_{,-} \\ f_{,+} & 0 & 0 & 0 \\ 0 & 0 & 0 & 0 \\ f_{,+} & 0 & 0 & 0 \end{bmatrix}$$

in which E_{g_0} is the band gap in the absence of any field, $P_{||}$ and $P_{\perp}$ are the momentum matrix elements parallel and perpendicular to the direction of crystal axis respectively, δ is the crystal field splitting constant, $\Delta_{\|}$ and $\Delta_{\perp}$ are the spin-orbit splitting constants parallel and perpendicular to the C-axis respectively, $f_{,\pm} \equiv (P_{\perp}/\sqrt{2})(k_x \pm ik_y)$ and $i = \sqrt{-1}$ .Thus, neglecting the contribution of the higher bands and the free electron term, the diagonalization of the above matrix leads to the dispersion relation of the conduction electrons in bulk specimens of nonlinear optical compounds a

ħ=h/2π, h is Planck's constant and $m_{\parallel}^{*}$ and $m_{\perp}^{*}$ are the longitudinal and transverse effective electron masses at the edge of the conduction band respectively.

$$\gamma(E) = f_1(E)k_s^2 + f_2(E)k_z^2 \tag{2}$$

where,

$$\gamma(E) \equiv E(E+E_{go})[(E+E_{go})(E+E_{go}+\Delta_{\parallel}) + \delta(E+E_{go}+\frac{2}{3}\Delta_{\parallel}) + \frac{2}{9}(\Delta_{\parallel}^2 - \Delta_{\perp}^2)],$$

E is the total energy of the electron as measured from the edge of the conduction band in the vertically upward direction in the absence of any quantization, $k_s^2 = k_x^2 + k_y^2$,

$$f_1(E) \equiv \frac{\hbar^2 E_{g_0}\left(E_{g_0}+\Delta_{\perp}\right)}{\left[2m_{\perp}^{*}\left(E_{g_0}+\frac{2}{3}\Delta_{\perp}\right)\right]}\left[\delta\left(E+E_{g_0}+\frac{1}{3}\Delta_{\parallel}\right) + \left(E+E_{g_0}\right)\left(E+E_{g_0}+\frac{2}{3}\Delta_{\parallel}\right) + \frac{1}{9}\left(\Delta_{\parallel}^2 - \Delta_{\parallel}^2\right)\right],$$

$$f_2(E) \equiv \frac{\hbar^2 E_{g_0}\left(E_{g_0}+\Delta_{\parallel}\right)}{\left[2m_{\parallel}^{*}\left(E_{g_0}+\frac{2}{3}\Delta_{\parallel}\right)\right]}\left[\left(E+E_{g_0}\right)\left(E+E_{g_0}+\frac{2}{3}\Delta_{\parallel}\right)\right]$$

For two dimensional quantization along x and y directions, the (1.2) assumes the form

$$k_z^2 = A_{11}(E, n_x, n_y) \tag{3}$$

where $A_{11}(E,n_x,n_y) = [f_2(E)]^{-1}[\gamma(E) - \phi_1(n_x,n_y)f_1(E)], \phi_1(n_x,n_y) = (\frac{n_x\pi}{d_x})^2 + (\frac{n_y\pi}{d_y})^2$

$n_x = (1,2,3.......), n_y = (1,2,3.......)$ are the size quantum numbers along x and y directions respectively and d_x and d_y are the nano-thickness along x and y directions respectively.

The quantized sub-band energy (E_{11}) is given by

$$\gamma(E_{11}) = f_1(E_{11})\phi_1(n_x,n_y) \tag{4}$$

The electron concentration per unit length can be expressed as

$$n_0 = \frac{2g_v}{\pi} \sum_{n_x=1}^{n_{x\max}} \sum_{n_y=1}^{n_{y\max}} [B_{11}(E_{F1D}, n_x, n_y) + B_{12}(E_{F1D}, n_x, n_y)] \tag{5}$$

where g_v is the valley degeneracy

$$B_{11}(E_{F1D}, n_x, n_y) = [A_{11}(E_{F1D}, n_x, n_y)]^{1/2}, B_{12}(E_{F1D}, n_x, n_y) = \sum_{r=1}^{r_0} Z_{1D}(r)[B_{11}(E_{F1D}, n_x, n_y)]$$

$$Z_{1D}(r) = 2(k_B T)^{2r}(1 - 2^{1-2r})\xi(2r)\frac{\partial^{2r}}{\partial E_{F1D}^{2r}},$$

,

k_B is the Boltzmann constant, T is the temperature, r is the set of real positive integers whose upper limit is r_0, $\xi(2r)$ is the Zeta function of order 2r [107] and E_{F1D} is the Fermi energy in the presence of 2D quantization as measured from the edge of the conduction band in the vertically upward direction in the absence of any quantization.

The current (I) due to Fowler-Nordheim {FN}field emission can be written as

$$I = \frac{1}{2} \sum_{n_x=1}^{n_{x\max}} \sum_{n_y=1}^{n_{y\max}} [\int_{E_{11}}^{\infty} [e.vn_1 t_{11}] \tag{6a}$$

where e is the magnitude of the electron charge, v is the velocity of the electron and given by $v = \frac{1}{\hbar}\frac{\partial E}{\partial k_z}$, $n_1 = N_{1D}(E).f(E)dE$, $N_{1D}(E)$ is the 1D density –of – states function per subbands and can be written as $N_{1D}(E) = \frac{2g_v}{\pi}\frac{\partial k_z}{\partial E}$, $f(E)$ is the Fermi-Dirac occupation probability factor and is given by $f(E) = [1 + \exp(\frac{E - E_{F1D}}{k_B T})]^{-1}$ and t_{11} is the transmission coefficient. From (1.6a), it appears that I is a function of the product of the carrier velocity, concentration and transmission coefficient. These three quantities in turn depend totally on the dispersion relation of the material.

As the basic **E-k** relation changes, all the aforementioned quantities will change and the current due to FN field emission will be different consequently. Thus, the field emission will change altogether in 1D, 2D and 3D quantization of the wave vector space encompassing the whole arena of quantized structures.

Thus from (6a) one can write

$$I = \frac{1}{2}\sum_{n_x=1}^{n_{x\max}}\sum_{n_y=1}^{n_{y\max}}[\int_{E_{11}}^{\infty}[\frac{e}{\hbar}\frac{\partial E}{\partial k_z}.\frac{2g_v}{\pi}\frac{\partial k_z}{\partial E}f(E)dE]t_{11}] \quad (6b)$$

The term $\frac{\partial E}{\partial k_z}$ from velocity and the term $\frac{\partial k_z}{\partial E}$ from the density –of – states function per sub-bands cancel each other leaving the constant pre-factor.

Therefore (6b) assumes the form

$$I = \frac{e.g_v}{\hbar\pi}\sum_{n_x=1}^{n_{x\max}}\sum_{n_y=1}^{n_{y\max}}[\int_{E_{11}}^{\infty}[f(E)dE].t_{11} \quad (6c)$$

The term t_{11} in this case can be expressed by using the method as given in [108] as

$$t_{11} = \exp(-\beta_{11}) \quad (7)$$

in which

$$\beta_{11} = \frac{4[A_{11}(V_0,n_x,n_y)]^{3/2}}{3eF_{sz}[A'_{11}(V_0,n_x,n_y)]} \quad (8)$$

F_{sz} is the surface electric field along z-axis, $A_{11}(V_0,n_x,n_y) = [f_2(V_0)]^{-1}[\gamma(V_0) - \phi_1(n_x,n_y)f_1(V_0)]$, V_0 is equal to the addition of the Fermi energy in the corresponding case and the work function ϕ_w of the material

$$f_2(V_0)\equiv\frac{\hbar^2 E_{g_0}\left(E_{g_0}+\Delta_{\|}\right)}{\left[2m_{\|}^*\left(E_{g_0}+\frac{2}{3}\Delta_{\|}\right)\right]}\left[\left(V_0+E_{g_0}\right)\left(V_0+E_{g_0}+\frac{2}{3}\Delta_{\|}\right)\right],$$

$$\gamma(V_0)\equiv V_0(V_0+E_{g_0})\left[\left(V_0+E_{g_0}\right)\left(V_0+E_{g_0}+\Delta_{\|}\right)+\delta\left(V_0+E_{g_0}+\frac{2}{3}\Delta_{\|}\right)+\frac{2}{9}\left(\Delta_{\|}^2-\Delta_{\perp}^2\right)\right],$$

$$f_1(V_0)\equiv\frac{\hbar^2 E_{g_0}\left(E_{g_0}+\Delta_{\perp}\right)}{\left[2m_{\perp}^*\left(E_{g_0}+\frac{2}{3}\Delta_{\perp}\right)\right]}\left[\delta\left(V_0+E_{g_0}+\frac{1}{3}\Delta_{\|}\right)+\left(V_0+E_{g_0}\right)\left(V_0+E_{g_0}+\frac{2}{3}\Delta_{\|}\right)+\frac{1}{9}\left(\Delta_{\|}^2-\Delta_{\|}^2\right)\right],$$

$$A_{11}'\left(V_0,n_x,n_y\right)=\left[\frac{-A_{11}\left(V_0,n_x,n_y\right)f_2'\left(V_0\right)}{f_2\left(V_0\right)}\right]\left[f_2\left(V_0\right)\right]^{-1}\left[\gamma'\left(V_0\right)-f_1'\left(V_0\right)\phi_1\left(n_x,n_y\right)\right]$$

$$f_2'\left(V_0\right)=\left[\left[\hbar^2E_{g0}\left(E_{g0}+\Delta_{\|}\right)\right]\left[2m_{\|}^*\left(E_{g0}+\frac{2}{3}\Delta_{\|}\right)\right]^{-1}\left[2V_0+2E_{g0}+\frac{2}{3}\Delta_{\|}\right]\right]$$

$$f_1'\left(V_0\right)=\left[\left[\hbar^2E_{g0}\left(E_{g0}+\Delta_{\perp}\right)\right]\left[2m_{\perp}^*\left(E_{g0}+\frac{2}{3}\Delta_{\|}\right)\right]^{-1}\left[2V_0+2E_{g0}+\frac{2}{3}\Delta_{\|}\right]\right]$$

and

$$\gamma'\left(V_0\right)=\left[\frac{\gamma\left(V_0\right)\left(2V_0+E_{g0}\right)}{V_0\left(V_0+E_{g0}\right)}+\left(V_0\left(V_0+E_{g0}\right)\right)\left[2V_0+2E_{g0}+\Delta_{\|}+\delta\right]\right]$$

Thus we can write

$$I=\frac{2g_v ek_BT}{h}\sum_{n_x=1}^{n_{x\max}}\sum_{n_y=1}^{n_{y\max}}F_0(\eta_{11})t_{11} \tag{9}$$

where $\eta_{11}=\frac{E_{F1D}-E_{11}}{k_BT}$, $F_0(\eta_{11})$ is the special case of the one parameter Fermi-Dirac integral of order j which can be written as

$$F_j(\eta)=\left(\frac{1}{\Gamma(j+1)}\right)\int_0^\infty\frac{x^j\,dx}{1+\exp(x-\eta)},\qquad j>-1 \tag{10}$$

for all *j*, analytically continued as a complex contour integral around the negative x-axis

$$F_j(\eta)=\left(\frac{\Gamma(-j)}{2\pi\sqrt{-1}}\right)\int_{-\infty}^{+0}\frac{x^j dx}{1+\exp(-x-\eta)} \quad (11)$$

where η is the dimensionless x independent variable,

$$\Gamma(j+1)=j\Gamma(j),\Gamma\left(\frac{1}{2}\right)=\sqrt{\pi}\ and\ \Gamma(0)=1$$

Therefore the field emitted current is given by

$$I=\frac{2g_v ek_B T}{h}\sum_{n_x=1}^{n_{x_{\max}}}\sum_{n_y=1}^{n_{y_{\max}}}F_0(\eta_{11})\exp(-\beta_{11}) \quad (12)$$

The FNFE from Quantum Wires of III-V Semiconductors

The dispersion relation of the conduction electrons of III-V compounds are described by the models of Kane (both three and two bands) [57], Stillman et al. [58], Newson and Kurobe [59] and Palik et al. [60] respectively. For the purpose of complete and coherent presentation, the FNFE from QWs of III-V semiconductors have also been investigated in accordance with the aforementioned different dispersion relations for the purpose of relative comparison as follows:

The Three Band Model of Kane

Under the conditions, $\delta=0$, $\Delta_{\|}=\Delta_{\perp}=\Delta$ (isotropic spin orbit splitting constant) and $m_{\|}^{*}=m_{\perp}^{*}=m_c$ (isotropic effective electron mass at the edge of the conduction band), (1.2) gets simplified into the form

$$\frac{\hbar^2 k^2}{2m_c} = I_{11}(E), I_{11}(E) \equiv \frac{E\left(E+E_{g_0}\right)\left(E+E_{g_0}+\Delta\right)\left(E_{g_0}+\frac{2}{3}\Delta\right)}{E_{g_0}\left(E_{g_0}+\Delta\right)\left(E+E_{g_0}+\frac{2}{3}\Delta\right)} \tag{13}$$

which is known as the three band model of Kane [57] and is often used to study the electronic properties of III-V materials.

The 1-D E-k_z relation can be expressed as

$$k_z^2 = A_{12}(E, n_x, n_y) \tag{14}$$

$$where \quad A_{12}(E, n_x, n_y) = \frac{2m_c}{\hbar^2} I_{11}(E) - \phi_1(n_x, n_y)$$

The quantized sub-band energy (E_{12}) is given by

$$I_{11}(E_{12}) = [\frac{\hbar^2}{2m_c}]\phi_1(n_x, n_y) \tag{15}$$

The electron concentration per unit length can be written as

$$n_0 = \frac{2g_v}{\pi} \sum_{n_x=1}^{n_{x\max}} \sum_{n_y=1}^{n_{y\max}} [B_{13}(E_{F1D}, n_x, n_y) + B_{14}(E_{F1D}, n_x, n_y)] \tag{16}$$

$$\text{where} \quad B_{13}(E_{F1D}, n_x, n_y) = [A_{12}(E_{F1D}, n_x, n_y)]^{1/2}$$

and

$$B_{14}(E_{F1D}, n_x, n_y) = \sum_{r=1}^{r_0} Z_{1D}(r)[B_{13}(E_{F1D}, n_x, n_y)]$$

The field emitted current can be expressed as

$$I = \frac{2g_v e k_B T}{h} \sum_{n_x=1}^{n_{x\max}} \sum_{n_y=1}^{n_{y\max}} F_0(\eta_{12}) t_{12} \tag{17}$$

$$where, \eta_{12} = \frac{E_{F1D} - E_{12}}{k_B T}, t_{12} = \exp(-\beta_{12}), \beta_{12} = \frac{4}{3}[A_{12}(V_0, n_x, n_y)]^{3/2} \cdot [eF_{sz} A'_{12}(V_0)]^{-1},$$

$$A_{12}(V_0, n_x, n_y) = [\frac{2m_c}{\hbar^2} I_{11}(V_0) - \phi_1(n_x, n_y)], I_{11}(V_0) \equiv \frac{V_0\left(V_0+E_{g_0}\right)\left(V_0+E_{g_0}+\Delta\right)\left(E_{g_0}+\frac{1}{3}\Delta\right)}{E_{g_0}\left(E_{g_0}+\Delta\right)\left(V_0+E_{g_0}+\frac{2}{3}\Delta\right)}$$

$$and \ A'_{12}(V_0) = [\frac{2m_c}{\hbar^2} I_{11}(V_0)[\frac{1}{V_0} + \frac{1}{V_0+E_{g0}} + \frac{1}{V_0+E_{g0}+\Delta} - \frac{1}{V_0+E_{g0}+(2/3)\Delta}]]$$

Therefore the field emitted current is given by

$$I = \frac{2g_v e k_B T}{h} \sum_{n_x=1}^{n_{x\max}} \sum_{n_y=1}^{n_{y\max}} F_0(\eta_{12}) \exp(-\beta_{12}) \tag{18}$$

Two Band Model of Kane

Under the inequalities $\Delta >> E_{g_0}$ or $\Delta << E_{g_0}$, (1.13) assumes the form

$$E(1+\alpha E) = (\hbar^2 k^2 / 2m_c\ ), \alpha \equiv 1/E_{g_0} \tag{19}$$

The (19) is known as the two-band model of Kane where α is known as band non-parabolicity parameter and should be as such for studying the electronic properties of the semiconductors whose band-structures obey the above inequalities [57].

The 1-D E-k_z relation can be expressed as

$$k_z^2 = A_{13}(E, n_x, n_y) \tag{20}$$

$$\text{where } A_{13}(E, n_x, n_y) = [\frac{2m_c}{\hbar^2}\{E(1+\alpha E)\} - \phi_1(n_x, n_y)]$$

The quantized sub-band energy (E_{12}) is given

$$E_{13}(1+\alpha E_{13}) = [\frac{\hbar^2}{2m_c}]\phi_1(n_x, n_y) \tag{21}$$

The electron concentration per unit length can be written as

$$n_0 = \frac{2g_v}{\pi} \sum_{n_x=1}^{n_{x\max}} \sum_{n_y=1}^{n_{y\max}} [B_{15}(E_{F1D}, n_x, n_y) + B_{16}(E_{F1D}, n_x, n_y)] \tag{22}$$

$$\text{where } B_{15}(E_{F1D}, n_x, n_y) = [A_{13}(E_{F1D}, n_x, n_y)]^{1/2}$$

and

$$B_{16}(E_{F1D}, n_x, n_y) = \sum_{r=1}^{r_0} Z_{1D}(r)[B_{15}(E_{F1D}, n_x, n_y)]$$

The field emitted current can be expressed as

$$I = \frac{2g_v e k_B T}{h} \sum_{n_x=1}^{n_{x\max}} \sum_{n_y=1}^{n_{y\max}} F_0(\eta_{13}) t_{13} \tag{23}$$

$$\text{where } \eta_{13} = \frac{E_{F1D} - E_{13}}{k_B T}, t_{13} = \exp(-\beta_{13}), \beta_{13} = \frac{4}{3}[A_{13}(V_0, n_x, n_y)]^{3/2} \cdot [eF_{sz} A'_{13}(V_0)]^{-1}$$

$$\text{and } A'_{13}(V_0) = \frac{2m_c}{\hbar^2}(1 + 2\alpha V_0)$$

Therefore the field emitted current assumes the from

$$I = \frac{2g_v e k_B T}{h} \sum_{n_x=1}^{n_{x\max}} \sum_{n_y=1}^{n_{y\max}} F_0(\eta_{13}) \exp(-\beta_{13}) \tag{24}$$

Parabolic Energy Bands

The expressions for the electron concentration per unit length and the FNFE from QWs having parabolic energy bands can, respectively, be written as

$$n_0 = \frac{2g_v \sqrt{2\pi m_c k_B T}}{h} \sum_{n_x=1}^{n_{x\max}} \sum_{n_y=1}^{n_{y\max}} F_{-1/2}(\eta_{14}) \tag{25}$$

$$I = \frac{2g_v e k_B T}{h} \sum_{n_x=1}^{n_{x\max}} \sum_{n_y=1}^{n_{y\max}} F_0(\eta_{14}) \exp[-\frac{4}{3eF_{sz}}(\frac{2m_c}{\hbar^2})^{1/2} V_0^{3/2} \{1 - \frac{\hbar^2 \phi_1(n_x, n_y)}{2m_c V_0}\}^{3/2}] \tag{26}$$

$$\text{where } \eta_{14} = [E_{F1D} - \frac{\hbar^2 \phi_1(n_x, n_y)}{2m_c}][k_B T]^{-1}$$

Converting the summations over n_x and n_y to the corresponding integrations, (26) gets transformed as

$$J = \frac{g_v e^2 F_{sz}^2}{8\pi h \phi_w} \exp\{-\frac{4}{3eF_{sz}}(\frac{2m_c}{\hbar^2})^{1/2} \phi_w^{3/2}\} \tag{27}$$

The (27) is the well-known expression of the FNFE from bulk semiconductors having parabolic energy bands [109].

The Model of Stillman et al.

In accordance with the model of Stillman et al. [58], the electron dispersion law of III-V materials assumes the form

$$E = \bar{t}_{11}k^2 - \bar{t}_{12}k^4 \tag{28}$$

where

$$\bar{t}_{11} \equiv \frac{\hbar^2}{2m_c}, \bar{t}_{12} \equiv (1-\frac{m_c}{m_0})^2(\frac{\hbar^2}{2m_c})^2[(3E_{g_0}+4\Delta+\frac{2\Delta^2}{E_{g_0}}).\{(E_{g_0}+\Delta)(2\Delta+3E_{g_0})\}^{-1}]$$

and

m_0 is the free electron mass.

The (1.28) can be expressed as

$$\frac{\hbar^2k^2}{2m_c} = I_{12}(E) \tag{29}$$

where,

$$I_{12}(E) \equiv a_{11}[1-(1-a_{12}E)^{1/2}], a_{11} \equiv (\frac{\hbar^2\bar{t}_{11}}{4m_c\,\bar{t}_{12}}) and\ a_{12} \equiv \frac{4\bar{t}_{12}}{\bar{t}_{11}^{\,2}}$$

The 1-D E-k_z relation can be written as

$$k_z^2 = A_{14}(E,n_x,n_y) \tag{30}$$

where

$$A_{14}(E,n_x,n_y) = [\frac{2m_c}{\hbar^2}\{I_{12}(E)\} - \phi_1(n_x,n_y)]$$

The quantized sub-band energy (E_{14}) is given by

$$I_{12}(E_{14}) = [\frac{\hbar^2}{2m_c}]\phi_1(n_x,n_y) \tag{31}$$

The electron concentration per unit length can be expressed as

$$n_0 = \frac{2g_v}{\pi} \sum_{n_x=1}^{n_{x\max}} \sum_{n_y=1}^{n_{y\max}} [B_{17}(E_{F1D}, n_x, n_y) + B_{18}(E_{F1D}, n_x, n_y)] \tag{32}$$

where

$$B_{17}(E_{F1D}, n_x, n_y) = [A_{14}(E_{F1D}, n_x, n_y)]^{1/2}$$

and

$$B_{18}(E_{F1D}, n_x, n_y) = \sum_{r=1}^{r_0} Z_{1D}(r)[B_{17}(E_{F1D}, n_x, n_y)]$$

The field emitted current can be written as

$$I = \frac{2g_v e k_B T}{h} \sum_{n_x=1}^{n_{x\max}} \sum_{n_y=1}^{n_{y\max}} F_0(\eta_{15}) \exp(-\beta_{15}) \tag{33}$$

where $\eta_{15} = \dfrac{E_{F1D} - E_{14}}{k_B T}, \beta_{15} = \dfrac{4}{3}[A_{14}(V_0, n_x, n_y)]^{3/2} \cdot [eF_{sz} A'_{14}(V_0)]^{-1}$,

$$A_{14}(V_0, n_x, n_y) = [\frac{2m_c}{\hbar^2}\{I_{12}(V_0)\} - \phi_1(n_x, n_y)], A'_{14}(V_0) = \frac{2m_c}{\hbar^2} I'_{12}(V_0)$$

and

$$I'_{12}(V_0) = (\frac{a_{11}a_{12}}{2})(1 - a_{12}V_0)^{-1/2}$$

The Model of Newson and Kurobe

In accordance with the model of Newson and Kurobe [59], the electron dispersion law in this case assumes the form

$$E = a_{13}k_z^4 + [\frac{\hbar^2}{2m_c} + a_{14}k_s^2]k_z^2 + \frac{\hbar^2}{2m_c}k_s^2 + a_{14}k_x^2 k_y^2 + a_{13}(k_x^4 + k_y^4) \tag{34}$$

where a_{13} is the non-parabolicity constant, $a_{14}\left(\equiv 2a_{13} + a_{15}\right)$ and a_{15} is known as the warping constant.

The 1-D E-k_z relation can be expressed as

$$k_z^2 = A_{15}(E, n_x, n_y) \tag{35}$$

where $A_{15}(E, n_x, n_y) = (2a_{13})^{-1}[-\overline{L_1}(n_x, n_y) + [\{\overline{L_1}(n_x, n_y)\}^2 - 4a_{13}[\overline{L_2}(n_x, n_y) - E)]^{1/2}]$,

$$\overline{L_1}(n_x, n_y) = \frac{\hbar^2}{2m_c} + a_{14}[(\frac{n_x\pi}{d_x})^2 + (\frac{n_y\pi}{d_y})^2]$$

and

$$\overline{L_2}(n_x, n_y) = [\frac{\hbar^2}{2m_c}\phi_1(n_x, n_y) + a_{14}(\frac{n_x\pi}{d_x} \cdot \frac{n_y\pi}{d_y})^2 + a_{13}[(\frac{n_x\pi}{d_x})^4 + (\frac{n_y\pi}{d_y})^4]]$$

The quantized sub-band energy (E_{16}) is given b

$$E_{16} = \overline{L_2}(n_x, n_y) \tag{36}$$

The electron concentration per unit length can be written as

$$n_0 = \frac{2g_v}{\pi} \sum_{n_x=1}^{n_{x\max}} \sum_{n_y=1}^{n_{y\max}} [B_{19}(E_{F1D}, n_x, n_y) + B_{20}(E_{F1D}, n_x, n_y)] \tag{37}$$

where $B_{19}(E_{F1D}, n_x, n_y) = [A_{15}(E_{F1D}, n_x, n_y)]^{1/2}$

and

$$B_{20}(E_{F1D}, n_x, n_y) = \sum_{r=1}^{r_0} Z_{1D}(r)[B_{19}(E_{F1D}, n_x, n_y)] \tag{38}$$

The field emitted current assumes the form

$$I = \frac{2g_v e k_B T}{h} \sum_{n_x=1}^{n_{x\max}} \sum_{n_y=1}^{n_{y\max}} F_0(\eta_{16}) \exp(-\beta_{16}) \tag{39}$$

$$where,\ \eta_{16} = \frac{E_{F1D} - E_{16}}{k_B T}, \beta_{16} = \frac{4}{3}[A_{15}(V_0, n_x, n_y)]^{3/2} \cdot [eF_{sz} A'_{15}(V_0, n_x, n_y)]^{-1}$$

and

$$A'_{15}(V_0, n_x, n_y) = [\{\overline{L_1}(n_x, n_y)\}^2 - 4a_{13}[\overline{L_2}(n_x, n_y) - V_0)]^{-1/2}]$$

Model of Palik et al.

The energy spectrum of the conduction electrons in III-V semiconductors up to the fourth order in effective mass theory, taking into account the interactions of heavy hole, light hole and the split-off holes can be expressed in accordance with the model of Palik et al. [60] as

$$E = \frac{\hbar^2 k^2}{2m_c} - \bar{B}_{11} k^4 \tag{40}$$

where

$$\bar{B}_{11} = [\frac{\hbar^4}{4E_{g_0}(m_c)^2}][\frac{1+\frac{x_{11}^2}{2}}{1+\frac{x_{11}}{2}}](1-y_{11})^2, x_{11} = [1+(\frac{\Delta}{E_{g0}})]^{-1} and\ y_{11} = \frac{m_c}{m_o}$$

The (40) gets simplified as

$$\frac{\hbar^2 k^2}{2m_c} = I_{13}(E) \tag{41}$$

$$where\ I_{13}(E) = \bar{b}_{12}[\bar{a}_{12} - ((\bar{a}_{12})^2 - 4E\bar{B}_{11})^{1/2}], \bar{a}_{12} = (\frac{\hbar^2}{2m_c}) and\ \bar{b}_{12} = [\frac{\bar{a}_{12}}{2\bar{B}_{11}}]$$

The 1-D E-k_z relation can be written as

$$k_z^2 = A_{16}(E, n_x, n_y) \tag{42}$$

$$where\ A_{16}(E, n_x, n_y) = [\frac{2m_c}{\hbar^2}\{I_{13}(E)\} - \phi_1(n_x, n_y)]$$

The electron concentration per unit length can be expressed as

$$n_0 = \frac{2g_v}{\pi} \sum_{n_x=1}^{n_{x_{\max}}} \sum_{n_y=1}^{n_{y_{\max}}} [B_{21}(E_{F1D}, n_x, n_y) + B_{22}(E_{F1D}, n_x, n_y)] \tag{43}$$

$$where\ B_{21}(E_{F1D}, n_x, n_y) = [A_{16}(E_{F1D}, n_x, n_y)]^{1/2}\ and\ B_{22}(E_{F1D}, n_x, n_y) = \sum_{r=1}^{r_0} Z_{1D}(r)[B_{21}(E_{F1D}, n_x, n_y)]$$

The field emitted current can be written as

$$I = \frac{2g_v e k_B T}{h} \sum_{n_x=1}^{n_{x_{\max}}} \sum_{n_y=1}^{n_{y_{\max}}} F_0(\eta_{16}) \exp(-\beta_{16}) \tag{44}$$

$$where\ \eta_{16} = \frac{E_{F1D} - E_{17}}{k_B T}$$

and E_{17} is the root of

$$\frac{\hbar^2}{2m_c}\phi_1(n_x, n_y) = I_{13}(E_{17}) \tag{45}$$

$$where\ \beta_{16} = \frac{4}{3}[A_{16}(V_0, n_x, n_y)]^{3/2} \cdot [eF_{sz} A'_{16}(V_0)]^{-1}, A'_{16}(V_0) = \frac{2m_c}{\hbar^2} I'_{13}(V_0)$$

$$and\ I'_{13}(V_0) = 2\bar{b}_{12}\bar{B}_{11}[(\bar{a}_{12})^2 - 4V_0\bar{B}_{11}]^{-1/2}$$

The FNFE from Quantum Wires of II-VI Semiconductors

The carrier energy spectra in bulk specimens of II-VI compounds in accordance with Hopfield model [72] can be written as

$$E = A_o k_s^2 + B_o k_z^2 \pm C_o k_s \tag{46}$$

where $A_0 \equiv \hbar^2/2m_\perp^*$, $B_0 \equiv \hbar^2/2m_\parallel^*$, and C_0 represents the splitting of the two-spin states by the spin orbit coupling and the crystalline field.

The 1D dispersion relations for quantum wires of II-VI semiconductors can be expressed as

$$E = B_0 k_z^2 + G_{3,\pm}(n_x, n_y) \tag{47}$$

$$\text{where } G_{3,\pm}(n_x, n_y) \equiv [A_0\{(\frac{\pi n_x}{d_x})^2 + (\frac{\pi n_y}{d_y})^2\} \pm C_0\{(\frac{\pi n_x}{d_x})^2 + (\frac{\pi n_y}{d_y})^2\}^{1/2}]$$

The FNFE in this case is given by

$$I = \frac{eg_v k_B T}{h} \sum_{n_x=1}^{n_{x\max}} \sum_{n_y=1}^{n_{y\max}} [\{F_0\{(k_B T)^{-1}[E_{F1D} - [G_{3,+}(n_x, n_y)]]\}\exp(-\bar{\beta}_{016,+})\} +$$

$$F_0\{(k_B T)^{-1}[E_{F1D} - [G_{3,-}(n_x, n_y)]]\}\exp(-\bar{\beta}_{016,-})] \tag{48}$$

$$\text{and } \bar{\beta}_{016,\pm} = \frac{4}{3}[(V_0 - G_{3,\pm}(n_x, n_y))]^{3/2} \cdot [eF_{sz}\hbar]^{-1}\sqrt{2m_\parallel^*}$$

The 1D electron statistics can be written as

$$n_{1D} = \frac{g_v}{\pi\sqrt{B_0}} \sum_{n_x=1}^{n_{x\max}} \sum_{n_y=1}^{n_{y\max}} [t_7(E_{F1D}, n_x, n_y) + t_8(E_{F1D}, n_x, n_y)] \tag{49}$$

$$\text{where } t_7(E_{F1D}, n_x, n_y) \equiv [E_{F1D} - [G_{3,+}(n_x, n_y)]]^{1/2} + [E_{F1D} - [G_{3,-}(n_x, n_y)]]^{1/2}$$

The FNFE from Quantum Wires of Bismuth

The McClure and Choi Model

The dispersion relation of the carriers in Bi can be written, following the McClure and Choi [81], as

$$E(1+\alpha E)=\frac{p_x^{\ 2}}{2m_1}+\frac{p_y^{\ 2}}{2m_2}+\frac{p_z^{\ 2}}{2m_3}+\frac{p_y^{\ 2}}{2m_2}\alpha E\{1-(\frac{m_2}{m_2'})\}+\frac{p_y^{\ 4}\alpha}{4m_2m_2'}-\frac{\alpha p_x^{\ 2}p_y^{\ 2}}{4m_1m_2}-\frac{\alpha p_y^{\ 2}p_z^{\ 2}}{4m_2m_3} \tag{50}$$

where $p_i \equiv \hbar k_i$, $i=x,y,z$, m_1, m_2 and m_3 are the effective carrier masses at the band-edge along x, y and z directions respectively and m_2' is the effective-mass tensor component at the top of the valence band (for electrons) or at the bottom of the conduction band (for holes).

The 1D dispersion relation of the carriers in Bi in this case assumes the form

$$E(1+\alpha E)=\left\{\frac{\hbar^2k_x^2}{2m_1}\left[1-\frac{\alpha\hbar^2}{2m_2}\left(\frac{\pi n_y}{d_y}\right)^2\right]+G_{12}+\frac{\hbar^2}{2m_2}\alpha E\left\{1-\left(\frac{m_2}{m_2'}\right)\right\}\left(\frac{\pi n_y}{d_y}\right)^2\right\} \tag{51}$$

$$\text{where } G_{12}\equiv\left\{\frac{\hbar^2}{2m_2}\left(\frac{\pi n_y}{d_y}\right)^2+\frac{\hbar^2}{2m_3}\left(\frac{\pi n_z}{d_z}\right)^2+\frac{\alpha\hbar^4}{4m_2m_2'}\left(\frac{\pi n_y}{d_y}\right)^4-\frac{\alpha}{4m_2m_3}\left(\frac{\hbar^2n_yn_z\pi^2}{d_yd_z}\right)^2\right\}$$

Using (1.51), the 1D electron statistics can be expressed as

$$n_{1D}=\frac{2g_v}{\pi}\frac{\sqrt{2m_1}}{\hbar}\sum_{n_y=1}^{n_{y\max}}\sum_{n_z=1}^{n_{z\max}}[t_{27}(E_{F1D},n_y,n_z)+t_{28}(E_{F1D},n_y,n_z)] \tag{52}$$

where

$$t_{27}\left(E_{F1D},n_y,n_z\right)\equiv\left\{\left[1-\frac{\alpha\hbar^2}{2m_2}\left(\frac{\pi n_y}{d_y}\right)^2\right]^{-1/2}\left[E_{F1D}\left(1+\alpha E_{1DF}\right)-G_{12}-\frac{\hbar^2}{2m_2}\alpha E\left\{1-\left(\frac{m_2}{m_2'}\right)\right\}\left(\frac{\pi\hbar n_y}{d_y}\right)^2\right]^{1/2}\right\}$$

$$\text{and } t_{28}\left(E_{F1D},n_y,n_z\right)\equiv\sum_{r=1}^{s_0}Z_{1D}(r)\left[t_{27}\left(E_{F1D},n_y,n_z\right)\right]$$

The field emitted current is given by

$$I = \frac{2g_v e k_B T}{h} \sum_{n_z=1}^{n_{z\max}} \sum_{n_y=1}^{n_{y\max}} F_0(\eta_{17}) \exp(-\beta_{17}) \tag{53}$$

where $\eta_{17} = \dfrac{E_{F1D} - E_{19}}{k_B T}$

and E_{19} is the root of

$$E_{19}\left(1+\alpha E_{19}\right) = \left\{G_{12} + \frac{\hbar^2}{2m_2}\alpha E_{19}\left\{1-\left(\frac{m_2}{m_2'}\right)\right\}\left(\frac{\pi n_y}{d_y}\right)^2\right\} \tag{54}$$

$$\beta_{17} = \frac{4}{3}[A_{17}(V_0, n_z, n_y)]^{3/2} \cdot [eF_{sz} A_{17}'(V_{0,} n_z, n_y)]^{-1}$$

$$\left[1 - \frac{\alpha\hbar^2}{2m_2}\left(\frac{\pi n_y}{d_y}\right)^2\right]^{-1}\left[V_0\left(1+\alpha V_0\right) - G_{12} - \left\{\frac{\hbar^2}{2m_2}\alpha V_0\left\{1-\left(\frac{m_2}{m_2'}\right)\right\}\left(\frac{\pi n_y}{d_y}\right)^2\right\}\right]\left(\frac{\hbar^2}{2m_1}\right)^{-1} = A_{17}\left(V_{0,} n_{y,} n_z\right),$$

and

$$\left[1 - \frac{\alpha\hbar^2}{2m_2}\left(\frac{\pi n_y}{d_y}\right)^2\right]^{-1}\left[\left(1+2\alpha V_0\right) - \left\{\frac{\hbar^2}{2m_2}\alpha\left\{1-\left(\frac{m_2}{m_2'}\right)\right\}\left(\frac{\pi n_y}{d_y}\right)^2\right\}\right]\left(\frac{\hbar^2}{2m_1}\right)^{-1} = A_{17}'\left(V_{0,} n_{y,} n_z\right)$$

The Hybrid Model

The dispersion relation of the carriers in bulk specimens of Bi in accordance with the Hybrid model can be written as [80]

$$E\left(1+\alpha E\right) = \frac{\theta_0\left(E\right)\left(\hbar k_y\right)^2}{2M_2} + \frac{\alpha\gamma_0 \hbar^4 k_y^4}{4M_2^2} + \frac{\hbar^2 k_x^2}{2m_1} + \frac{\hbar^2 k_z^2}{2m_3} \tag{55}$$

$$\text{and } \theta_0(E) \equiv [1 + \alpha E(1-\gamma_0) + \bar{\delta}_0],\ \gamma_0 \equiv \frac{M_2}{m_2},\ \bar{\delta}_0 \equiv \frac{M_2}{M_2'}$$

and the other notations are defined in [80].

The 1D dispersion relation in this case assumes the form

$$E\left(1+\alpha E\right) = \frac{\hbar^2 k_x^2}{2m_1} + G_{14} + \frac{\hbar^2}{2M_2}\left(\frac{\pi n_y}{d_y}\right)^2 \alpha E\left(1-\gamma_0\right) \tag{56}$$

$$\text{where } G_{14} = \left[\frac{\hbar^2}{2m_3}\left(\frac{\pi n_z}{d_z}\right)^2 + \frac{\hbar^2}{2M_2}\left(\frac{\pi n_y}{d_y}\right)^2\left(1+\bar{\delta}_0\right) + \frac{\alpha\gamma_0\hbar^4}{4M_2^2}\left(\frac{\pi n_y}{d_y}\right)^4\right]$$

The use of (1.56) leads to the expression for the electron concentration per unit length as

$$n_{1D} = \frac{2g_v}{\pi}\frac{\sqrt{2m_1}}{\hbar}\sum_{n_y=1}^{n_{y\max}}\sum_{n_z=1}^{n_{z\max}}\left[t_{31}\left(E_{F1D},n_y,n_z\right)+t_{32}\left(E_{F1D},n_y,n_z\right)\right] \tag{57}$$

$$\text{where } t_{31}\left(E_{F1D},n_y,n_z\right) \equiv \left[E_{F1D}\left(1+\alpha E_{1DF}\right)-G_{14}-\frac{\hbar^2}{2M_2}\left(\frac{\pi n_y}{d_y}\right)^2 \alpha E_{F1D}\left(1-\gamma_0\right)\right]^{1/2}$$

$$\text{and } t_{32}\left(E_{F1D},n_y,n_z\right) \equiv \sum_{r=1}^{s_o} Z_{1D}(r)\left[t_{31}\left(E_{F1D},n_y,n_z\right)\right]$$

The field emitted current is given by

$$I = \frac{2g_v e k_B T}{h}\sum_{n_z=1}^{n_{z\max}}\sum_{n_y=1}^{n_{y\max}} F_0(\eta_{18})\exp(-\beta_{18}) \tag{58}$$

$$\text{where } \eta_{18} = \frac{E_{F1D}-E_{20}}{k_B T}$$

and E_{20} is the root of

$$E_{20}\left(1+\alpha E_{20}\right) = G_{14}+\frac{\hbar^2}{2M_2}\left(\frac{\pi n_y}{d_y}\right)^2 \alpha E_{20}\left(1-\gamma_0\right) \tag{59}$$

$$\beta_{18} = \frac{4}{3}[A_{18}(V_0,n_z,n_y)]^{3/2}\cdot[eF_{sz}A'_{18}(V_0,n_z,n_y)]^{-1}$$

$$\left[V_0\left(1+\alpha V_0\right)-G_{14}-\left\{\frac{\hbar^2}{2M_2}\alpha V_0\{1-\gamma_0\}\left(\frac{\pi n_y}{d_y}\right)^2\right\}\right]\left(\frac{\hbar^2}{2m_1}\right)^{-1} = A_{18}\left(V_0,n_y,n_z\right),$$

$$\text{and } \left[\left(1+2\alpha V_0\right)-\left\{\frac{\hbar^2}{2M_2}\alpha\{1-\gamma_0\}\left(\frac{\pi n_y}{d_y}\right)^2\right\}\right]\left(\frac{\hbar^2}{2m_1}\right)^{-1} = A_{18}'\left(V_0,n_y,n_z\right)$$

The Cohen Model

In accordance with the Cohen model [79], the dispersion law of the carriers in Bi is given by

$$E\left(1+\alpha E\right) = \frac{p_x^{\,2}}{2m_1}+\frac{p_z^{\,2}}{2m_3}-\frac{\alpha E p_y^{\,2}}{2m_2'}+\frac{p_y^{\,2}(1+\alpha E)}{2m_2}+\frac{\alpha p_y^{\,4}}{4m_2 m_2'} \tag{60}$$

The 1D carrier dispersion law in this case can be written as

$$\alpha E^2 + El_7 - G_{15} = \frac{\hbar^2 k_x^2}{2m_1} \tag{61}$$

$$where,\ l_7 = [1 - \frac{\alpha\hbar^2}{2m_2}(\frac{\pi n_y}{d_y})^2 + \frac{\alpha\hbar^2}{2m_2'}(\frac{\pi n_y}{d_y})^2]$$

$$and\ G_{15} = [\frac{\hbar^2}{2m_3}(\frac{\pi n_z}{d_z})^2 + \frac{\hbar^2}{2m_2}(\frac{\pi n_y}{d_y})^2 + \frac{\alpha\hbar^4}{4m_2 m_2'}(\frac{\pi n_y}{d_y})^4]$$

The 1D electron concentration per unit length assumes the form

$$n_{1D} = \frac{2g_v}{\pi}\frac{\sqrt{2m_1}}{\hbar}\sum_{n_y=1}^{n_{y_{\max}}}\sum_{n_z=1}^{n_{z_{\max}}}\left[t_{35}\left(E_{F1D}, n_y, n_z\right) + t_{36}\left(E_{F1D}, n_y, n_z\right)\right] \tag{62}$$

where

$$t_{35}\left(E_{F1D}, n_y, n_z\right) = \left[\alpha E_{F1D}^2 + E_{F1D} l_7 - G_{15}\right]^{1/2}$$

and

$$t_{36}\left(E_{F1D}, n_y, n_z\right) = \sum_{r=1}^{s_o} Z_{1D}(r)\left[t_{35}\left(E_{F1D}, n_y, n_z\right)\right].$$

The field emitted current is given by

$$I = \frac{2g_v e k_B T}{h}\sum_{n_z=1}^{n_{z_{\max}}}\sum_{n_y=1}^{n_{y_{\max}}} F_0(\eta_{19})\exp(-\beta_{19}) \tag{63}$$

$$where\ \eta_{19} = \frac{E_{F1D} - E_{21}}{k_B T}$$

E_{21} is the root of

$$E_{21}\left(l_7 + \alpha E_{21}\right) = G_{15} \tag{64}$$

$$\beta_{19} = \frac{4}{3}[A_{21}(V_0, n_z, n_y)]^{3/2} \cdot [eF_{sz} A_{21}'(V_{0,} n_z, n_y)]^{-1}, \left[V_0\left(l_7 + \alpha V_0\right) - G_{15}\right]\left(\frac{\hbar^2}{2m_1}\right)^{-1} = A_{21}\left(V_{0,} n_{y,} n_z\right)$$

$$and\ \left[\left(l_7 + 2\alpha V_0\right)\right]\left(\frac{\hbar^2}{2m_1}\right)^{-1} = A_{21}'\left(V_{0,} n_{y,} n_z\right)$$

The Lax Model

The electron energy spectra in bulk specimens of Bi in accordance with the Lax model can be written as [75]

$$E(1+\alpha E)=\frac{p_x^{\ 2}}{2m_1}+\frac{p_y^{\ 2}}{2m_2}+\frac{p_z^{\ 2}}{2m_3} \tag{65}$$

The 1D dispersion relation in this case can be expressed as

$$E(1+\alpha E)=\frac{\hbar^2 k_x^{\ 2}}{2m_1}+G_{16} \tag{66}$$

$$where\ G_{16}=\frac{\hbar^2}{2m_2}(\frac{\pi n_y}{d_y})^2+\frac{\hbar^2}{2m_3}(\frac{\pi n_z}{d_z})^2$$

The 1D electron statistics is given by

$$n_{1D}=\frac{2g_v}{\pi}\frac{\sqrt{2m_1}}{\hbar}\sum_{n_y=1}^{n_{y_{\max}}}\sum_{n_z=1}^{n_{z_{\max}}}[t_{37}(E_{F1D},n_y,n_z)+t_{38}(E_{F1D},n_y,n_z)] \tag{67}$$

$$where\ t_{37}(E_{F1D},n_y,n_z)=[E_{F1D}(1+\alpha E_{F1D})-G_{16}]^{1/2}$$

$$and\ t_{38}(E_{F1D},n_y,n_z)=\sum_{r=1}^{s_Q} Z_{1D}(r)[t_{37}(E_{F1D},n_y,n_z)]$$

The field emitted current in this case assumes the form

$$I=\frac{2g_v e k_B T}{h}\sum_{n_z=1}^{n_{z_{\max}}}\sum_{n_y=1}^{n_{y_{\max}}}F_0(\eta_{20})\exp(-\beta_{20}) \tag{68}$$

where, $\eta_{20}=\dfrac{E_{F1D}-E_{22}}{k_B T}$,

E_{22} is the root of

$$E_{22}(l_7+\alpha E_{22})=G_{16} \tag{69}$$
$$\beta_{20}=\frac{4}{3}[A_{22}(V_0,n_z,n_y)]^{3/2}\cdot[eF_{sz}A'_{22}(V_0)]^{-1}$$
$$[V_0(1+\alpha V_0)-G_{16}](\frac{\hbar^2}{2m_1})^{-1}=A_{22}(V_0,n_z,n_y)\ and\ [(1+2\alpha V_0)](\frac{\hbar^2}{2m_1})^{-1}=A_{22}'(V_0)$$

It may be noted that under the conditions $\alpha\to 0, M_2'\to\infty$ and isotropic effective electron mass at the edge of the conduction band, m all models of Bismuth convert into isotropic parabolic energy bands. Thus under the aforementioned conditions and the conversion of the summations over the quantum numbers to the corresponding integrations, all the equations for FNFE for Bismuth lead to the well known expression of the FNFE as given by (1.27).

The FNFE from Quantum Wires of IV-VI Semiconductors

The dispersion relation of the conduction electrons in IV-VI semiconductors can be expressed in accordance with Dimmock [110] as

$$[\bar{\varepsilon}-\frac{E_{g_0}}{2}-\frac{\hbar^2 k_s^{\ 2}}{2m_t^-}-\frac{\hbar^2 k_z^{\ 2}}{2m_l^-}][\bar{\varepsilon}+\frac{E_{g_0}}{2}+\frac{\hbar^2 k_s^{\ 2}}{2m_t^-}+\frac{\hbar^2 k_z^{\ 2}}{2m_l^-}]=P_\perp^{\ 2}k_s^{\ 2}+P_\parallel^2 k_z^{\ 2} \tag{70}$$

where $\bar{\varepsilon}$ is the energy as measured from the center of the band gap E_{g_0}, $m_t^{\pm}$ and $m_l^{\pm}$ represent the contributions to the transverse and longitudinal effective masses of the external L_6^+ and L_6^- bands arising from the $\vec{k}.\vec{p}$ perturbations with the other bands taken to the second order.

Using $\varepsilon=E+\left(E_{g_0}/2\right)$, $P_\perp^2=\frac{\hbar^2 E_{g_0}}{2m_t^*}$, $P_\parallel^2=\frac{\hbar^2 E_{g_0}}{2m_l^*}$ (m_t^* and m_l^* are the transverse and longitudinal effective electron masses at $k=0$) in (1.70), we can write

$$[E-\frac{\hbar^2 k_s^2}{2m_t^-}-\frac{\hbar^2 k_z^2}{2m_l^-}][1+\alpha E+\alpha\frac{\hbar^2 k_s^2}{2m_t^+}+\alpha\frac{\hbar^2 k_z^2}{2m_l^+}]=\frac{\hbar^2 k_s^2}{2m_t^*}+\frac{\hbar^2 k_z^2}{2m_l^*} \tag{71}$$

The 1D dispersion relation can be expressed from (71) as

$$k_z^2 = A_{23}(E,n_x,n_y) \tag{72}$$

$$where,\ A_{23}(E,n_x,n_y) = (2h_4)^{-1}[h_6(E,n_x,n_y) - [h_6^2(E,n_x,n_y) + 4h_4h_7(E,n_x,n_y)]^{1/2}],\ h_4 = [\frac{\alpha\hbar^4}{4x_3x_6}]$$

$$h_6(E,n_x,n_y) = [\frac{\alpha E\hbar^2}{2x_6} - \frac{\alpha\hbar^2}{2x_6}[(\frac{\pi n_x}{d_x})^2\frac{\hbar^2}{2x_1} + (\frac{\pi n_y}{d_y})^2\frac{\hbar^2}{2x_2}] -$$

$$x_3 = \frac{3m_t^- m_l^-}{2m_l^- + m_t^-},\ x_6 = \frac{3m_t^+ m_l^+}{2m_l^+ + m_t^+}$$

$$\frac{\alpha\hbar^2}{2x_3}[(\frac{\pi n_x}{d_x})^2\frac{\hbar^2}{2x_4} + (\frac{\pi n_y}{d_y})^2\frac{\hbar^2}{2x_5}] - \frac{\hbar^2}{2m_3} - \frac{(1+\alpha E)\hbar^2}{2x_3}],$$

$$x_2 = \frac{m_t^- + 2m_l^-}{3},\ x_4 = m_t^+,\ x_5 = \frac{m_t^+ + 2m_l^+}{3},\ m_3 = \frac{3m_t^* m_l^*}{m_t^* + 2m_l^*}$$

$$h_7(E,n_x,n_y) = [E(1+\alpha E) + \alpha E[(\frac{\pi n_x}{d_x})^2\frac{\hbar^2}{2x_4} + (\frac{\pi n_y}{d_y})^2\frac{\hbar^2}{2x_5}]$$

$$-(1+\alpha E)[(\frac{\pi n_x}{d_x})^2\frac{\hbar^2}{2x_1} + (\frac{\pi n_y}{d_y})^2\frac{\hbar^2}{2x_2}] - \alpha[(\frac{\pi n_x}{d_x})^2\frac{\hbar^2}{2x_1} + (\frac{\pi n_y}{d_y})^2\frac{\hbar^2}{2x_2}][(\frac{\pi n_x}{d_x})^2\frac{\hbar^2}{2x_4} + (\frac{\pi n_y}{d_y})^2\frac{\hbar^2}{2x_5}]$$

$$-[(\frac{\pi n_x}{d_x})^2\frac{\hbar^2}{2m_1} + (\frac{\pi n_y}{d_y})^2\frac{\hbar^2}{2m_2}]],,\ m_1 = m_t^*,\ and\ m_2 = \frac{m_t^* + 2m_l^*}{3}$$

The electron concentration is given by

$$n_0 = \frac{2g_v}{\pi}\sum_{n_x=1}^{n_{x\max}}\sum_{n_y=1}^{n_{y\max}}[B_{32}(E_{F1D},n_x,n_y) + B_{33}(E_{F1D},n_x,n_y)] \tag{73}$$

$$where\ B_{32}(E_{F1D},n_x,n_y) = [A_{23}(E_{F1D},n_x,n_y)]^{1/2}$$

$$and\ B_{33}(E_{F1D},n_x,n_y) = \sum_{r=1}^{r_0} Z_{1D}(r)[B_{32}(E_{F1D},n_x,n_y)]$$

The field emitted current can be written as

$$I = \frac{2g_v e k_B T}{h}\sum_{n_x=1}^{n_{x\max}}\sum_{n_y=1}^{n_{y\max}} F_0(\eta_{22})\exp(-\beta_{22}) \tag{74}$$

$$where,\ \eta_{22} = \frac{E_{F1D} - E_{22}}{k_B T}$$

The sub- band energy E_{22} assumes the form

$$E_{22} = (2\alpha)^{-1}[-\zeta_{70}(n_x,n_y) + [\zeta_{70}^2(n_x,n_y) + 4\alpha\zeta_{71}(n_x,n_y)]^{1/2}] \tag{75}$$

in which

$$\zeta_{70}(n_x,n_y) = [1 + \alpha[(\frac{\pi n_x}{d_x})^2 \frac{\hbar^2}{2x_4} + (\frac{\pi n_y}{d_y})^2 \frac{\hbar^2}{2x_5}] - \alpha[(\frac{\pi n_x}{d_x})^2 \frac{\hbar^2}{2x_1} + (\frac{\pi n_y}{d_y})^2 \frac{\hbar^2}{2x_2}]]$$

$$\zeta_{71}(n_x,n_y) = [[(\frac{\pi n_x}{d_x})^2 \frac{h^2}{2x_1} + (\frac{\pi n_y}{d_y})^2 \frac{h^2}{2x_2}] + \alpha[(\frac{\pi n_x}{d_x})^2 \frac{h^2}{2x_1} + (\frac{\pi n_y}{d_y})^2 \frac{h^2}{2x_2}][(\frac{\pi n_x}{d_x})^2 \frac{h^2}{2x_4} + (\frac{\pi n_y}{d_y})^2 \frac{h^2}{2x_5}]$$
$$+[(\frac{\pi n_x}{d_x})^2 \frac{\hbar^2}{2m_1} + (\frac{\pi n_y}{d_y})^2 \frac{\hbar^2}{2m_2}]],$$

$$\beta_{22} = \frac{4}{3}[A_{23}(V_0,n_x,n_y)]^{3/2} \cdot [eF_{sz} A'_{23}(V_0,n_x,n_y)]^{-1},$$

$$A'_{23}(V_0,n_x,n_y) = (2h_4)^{-1}[h'_6 - \frac{h_6(V_0,n_x,n_y)h'_6 + 2h_4 h'_7(V_0,n_x,n_y)}{[h_6^2(V_0,n_x,n_y) + 4h_4 h_7(V_0,n_x,n_y)]^{1/2}}], h'_6 = \frac{\alpha\hbar^2}{2}(\frac{1}{x_6} - \frac{1}{x_3})$$

and

$$h'_7(V_0,n_x,n_y) = [1 + 2\alpha V_0 + \alpha[\frac{h^2}{2x_4}(\frac{\pi n_x}{d_x})^2 + (\frac{\pi n_y}{d_y})^2 \frac{h^2}{2x_5}] - \alpha[\frac{h^2}{2x_1}(\frac{\pi n_y}{d_y})^2 + (\frac{\pi n_y}{d_y})^2 \frac{h^2}{2x_2}]]$$

The FNFE from Quantum Wires of Stressed Semiconductors

The dispersion relation of the conduction electrons in bulk specimens of stressed semiconductors can be written as [111]

$$\frac{k_x^2}{[a^*(E)]^2} + \frac{k_y^2}{[b^*(E)]^2} + \frac{k_z^2}{[c^*(E)]^2} = 1 \tag{76}$$

$$where\ [a^*(E)]^2 = \frac{K_0(E)}{M_0(E) + \frac{1}{2}N_0(E)}, K_0(E) = [E - C_1\varepsilon - \frac{2C_2^2\varepsilon_{xy}^2}{3E'_{g_0}(E)}](\frac{3E'_{g_0}(E)}{2B_2^2})$$

C_1 is the conduction band deformation potential constant, ε is the trace of the strain tensor $\hat{\varepsilon}$ which can be written as $\hat{\varepsilon} = \begin{bmatrix} \varepsilon_{xx} & \varepsilon_{xy} & 0 \\ \varepsilon_{xy} & \varepsilon_{yy} & 0 \\ 0 & 0 & \varepsilon_{zz} \end{bmatrix}$, C_2 is a constant which describes the strain interaction between the conduction and valance bands, $E'_{g_0}(E) = E_{g_0} + E - C_1\varepsilon$, E_{g_0} is the band gap in the absence of stress, B_2 is the momentum matrix element,

$$M_0(E) = [1 - \frac{(\bar{a}_0 + C_1)\varepsilon}{E'_{g_0}(E)} + \frac{3\bar{b}_0\varepsilon_{xx}}{2E'_{g_0}(E)} - \frac{\bar{b}_0\varepsilon}{2E'_{g_0}(E)}], \bar{a}_0 = -\frac{1}{3}(\bar{b}_0 + 2\bar{m}), \bar{b}_0 = \frac{1}{3}(\bar{l} - \bar{m}), \bar{l}, \bar{m}, \bar{n}$$

are the matrix elements of the strain perturbation operator,

$$N_0(E) = (\bar{d}_0\sqrt{3})\frac{\varepsilon_{xy}}{E'_{g_0}(E)}$$

$$[b^*(E)]^2 = \frac{K_0(E)}{M_0(E) - \frac{1}{2}N_0(E)} [c^*(E)]^2 = \frac{K_0(E)}{L_0(E)}$$

$$and\ L_0(E) = [1 - \frac{(\bar{a}_0 + C_1)\varepsilon}{E'_{g_0}(E)} + \frac{3\bar{b}_0\varepsilon_{zz}}{2E'_{g_0}(E)} - \frac{\bar{b}_0\varepsilon}{2E'_{g_0}(E)}],$$

The 1D dispersion relation of the carriers in stressed materials in this case can be written as

$$k_z^2 = A_{24}(E, n_x, n_y) \tag{77}$$

$$where\ A_{24}(E, n_x, n_y) = [c^*(E)]^2[1 - (\frac{\pi n_x}{d_x})^2[a^*(E)]^{-2} - (\frac{\pi n_y}{d_y})^2[b^*(E)]^{-2}]$$

The sub-band energy E_{23} assumes the form

$$[(\frac{\pi n_x}{d_x})^2[a^*(E)]^{-2} + (\frac{\pi n_y}{d_y})^2[b^*(E)]^{-2}] = 1 \tag{78}$$

Using (1.77), the 1D electron statistics can be expressed as

$$n_{1D} = \frac{2g_v}{\pi}\sum_{n_y=1}^{n_{y\max}}\sum_{n_z=1}^{n_{z\max}}\left[B_{34}\left(E_{F1D}, n_y, n_z\right) + B_{35}\left(E_{F1D}, n_y, n_z\right)\right] \tag{79}$$

$$\text{where } B_{34}\left(E_{F1D}, n_y, n_z\right) = \sqrt{A_{24}(E_{FID}, n_x, n_y)}$$

$$and\ B_{35}\left(E_{F1D}, n_x, n_y\right) = \sum_{r=1}^{s_o} Z_{1D}(r)\left[B_{35}\left(E_{F1D}, n_x, n_y\right)\right]$$

The field emitted current is given by

$$I=\frac{2g_v e k_B T}{h}\sum_{n_x=1}^{n_{x\max}}\sum_{n_y=1}^{n_{y\max}}F_0(\eta_{23})\exp(-\beta_{23}) \tag{80}$$

where $\eta_{23}=\dfrac{E_{F1D}-E_{23}}{k_B T}$, $\beta_{23}=\dfrac{4}{3}[A_{24}(V_0,n_x,n_y)]^{3/2}\cdot[eF_{sz}A'_{24}(V_0,n_x,n_y)]^{-1}$,

$$A_{24}'(V_0,n_x,n_y)=[\frac{K_0'(V_0)}{L_0(V_0)}-\frac{K_0(V_0)L_0'(V_0)}{L_0^2(V_0)}-(\frac{n_x\pi}{d_x})^2\frac{[M_0'(V_0)+\frac{1}{2}N_0'(V_0)]}{L_0(V_0)}+$$
$$(\frac{n_x\pi}{d_x})^2\cdot\frac{L_0'(V_0)}{L_0^2(V_0)}[M_0(V_0)+\frac{1}{2}N_0(V_0)]+(\frac{n_y\pi}{d_y})^2\cdot\frac{L_0'(V_0)}{L_0^2(V_0)}[M_0(V_0)-\frac{1}{2}N_0(V_0)]$$
$$-(\frac{n_y\pi}{d_y})^2\cdot\frac{1}{L_0(V_0)}[M_0'(V_0)-\frac{1}{2}N_0'(V_0)]],$$

$$K_0'(V_0)=[[1+\frac{2C_2^2\varepsilon_{xy}^2}{3\{E'_{g_0}(V_0)\}^2}](\frac{3E'_{g_0}(V_0)}{2B_2^2})+[V_0-C_1\varepsilon-\frac{2C_2^2\varepsilon_{xy}^2}{3E'_{g_0}(E)}](\frac{3}{2B_2^2})],$$

$$M_0'(V_0)=[\frac{(\bar{a}_0+C_1)\varepsilon}{(E'_{g_0}(V_0))^2}-\frac{3\bar{b}_0\varepsilon_{xx}}{2(E'_{g_0}(V_0))^2}+\frac{\bar{b}_0\varepsilon}{2(E'_{g_0}(V_0))^2}]$$

and

$$N_0'(V_0)=-(\bar{d}_0\sqrt{3})\frac{\varepsilon_{xy}}{(E'_{g_0}(V_0))^2}$$

The FNFE from Quantum Wires of Tellurium

The dispersion relation of the conduction electrons in Te can be expressed as [112]

$$E=\psi_1k_z^2+\psi_2k_s^2\pm[\psi_3^2k_z^2+\psi_4^2k_s^2]^{1/2} \tag{81}$$

where, $\psi_1=6.7\times10^{-16}\,mev.m^2$, $\psi_2=4.2\times10^{-16}\,mev.m^2$, $\psi_3=6\times10^{-8}\,mev.m$ and $\psi_4=3.6\times10^{-8}\,mev.m$

From (1.81), the1-D dispersion relation can be written as

$$k_z^2=A_{25,\pm}(E,n_x,n_y) \tag{82}$$

where

$$A_{25,\pm}(E,n_x,n_y)=[(\psi_5(E)-\psi_6k_s^2\pm\psi_7[\psi_8(E)-k_s^2]^{1/2}],\ k_s^2=\phi_1(n_x,n_y),\ \psi_5(E)=[\frac{E}{\psi_1}+\frac{E}{2\psi_1^2}],$$

$$\psi_6=\frac{\psi_2}{\psi_1},\ \psi_7=(2\psi_1^2)^{-1}[4\psi_3^2\psi_1\psi_2-4\psi_1^2\psi_4^2]^{1/2}\ and\ \psi_8(E)=[\frac{\psi_4^4+4E\psi_3^2\psi_1}{4\psi_3^2\psi_1\psi_2-4\psi_1^2\psi_4^2}]$$

The sub-band energies are given by

$$E_{26,\pm}=\psi_2\phi_1(n_x,n_y)\pm\psi_4(\phi_1(n_x,n_y))^{1/2} \tag{83}$$

The electron concentration per unit length can be expressed as

$$n_0 = \frac{g_v}{\pi} \sum_{n_x=1}^{n_{x\max}} \sum_{n_y=1}^{n_{y\max}} [B_{36,\pm}(E_{F1D}, n_x, n_y) + B_{37,\pm}(E_{F1D}, n_x, n_y)] \tag{84}$$

$$where\ B_{36,\pm}(E_{F1D}, n_x, n_y) = \sqrt{A_{25,+}(E_{F1D}, n_x, n_y)} + \sqrt{A_{25,-}(E_{F1D}, n_x, n_y)}$$

$$and\ B_{37,\pm}(E_{F1D}, n_x, n_y) = \sum_{r=1}^{s_o} Z_{1D}(r)[B_{36,+}(E_{F1D}, n_x, n_y) + B_{36,-}(E_{F1D}, n_x, n_y)]$$

The field emitted current assumes the form +

$$I = \frac{g_v e k_B T}{h} \sum_{n_x=1}^{n_{x\max}} \sum_{n_y=1}^{n_{y\max}} [F_0(\eta_{24,+})\exp(-\beta_{24,+}) + F_0(\eta_{24,-})\exp(-\beta_{24,-})] \tag{85}$$

$$where\ \eta_{24,\pm} = \frac{E_{F1D} - E_{26,\pm}}{k_B T},,\beta_{24,\pm} = \frac{4}{3}[A_{25,\pm}(V_0, n_x, n_y)]^{3/2} \cdot [eF_{sz} A'_{25,\pm}(V_0, n_x, n_y)]^{-1},$$

$$A_{25,\pm}(V_0, n_x, n_y) = [\frac{1}{\psi_1} \pm \frac{\psi_7}{2}[\psi_8(V_0) - \phi_1(n_x, n_y)]^{-1/2} \psi'_8(V_0)] and\ \psi'_8(V_0) = [\frac{4\psi_3^2\psi_1}{4\psi_3^2\psi_1\psi_2 - 4\psi_1^2\psi_4^2}]$$

The FNFE from Quantum Wires of Gallium Phosphide

The energy spectrum of the conduction electrons in n-GaP can be written as [113]

$$E = \frac{\hbar^2 k_s^2}{2m_\perp^*} + \frac{\hbar^2}{2m_\parallel^*}[\bar{A}' k_s^2 + k_z^2] - [\frac{\hbar^4 k_0^2}{m_\parallel^{*2}}(k_s^2 + k_z^2) + |V_G|^2]^{1/2} + |V_G| \tag{86}$$

where, k_0 and $|V_G|$ are constants of the energy spectrum and $\bar{A}' = 1$. The 1D dispersion relations assumes the form

$$k_z^2 = A_{26}(E, n_x, n_y) \tag{87}$$

where,

$$A_{26}(E, n_x, n_y) = [t_1 E + t_2 - t_3\phi_1(n_x, n_y) - [t_4\phi_1(n_x, n_y) + t_6 E + t_5]^{1/2}], t_1 = (b_0)^{-1}, b_0 = (\hbar^2 / 2m_\parallel^*),$$

$$t_2 = [-2b_0 D_0 + C][2b_0^2]^{-1}, D_0 = |V_g|, C = (\hbar^2 k_0 / m_\parallel^*)^2, t_3 = (a_0 / b_0), a_0 = [(\hbar^2 / 2m_\perp^*) + (A\hbar^2 / 2m_\parallel^*)],$$

$$t_4 = (\bar{a}_1 / (4b_0^4)), \bar{a} = 4b_0 C(b_0 - C), t_6 = (\bar{a}_3 / (4b_0^4)), \bar{a}_3 = 4Cb_0,, t_5 = (\bar{a}_2 / (4b_0^4))$$

$$and\ \bar{a}_2 = [C^2 + 4b_0^2 D_0^2 - 4b_0 C D_0]$$

The sub-band energy E_{27} can be written as

$$E_{27} = a_0\phi_1(n_x, n_y) - (C\phi_1(n_x, n_y) + D_0^2)^{1/2} + D_0 \tag{88}$$

The electron concentration per unit length can be expressed as

$$n_0 = \frac{2g_v}{\pi}\sum_{n_x=1}^{n_{x\max}}\sum_{n_y=1}^{n_{y\max}}[B_{38}(E_{F1D}, n_x, n_y) + B_{39}(E_{F1D}, n_x, n_y)] \tag{89}$$

$$where,\ B_{38}(E_{F1D}, n_x, n_y) = \sqrt{A_{26}(E_{FID}, n_x, n_y)}$$

$$and\ B_{39}(E_{F1D}, n_x, n_y) = \sum_{r=1}^{s_o} Z_{1D}(r)[B_{38}(E_{F1D}, n_x, n_y)]$$

The field emitted current assumes the form

$$I = \frac{2g_v e k_B T}{h}\sum_{n_x=1}^{n_{x\max}}\sum_{n_y=1}^{n_{y\max}}[F_0(\eta_{26})\exp(-\beta_{26})] \tag{90}$$

$$where\ \eta_{26} = \frac{E_{F1D} - E_{27}}{k_B T},, \beta_{26} = \frac{4}{3}[A_{26}(V_0, n_x, n_y)]^{3/2} \cdot [eF_{sz}A'_{26}(V_0, n_x, n_y)]^{-1}$$

$$and\ A'_{26}(V_0, n_x, n_y) = [t_1 - \frac{1}{2}t_6[t_4\phi_1(n_x, n_y) + t_6V_0 + t_5]^{-1/2}]$$

The FNFE from Quantum Wires of Platinum Antimonide

The dispersion relation for n-$PtSb_2$ is given by [114]

$$[E + \overline{\lambda_0}\frac{(\bar{a})^2}{4}k^2 - \bar{l}\frac{(\bar{a})^2}{4}k_s^2][E + \overline{\delta_0} - \bar{\nu}\frac{(\bar{a})^2}{4}k^2 - \bar{n}\frac{(\bar{a})^2}{4}k_s^2] = \frac{I(\bar{a})^4}{16}k^4 \tag{91}$$

where $\overline{\lambda_0}, \bar{a}, \bar{l}, \overline{\delta_0}, \bar{\nu}, \bar{n}$ and I are system constants.

The 1D dispersion relations can be written as

$$k_x^2 = [2\overline{A_9}]^{-1}[-\overline{A_{10}}(E, n_z) + [\overline{A_{10}}^2(E, n_z) + 4(\overline{A_9})\overline{A_{11}}(E, n_z)]]^{1/2} - (\frac{n_y\pi}{d_Y})^2 \tag{92}$$

$$where\ \overline{A_9} = (I_1 + \omega_1\omega_3), I_1 = I\frac{(\bar{a})^4}{16}$$

$$\overline{A_{10}}(E, n_z) = [\omega_3 E + \omega_1[E + \overline{\delta_0} - \omega_4(\frac{n_z\pi}{d_z})^2] + \omega_2\omega_3(\frac{n_z\pi}{d_z})^2 + 2I_1(\frac{n_z\pi}{d_z})^2],$$

$$\omega_1 = \frac{(\bar{a})^2}{4}[(\overline{\lambda_0}) - \bar{l}], \omega_2 = (\overline{\lambda_0}\frac{(\bar{a})^2}{4}), \omega_3 = \frac{(\bar{a})^2}{4}(\bar{n} + \bar{\nu}), \omega_4 = \frac{(\bar{a})^2}{4}\bar{\nu}$$

and

$$\overline{A_{11}}(E,n_z)=[E[E+\overline{\delta_0}-\omega_4(\frac{n_z\pi}{d_z})^2]+\omega_2(\frac{n_z\pi}{d_z})^2[E+\overline{\delta_0}-\omega_4(\frac{n_z\pi}{d_z})^2]-I_1(\frac{n_z\pi}{d_z})^4]$$

The (92) can be expressed as

$$k_x^2 = A_{27}\left(E,n_y,n_z\right) \tag{93}$$

$$\text{where, } A_{27}\left(E,n_y,n_z\right)=\left[2\overline{A_9}\right]^{-1}\left[-\overline{A_{10}}\left(E,n_z\right)+\left[\overline{A_{10}}^2\left(E,n_z\right)+4\left(\overline{A_9}\right)\overline{A_{11}}\left(E,n_z\right)\right]\right]^{1/2}-\left(\frac{n_y\pi}{d_Y}\right)^2$$

The electron concentration per unit length can be written as

$$n_0=\frac{2g_v}{\pi}\sum_{n_x=1}^{n_{x\max}}\sum_{n_y=1}^{n_{y\max}}[B_{40}(E_{F1D},n_x,n_y)+B_{41}(E_{F1D},n_x,n_y)] \tag{94}$$

$$\text{where, } B_{40}(E_{F1D},n_x,n_y)=\sqrt{A_{27}(E_{FID},n_x,n_y)}$$

$$\text{and } B_{41}(E_{F1D},n_x,n_y)=\sum_{r=1}^{s_o}Z_{1D}(r)[B_{40}(E_{F1D},n_x,n_y)]$$

The field emitted current assumes the form

$$I=\frac{2g_v e k_B T}{h}\sum_{n_x=1}^{n_{x\max}}\sum_{n_y=1}^{n_{y\max}}[F_0(\eta_{27})\exp(-\beta_{27})] \tag{95}$$

$\eta_{27}=\dfrac{E_{F1D}-E_{28}}{k_BT}$, E_{28} is the root of the equation

$$[2\overline{A_9}]^{-1}[-\overline{A_{10}}(E_{28},n_z)+[\overline{A_{10}}^2(E_{28},n_z)+4(\overline{A_9})\overline{A_{11}}(E_{28},n_z)]]^{1/2}-(\frac{n_y\pi}{d_Y})^2=0 \tag{96}$$

$$\beta_{27}=\frac{4}{3}[A_{27}(V_0,n_x,n_y)]^{3/2}\cdot[eF_{sz}A_{27}'(V_0,n_x,n_y)]^{-1}$$

$$A_{27}'(E,n_y,n_z)=[2\overline{A_9}]^{-1}[-(\overline{A_{10}})'+[\overline{A_{10}}^2(V_0,n_z)+4(\overline{A_9})\overline{A_{11}}(V_0,n_z)]^{-1/2}.$$

$$.[\overline{A_{10}}(V_0,n_z)(\overline{A_{10}})'+2(\overline{A_9})(\overline{A_{11}}(V_0,n_z))']],\ A_{10}'=(\omega_1+\omega_3)$$

and

$$(\overline{A_{11}}(V_0,n_z))'=[2V_0+\overline{\delta_0}-\omega_4(\frac{n_z\pi}{d_z})^2+\omega_2(\frac{n_z\pi}{d_z})^2]$$

The FNFE from Quantum Wires of Bismuth Telluride

The dispersion relation of the conduction electron in Bi_2Te_3 can be written as [115]

$$E(1+\alpha E)=\bar{\omega}_1 k_x^{\;2}+\bar{\omega}_2 k_y^{\;2}+\bar{\omega}_3 k_z^{\;2}+\bar{\omega}_4 k_x k_y \tag{97}$$

where $\bar{\omega}_1=\frac{\hbar^2}{2m_0}\bar{\alpha}_{11},\bar{\omega}_2=\frac{\hbar^2}{2m_0}\bar{\alpha}_{22},\bar{\omega}_3=\frac{\hbar^2}{2m_0}\bar{\alpha}_{33},\bar{\omega}_4=\frac{\hbar^2}{2m_0}\bar{\alpha}_{23}$ in which $\bar{\alpha}_{11},\bar{\alpha}_{22},\bar{\alpha}_{33}$ and $\bar{\alpha}_{23}$ are system constants.

The 1D electron energy spectrum assumes the form

$$k_x^{\;2}=A_{28}(E,n_y,n_z) \tag{98}$$

$$where\ A_{28}(E,n_y,n_z)=[E(1+\alpha E)-\bar{\omega}_2(\frac{n_y\pi}{d_y})^2-\bar{\omega}_3(\frac{n_z\pi}{d_z})^2-\bar{\omega}_4(\frac{n_y\pi}{d_y})(\frac{n_z\pi}{d_z})](\bar{\omega}_1)^{-1}$$

The subband energy (E_{30}) can be expressed as

$$E_{30}=(2\alpha)^{-1}[-1+\sqrt{1+4\alpha\theta_{32}(n_y,n_z)}] \tag{99}$$

$$where,\ \theta_{32}(n_y,n_z)=[\bar{\omega}_2(\frac{n_y\pi}{d_y})^2+\bar{\omega}_3(\frac{n_z\pi}{d_z})^2+\bar{\omega}_4(\frac{n_y\pi}{d_y})(\frac{n_z\pi}{d_z})]$$

The electron concentration per unit length is given by

$$n_0=\frac{2g_v}{\pi}\sum_{n_z=1}^{n_{z\max}}\sum_{n_y=1}^{n_{y\max}}[B_{42}(E_{F1D},n_z,n_y)+B_{43}(E_{F1D},n_z,n_y)] \tag{100}$$

$$where,\ B_{42}(E_{F1D},n_z,n_y)=\sqrt{A_{28}(E_{F1D},n_y,n_z)}$$

$$and\ B_{43}(E_{F1D},n_z,n_y)=\sum_{r=1}^{s_o}Z_{1D}(r)[B_{42}(E_{F1D},n_z,n_y)]$$

The field emitted current in this case can be written as

$$I = \frac{2g_v e k_B T}{h} \sum_{n_z=1}^{n_{z\max}} \sum_{n_y=1}^{n_{y\max}} [F_0(\eta_{28}) \exp(-\beta_{28})] \quad (101)$$

$$\eta_{28} = \frac{E_{F1D} - E_{30}}{k_B T}, \quad \beta_{28} = \frac{4}{3}[A_{28}(V_0, n_y, n_z)]^{3/2} \cdot [eF_{sz} A'_{28}(V_0)]^{-1}$$

$$\text{and } A_{28}'(V_0) = (\bar{\omega}_1)^{-1}[1 + 2\alpha V_0]$$

The FNFE from Quantum Wires of Germanium

It is well known that the conduction electrons of n-Ge obey two different types of dispersion laws since band non-parabolicity has been included in two different ways as given in the literature [116-117].

a) The energy spectrum of the conduction electrons in bulk specimens of n-Ge can be expressed in accordance with Cardona et al. [116] as

$$E = -\frac{E_{g_0}}{2} + \frac{\hbar^2 k_z^2}{2m_\parallel^*} + [\frac{E_{g_0}^2}{4} + E_{g_0} k_s^2 (\frac{\hbar^2}{2m_\perp^*})]^{1/2} \quad (102)$$

where in this case $m_\parallel^*$ and $m_\perp^*$ are the longitudinal and transverse effective masses along <111> direction at the edge of the conduction band respectively

The 1D electron energy spectrum assumes the form

$$k_y^2 = A_{29}(E, n_x, n_z) \quad (103)$$

$$\text{where } A_{29}(E, n_x, n_z) = [[\gamma_{15}(E, n_z) - (\frac{\hbar^2}{2m_1^*})(\frac{n_x \pi}{d_x})^2](2m_2^* / \hbar^2)]$$

$$\gamma_{15}(E, n_z) = [E(1+\alpha E) - (1+2\alpha E)(\frac{\hbar^2}{2m_3^*})(\frac{n_z \pi}{d_z})^2 + \alpha[(\frac{\hbar^2}{2m_3^*})(\frac{n_z \pi}{d_z})^2]^2],$$

$$m_1^* = m_\perp^*, \; m_2^* = \frac{m_\perp^* + 2m_\parallel^*}{3} \text{ and } m_3^* = \frac{3m_\perp^* m_\parallel^*}{m_\perp^* + 2m_\parallel^*}$$

The quantized energy levels (E_{31}) can be expressed through the equation

$$E_{31} = (2\alpha)^{-1}[-\rho_{91}(n_z) + \sqrt{\rho_{91}(n_z)^2 - 4\alpha\rho_{92}(n_z)}] \tag{104}$$
$$where\ \rho_{91}(n_z) = [1 - 2\alpha \frac{\hbar^2}{2m_3^*}(\frac{n_z\pi}{d_z})^2]$$
$$and\ \ \rho_{92}(n_z) = [\frac{\hbar^2}{2m_3^*}(\frac{n_z\pi}{d_z})^2 - \alpha[\frac{\hbar^2}{2m_3^*}(\frac{n_z\pi}{d_z})^2]^2]$$

The electron concentration per unit length is given by

$$n_0 = \frac{2g_v}{\pi}\sum_{n_x=1}^{n_{x_{\max}}}\sum_{n_z=1}^{n_{z_{\max}}}[B_{44}(E_{F1D}, n_x, n_z) + B_{45}(E_{F1D}, n_x, n_z)] \tag{105}$$

where

$$B_{44}(E_{F1D}, n_x, n_z) = \sqrt{A_{29}(E_{FID}, n_x, n_z)}$$

and

$$B_{45}(E_{F1D}, n_x, n_z) = \sum_{r=1}^{s_o} Z_{1D}(r)[B_{44}(E_{F1D}, n_x, n_z)]$$

The field emitted current assumes the form

$$I = \frac{2g_v e k_B T}{h}\sum_{n_x=1}^{n_{x_{\max}}}\sum_{n_z=1}^{n_{z_{\max}}}[F_0(\eta_{29})\exp(-\beta_{29})] \tag{106}$$
$$where,\ \eta_{29} = \frac{E_{F1D} - E_{31}}{k_B T},\ \beta_{29} = \frac{4}{3}[A_{29}(V_0, n_y, n_z)]^{3/2} \cdot [eF_{sz} A'_{29}(V_0)]^{-1}$$
$$and\ A'_{29}(V_0, n_x, n_z) = [[1 + 2\alpha V_0 - (\alpha\frac{\hbar^2}{m^*_3})(\frac{n_z\pi}{d_z})^2](2m^*_2 / \hbar^2)]$$

(b) The dispersion relation of the conduction electron in bulk specimens of n-Ge can be expressed in accordance with the model of Wang and Ressler [117] can be written as

$$E = \frac{\hbar^2 k_z^2}{2m_{\parallel}^*} + \frac{\hbar^2 k_s^2}{2m_{\perp}^*} - \bar{c}_1(\frac{\hbar^2 k_s^2}{2m_{\perp}^*})^2 - \bar{d}_1(\frac{\hbar^2 k_s^2}{2m_{\perp}^*})(\frac{\hbar^2 k_z^2}{2m_{\parallel}^*}) - \bar{e}_1(\frac{\hbar^2 k_z^2}{2m_{\parallel}^*})^2 \quad (107)$$

where $\bar{c}_1 = \bar{C}(2m_{\perp}^*/\hbar^2)^2, \bar{C} = 1.4\bar{A}, \bar{A} = \frac{1}{4}(\hbar^4/E_{g_0} m_{\perp}^{*2})(1 - \frac{m_{\perp}^*}{m_0})^2, \bar{d}_1 = \bar{d}(\frac{4m_{\perp}^* m_{\parallel}^*}{\hbar^4}), \bar{d} = 0.8\bar{A},$

$\bar{e}_1 = \bar{e}_0(2m_{\parallel}^*/\hbar^2)^2$ *and* $\bar{e}_0 = 0.005\bar{A}$

The 1D electron energy spectrum assumes the form

$$k_y^2 = A_{30}(E, n_x, n_z) \quad (108)$$

where $A_{30}(E, n_x, n_z) = [[I_{29}(E, n_z) - (\frac{\hbar^2}{2m_1^*})(\frac{n_x \pi}{d_x})^2](2m_2^* / \hbar^2)]$

$I_{29}(E, n_z) = [2\overline{C_1}]^{-1}[\overline{A_6}(n_z) + [\overline{A_6}^2(n_z) - 4\bar{C}_1 E + 4(\overline{C_1})\overline{A_5}(n_z)]^{\frac{1}{2}}]$

$\overline{A_5}(n_z) = (\frac{\hbar^2}{2m_3^*})(\frac{n_z \pi}{d_z})^2[1 - \bar{e}_1(\frac{\hbar^2}{2m_3^*})(\frac{n_z \pi}{d_z})^2]\sum_{i=1}^{n} X_i^2$ *and* $\overline{A_6}(n_z) = [1 - \bar{d}_1(\frac{\hbar^2}{2m_3^*})(\frac{n_z \pi}{d_z})^2]$

The quantized energy levels (E_{32}) can be expressed through the equation

$$E_{32} = \bar{A}_5(n_z) + (\frac{1}{4\bar{C}_1})[[\frac{\bar{C}_1 \hbar^2}{m_1^*}(\frac{n_x \pi}{d_x})^2]^2 - 2\bar{A}_6(n_z)\frac{\bar{C}_1 \hbar^2}{m_1^*}(\frac{n_x \pi}{d_x})^2] \quad (109)$$

The electron concentration per unit length is given by

$$n_0 = \frac{2g_v}{\pi} \sum_{n_x=1}^{n_{x_{\max}}} \sum_{n_z=1}^{n_{z_{\max}}} [B_{46}(E_{F1D}, n_x, n_z) + B_{47}(E_{F1D}, n_x, n_z)]$$

where, $B_{46}(E_{F1D}, n_x, n_z) = \sqrt{A_{30}(E_{FID}, n_x, n_z)}$ (110)

and $B_{47}(E_{F1D}, n_x, n_z) = \sum_{r=1}^{s_o} Z_{1D}(r)[B_{46}(E_{F1D}, n_x, n_z)]$

The field emitted current assumes the form

$$I = \frac{2g_v e k_B T}{h} \sum_{n_x=1}^{n_{x\max}} \sum_{n_z=1}^{n_{z\max}} [F_0(\eta_{30}) \exp(-\beta_{30})] \quad (111)$$

$$where\ \eta_{30} = \frac{E_{F1D} - E_{32}}{k_B T}, , \beta_{30} = \frac{4}{3}[A_{30}(V_0, n_y, n_z)]^{3/2} \cdot [eF_{sz} A'_{30}(V_0)]^{-1}, A'_{30}(V_0, n_x, n_z) = (2m^*_2 / \hbar^2) I'_{29}(V_0, n_z)$$

$$and\ I'_{29}(V_0, n_z) = [[\overline{A_6}^2(n_z) - 4\overline{C}_1 V_0 + 4(\overline{C}_1)\overline{A}_5(n_z)]^{-1/2}]$$

The FNFE from Quantum Wires of Gallium Antimonide

The dispersion relation of the conduction electrons in n-GaSb can be written as [118]

$$E = \frac{\hbar^2 k^2}{2m_0} - \frac{\overline{E}'_{go}}{2} + \frac{\overline{E}'_{go}}{2}[1 + \frac{2\hbar^2 k^2}{\overline{E}'_{go}}(\frac{1}{m_c} - \frac{1}{m_0})]^{\frac{1}{2}} \quad (112)$$

$$where\ \overline{E}'_{go} = [E_{go} + \frac{5.10^{-5} T^2}{2(112 + T)}]\ \ eV$$

The (112) can be expressed as

$$\frac{\hbar^2 k^2}{2m_c} = I_{16}(E) \quad (113)$$

where,

$$I_{16}(E) = [E + \overline{E}'_{g0} - (m_c / m_0)(\overline{E}'_{g0} / 2) - [(\overline{E}'_{g0} / 2)^2 +$$

$$[((\overline{E}'_{g0})^2 / 2)(1 - (m_c / m_0))] + [(\overline{E}'_{g0} / 2)(1 - (m_c / m_0))]^2 + E\overline{E}'_{g0}(1 - (m_c / m_0))]^{1/2}]$$

The 1D electron energy spectrum assumes the for

$$k_z^2 = A_{31}(E, n_x, n_y) \quad (114)$$

$$where\ A_{31}(E, n_x, n_y) = [[I_{16}(E) - \phi_1(n_x, n_y)](2m_c / \hbar^2)]$$

The quantized energy levels (E_{33}) can be expressed through the equation

$$I_{16}(E_{33}) = (\frac{\hbar^2}{2m_c})\phi_1(n_x, n_y) \quad (115)$$

The electron concentration per unit length is given by

$$n_0 = \frac{2g_v}{\pi} \sum_{n_x=1}^{n_{x_{\max}}} \sum_{n_y=1}^{n_{y_{\max}}} [B_{48}(E_{FID}, n_x, n_y) + B_{49}(E_{FID}, n_x, n_y)] \quad (116)$$

where $B_{48}(E_{FID}, n_x, n_y) = \sqrt{A_{31}(E_{FID}, n_x, n_y)}$

and $B_{49}(E_{FID}, n_x, n_y) = \sum_{r=1}^{s_o} Z_{1D}(r)[B_{48}(E_{FID}, n_x, n_y)]$

The field emitted current assumes the form

$$I = \frac{2g_v e k_B T}{h} \sum_{n_x=1}^{n_{x_{\max}}} \sum_{n_y=1}^{n_{y_{\max}}} [F_0(\eta_{31}) \exp(-\beta_{31})] \quad (117)$$

where, $\eta_{31} = \frac{E_{F1D} - E_{33}}{k_B T}$,, $\beta_{31} = \frac{4}{3}[A_{31}(V_0, n_x, n_y)]^{3/2} \cdot [eF_{sz} A'_{31}(V_0)]^{-1}$

and

$A'_{31}(V_0) = [1 - (m_c / m_0)(\bar{E}'_{g0} / 2)[(\bar{E}'_{g0} / 2)^2 +$
$[((\bar{E}'_{g0})^2 / 2)(1 - (m_c / m_0))] + [(\bar{E}'_{g0} / 2)(1 - (m_c / m_0))]^2 + V_0 \bar{E}'_{g0}(1 - (m_c / m_0))]^{-1/2}](2m_c / \hbar^2)$

The FNFE from Quantum Wires of II-V Materials

The dispersion relation of the holes are given by [119]

$$E = \theta_1 k_x^2 + \theta_2 k_y^2 + \theta_3 k_z^2 + \delta_4 k_x \mp [\{\theta_5 k_x^2 + \theta_6 k_y^2 + \theta_7 k_z^2 + \delta_5 k_x\}^2 + G_3^2 k_y^2 + \Delta_3^2]^{\frac{1}{2}} \pm \Delta_3 \quad (118)$$

where, k_x, k_y *and* k_z are expressed in the units of 10^{10} m$^-$

$\theta_1 = \frac{1}{2}(a_1 + b_1), \theta_2 = \frac{1}{2}(a_2 + b_2), \theta_3 = \frac{1}{2}(a_3 + b_3), \delta_4 = \frac{1}{2}(A + B),$
$\theta_5 = \frac{1}{2}(a_1 - b_1), \theta_6 = \frac{1}{2}(a_2 - b_2), \theta_7 = \frac{1}{2}(a_3 - b_3), \delta_5 = \frac{1}{2}(A - B),$
$a_i (i = 1, 2, 3, 4), b_i, A, B, G_3$ *and* Δ_3 are system constants

The 1D electron energy spectrum assumes the form

$$k_z^2 = A_{32,\pm}(E, n_x, n_y) \tag{119}$$

$$where,\ A_{32,\pm}(E,n_x,n_y) = \alpha_{4,\pm}(n_x,n_y) + \beta_4 E \pm [\beta_5 E^2 + E\beta_{6,\pm}(n_x,n_y) + \beta_{7,\pm}(n_x,n_y)]^{\frac{1}{2}}$$

$$\alpha_{4,\pm}(n_x,n_y) = [2(\theta_3^2 - \theta_7^2)]^{-1}[2\theta_7\alpha_2(n_x,n_y) - 2\alpha_{1,\mp}(n_x,n_y)\theta_3],$$
$$\alpha_{1,\mp}(n_x,n_y) = \theta_1(\tfrac{n_x\pi}{d_x})^2 + \theta_2(\tfrac{n_y\pi}{d_y})^2 + \delta_4(\tfrac{n_x\pi}{d_x}) \mp \Delta_3,$$
$$\alpha_2(n_x,n_y) = [\theta_5(\tfrac{n_x\pi}{d_x})^2 + \theta_6(\tfrac{n_y\pi}{d_y})^2 + \delta_5(\tfrac{n_x\pi}{d_x})],$$
$$\beta_4 = 2\theta_3[2(\theta_3^2 - \theta_7^2)]^{-1},\quad \beta_5 = [2(\theta_3^2 - \theta_7^2)]^{-2}[4\theta_7^2],$$
$$\beta_{6,\pm}(n_x,n_y) = [8\theta_3\theta_7\alpha_2(n_x,n_y) - 8\theta_7^2\alpha_{1,\mp}(n_x,n_y)][2(\theta_3^2 - \theta_7^2)]^{-1},$$
$$\beta_{7,\pm}(n_x,n_y) = [2(\theta_3^2 - \theta_7^2)]^{-2}[4\theta_7^2\alpha_2^2(n_x,n_y) - 8\theta_7\theta_3\ \alpha_2(n_x,n_y)\alpha_{1,\mp}(n_x,n_y) + 4\theta_3^2\alpha_3(n_y) +$$
$$4\theta_7^2\alpha_{1,\mp}(n_x,n_y) - 4\alpha_3(n_y)\theta_7^2]\ and\ \alpha_3(n_y) = G_3^2(\tfrac{n_y\pi}{d_y})^2 + \Delta_3^2$$

The quantized energy levels ($E_{34,\pm}$) can be expressed through the equation

$$E_{34,\pm} = \alpha_{1,\mp}(n_x,n_y) \pm [\alpha_2^2(n_x,n_y) + \alpha_3(n_y)]^{\frac{1}{2}} \tag{120}$$

The electron concentration per unit length is given by

$$n_0 = \frac{g_v}{\pi}\sum_{n_x=1}^{n_{x_{max}}}\sum_{n_y=1}^{n_{y_{max}}}[\bar{B}_{49}(E_{FID},n_x,n_y) + B_{50}(E_{FID},n_x,n_y)] \tag{121}$$
$$where\ \bar{B}_{49}(E_{FID},n_x,n_y) = [\sqrt{A_{32,+}(E_{FID},n_x,n_y)} + \sqrt{A_{32,-}(E_{FID},n_x,n_y)}]$$
$$and\ B_{50}(E_{FID},n_x,n_y) = \sum_{r=1}^{s_o} Z_{1D}(r)[B_{49}(E_{FID},n_x,n_y)]$$

The field emitted current assumes the form

$$I = \frac{g_v e k_B T}{h}\sum_{n_x=1}^{n_{x_{max}}}\sum_{n_y=1}^{n_{y_{max}}}[F_0(\eta_{32,+})\exp(-\beta_{32,+}) + F_0(\eta_{32,-})\exp(-\beta_{32,-})] \tag{122}$$
$$Where,\ \eta_{32,\pm} = \frac{E_{F1D} - E_{34,\pm}}{k_B T},\ \beta_{32,\pm} = \frac{4}{3}[A_{32,\pm}(V_0,n_x,n_y)]^{3/2}\cdot[eF_{sz}A'_{32,\pm}(V_0,n_x,n_y)]^{-1}$$
$$and\ A'_{32,\pm}(V_0,n_x,n_y) = [\beta_4 \pm \frac{1}{2}[2\beta_5 V_0 + \beta_{6,\pm}(n_x,n_y)].[\beta_5 V_0^2 + \beta_{6,\pm}(n_x,n_y)V_0 + \beta_{7,\pm}(n_x,n_y)]^{-\frac{1}{2}}]$$

RESULT AND DISCUSSIONS

Using (1) and (8) and taking the energy band constants as given in the Table 1.1, we have plotted the FNFE current from QWs of $CdGeAs_2$ (an example of nonlinear optical materials) as a function of d_y as shown by the dotted plot of Figure 1.1, in which the plot corresponds to the solid line represents the same for the two band model of Kane. Figure 1.2 exhibits the plot FNFE current from QWs of n-InSb as a function of film thickness in accordance with the three and two band models of Kane together with the models of Stillman, et al., Newson, et al., and Palik, et al., respectively for the purpose of assessing the influence of dispersion relations on the FNFE current from QWs of III-V semiconductors. Figure 1.3 exhibits the variation of the field emitted current with electric field which is normal to the two directions of quantizations in the wave-vector space of the material or the longitudinal direction which is along the axis of the wire for all cases Figure 1.2. Figure 1.4 presents the FNFE current as function of film thickness for two different values of alloy compositions for QWs of n-$Hg_{1-x}Cd_xTe$. Figure 1.5 shows the carrier concentration dependence of FNFE current from QWs of n-$Hg_{1-x}Cd_xTe$, n-InSb, n-InAs and n-GaAs respectively for the purpose of assessing the influence of different energy band constants on the field emitted current from QWs of III-V materials. In Figure 1.6, exhibits the film thickness dependence of FNFE current from QWs of n-$In_{1-x}Ga_xAs_yP_{1-y}$ in accordance with the three and two band models of Kane together with the models of Stillman, et al., Newson, et al. and Palik et al. respectively. Figure 1.7 shows the dependence of FNFE current on alloy composition from QWs of ternary and quaternary materials in accordance with the two band models of Kane.

The Figure 1.8 exhibits the film thickness dependence of FNFE current from QWs of II-Vi materials taking p-CdS as an example. The Figure 1.9 shows the FNFE current as a function of carrier concentration for the case of Figure 1.8. In Figure 1.10, we have plotted the FNFE current from QWs of Bismuth as a function of film thickness in accordance with the models

Table 1.1. The numerical values of the energy band constants of few materials

	Materials	Numerical Values of the Energy Band Constants
1	a) The conduction electrons of n-Cadmium Germanium Arsenide can be described by three types of band models	1. The values of the energy band constants in accordance with the generalized electron dispersion relation of nonlinear optical materials (as given by (1.2)) are as follows $E_{g_0} = 0.57eV$, $\Delta_{\parallel} = 0.30eV$, $\Delta_{\perp} = 0.36eV$, $m_{\parallel}^{*} = 0.034m_0$, $m_{\perp}^{*} = 0.039m_0$, $T = 4K$, $\delta = -0.21eV$, $g_v = 1$ [47,120], $\varepsilon_{sc} = 18.4\varepsilon_0$ [121] (ε_{sc} and ε_0 are the permittivity of the semiconductor material and free space respectively) and $W(electron\ affinity) = 4eV$ [122]. 2. In accordance with the three band model of Kane (as given by (1.13)), the spectrum constants are given by $\Delta = (\Delta_{\parallel} + \Delta_{\perp})/2 = 0.33eV, E_{g_0} = 0.57eV$, $m_c = (m_{\parallel}^{*} + m_{\perp}^{*})/2 = 0.0365m_0$ and $\delta = 0eV$. 3. In accordance with two band model of Kane (as given by (1.19)), the spectrum constants are given by $E_{g_0} = 0.57eV$ and $m_c = 0.0365m_0$.
	b) The conduction electrons of n-Cadmium Arsenide can be described by three types of band models	1. The values of the energy band constants in accordance with the generalized electron dispersion relation of nonlinear optical materials (as given by (1.2)) are as follows $\lvert E_{g_0} \rvert = 0.095eV$, $\Delta_{\parallel} = 0.27eV$, $\Delta_{\perp} = 0.25eV$, $m_{\parallel}^{*} = 0.00697m_0$, $m_{\perp}^{*} = 0.013933m_0$, $T = 4K$, $\delta = 0.085eV$, $g_v = 1$ [47,120] and $\varepsilon_{sc} = 16\varepsilon_0$ [122]. 2. In accordance with the three band model of Kane (as given by (1.13)), the spectrum constants are given by $\Delta = (\Delta_{\parallel} + \Delta_{\perp})/2 = 0.26eV$ $\lvert E_{g_0} \rvert = 0.095eV$, $m_c = (m_{\parallel}^{*} + m_{\perp}^{*})/2 = 0.020903m_0$ and $\delta = 0eV$. 3. In accordance with two band model of Kane (as given by (1.19)), the spectrum constants are given by $\lvert E_{g_0} \rvert = 0.095eV$, and $m_c = 0.020903m_0$.

	Materials	Numerical Values of the Energy Band Constants
2	n-Indium Arsenide	The values $E_{g_0} = 0.36eV$, $\Delta = 0.43eV$, $m_c = 0.026m_0$, $g_v = 1$, $\varepsilon_{sc} = 12.25\varepsilon_0$ [57] and $W = 5.06eV$ [123] are valid for three band model of Kane as given by (1.13).
3	n-Gallium Arsenide	The values $E_{g_0} = 1.55eV$, $\Delta = 0.35eV$, $m_c = 0.07m_0$, $g_v = 1$, $\varepsilon_{sc} = 12.9\varepsilon_0$ [57] and $W = 4.07eV$ [124] are valid for three band model of Kane as given by (1.13). The values $a_{13} = -1.97\times10^{-37}eVm^4$ and $a_{15} = -2.3\times10^{-34}eVm^4$ [59] are valid for the Newson and Kurobe model [59] as given by (1.34).
4	n-Gallium Aluminium Arsenide	$E_{g_0} = (1.424 + 1.266x + 0.26x^2)eV$, $\Delta = (0.34 - 0.5x)eV$, $m_c = [0.066 + 0.088x]m_0$, $\varepsilon_{sc} = [13.18 - 3.12x]\varepsilon_0$ [125] and $W = (3.64 - 0.14x)eV$ [126].
5	n-Mercury Cadmium Telluride	$E_{g_0} = (-0.302 + 1.93x + 5.35\times10^{-4}(1-2x)T - 0.810x^2 + 0.832x^3)eV$, $\Delta = (0.63 + 0.24x - 0.27x^2)eV, m_c = 0.1m_0E_{g_0}(eV)^{-1}, g_v = 1$, $\varepsilon_{sc} = [20.262 - 14.812x + 5.22795x^2]\varepsilon_0$ [127] and $W = (4.23 - 0.813(E_{g_0} - 0.083))eV$ [128].
6	n-Indium Gallium Arsenide Phosphide lattice matched to Indium Phosphide	$E_{g_0} = \left(1.337 - 0.73y + 0.13y^2\right)eV$, $\Delta = \left(0.114 + 0.26y - 0.22y^2\right)eV$, $y = (0.1896 - 0.4052x)/(0.1896 - 0.0123x), m_c = (0.08 - 0.039y)m_0$, $g_v = 1, \varepsilon_{sc} = [10.65 + 0.1320y]\varepsilon_0$ *and* [129] $W(x,y) = \left[5.06(1-x)y + 4.38(1-x)(1-y) + 3.64xy + 3.75\{x(1-y)\}\right]eV$
7	n-Indium Antimonide	$E_{g_0} = 0.2352eV$, $\Delta = 0.81eV$, $m_c = 0.01359m_0$, $g_v = 1$, $\varepsilon_{sc} = 15.56\varepsilon_0$ [57] and $W = 4.72eV$ [123].
8	n-Gallium Antimonide	The values of $E_{g_0} = 0.81eV$, $\Delta = 0.80eV$, $P = 9.48\times10^{-10}eVm$, $\bar{\varsigma}_0 = -2.1$, $\bar{v}_0 = -1.49$, $\bar{\omega}_0 = 0.42$, $g_v = 1$ [130] and $\varepsilon_{sc} = 15.85\varepsilon_0$ [131] are valid for the model of Seiler et al. [130] as given by (R1.5).).

Table 1.1. (Continued)

	Materials	Numerical Values of the Energy Band Constants
9	n-Cadmium Sulphide	$m_{\parallel}^{*}=0.7m_0$, $m_{\perp}^{*}=1.5m_0$, $C_0=1.4\times10^{-8}eVm$, $g_v=1$ [57], $\varepsilon_{sc}=15.5\varepsilon_0$ [132] and $W=4.5eV$ [123]
10	n-Lead Telluride	The values $m_t^{-}=0.070m_0$, $m_l^{-}=0.54m_0$, $m_t^{+}=0.010m_0$, $m_l^{+}=1.4m_0$, $P_{\parallel}=141meVnm$, $P_{\perp}=486meVnm$, $E_{g_0}=190meV$, $g_v=4$ [57], $\varepsilon_{sc}=33\varepsilon_0$ [57,133] and $W=4.6eV$ [134] are valid for the Dimmock model [110] as given by (1.70). The values $m_1=0.0239m_0$, $m_2=0.024m_0$, $m_2'=0.31m_0$, $m_3=0.24m_0$ [135] are valid for the Cohen model [79] as given by (1.60).
11	Stressed n-Indium Antimonide	The values $m_c=0.048m_o$, $E_{g_0}=0.081eV$, $B_2=9\times10^{-10}eVm$, $C_1=3eV$, $C_2=2eV$, $\bar{a}_0=-10eV$, $\bar{b}_0=-1.7eV$, $\bar{d}_0=-4.4eV$, $S_{xx}=0.6\times10^{-3}(kbar)^{-1}$, $S_{yy}=0.42\times10^{-3}(kbar)^{-1}$, $S_{zz}=0.39\times10^{-3}(kbar)^{-1}$, $S_{xy}=0.5\times10^{-3}(kbar)^{-1}$, $\varepsilon_{xx}=\sigma S_{xx}$, $\varepsilon_{yy}=\sigma S_{yy}$, $\varepsilon_{zz}=\sigma S_{zz}$, $\varepsilon_{xy}=\sigma S_{xy}$, σ is the stress in kilobar, $g_v=1$ [111] are valid for the model of Seiler et al. [111] as given by (1.76).
12	Bismuth	$E_{g_0}=0.0153eV$, $m_1=0.00194m_0$, $m_2=0.313m_0$, $m_3=0.00246m_0$, $m_2'=0.36m_0$, $g_v=3$, $g_s=2$ [136], $M_2=1.25m_0$, $M_2'=0.36m_0$ [137] and $W=4.34eV$.
13	Mercury Telluride	$m_v^{*}=0.028m_0$, $g_v=1$, $\varepsilon_{\infty}=15.2\varepsilon_0$ [138] and $W=5.5eV$ [139].

	Materials	**Numerical Values of the Energy Band Constants**
14	Platinum Antimonide	For valence bands, along <100> direction, $\bar{\lambda}_0 = (0.02/4)eV$, $\bar{l} = (-0.32/4)eV$, $\bar{v} = (0.39/4)eV$, $\bar{n} = (-0.65/4)eV$, $\bar{a} = 0.643nm$, $I = 0.30(eV)^2$, $\bar{\delta}_0 = 0.02eV$, $g_v = 6$ [114], $\varepsilon_{sc} = 30\varepsilon_0$ [140] and $\phi_w \approx 3.0eV$ [141]. For conduction bands, along <111> direction, $g_v = 8$ [140], $\bar{\lambda}_0 = (0.33/4)eV$, $\bar{l} = (1.09/4)eV$, $\bar{v} = (0.17/4)eV$ and $\bar{n} = (0.22/4)eV$
15	n-Gallium Phosphide	$m_{\parallel}^* = 0.92m_0, m_{\perp}^* = 0.25m_0, k_0 = 1.7\times10^{15}m^{-1}$, $\lvert V_G\rvert = 0.21eV, g_v = 6, g_s(spin\ \boldsymbol{deg}eneracy) = 2$ [113] and $W = 3.75eV$ [123].
16	Germanium	$E_{g_0} = 0.785eV$, $m_{\parallel}^* = 1.57m_0$, $m_{\perp}^* = 0.0807m_0$ [123] and $W = 4.14eV$ [124]
17	Tellurium	The values $\psi_1 = 6.7\times10^{-16}meVm^2$, $\psi_2 = 4.2\times10^{-16}meVm^2, \psi_3 = 6\times10^{-8}meVm$ and $\psi_4 = (3.6\times10^{-8}meVm)$ [112] are valid for the model of Bouat et al. [112] as given by (1.81).
18	Graphite	The values $\bar{\Delta} = -0.0002eV, \bar{\gamma}_1 = 0.392eV, \bar{\gamma}_5 = 0.194eV$, $c_6 = 0.674nm$, $\bar{\gamma}_2 = -0.019eV, a_6 = 0.246nm$, $\bar{\gamma}_0 = 3eV, \bar{\gamma}_4 = 0.193eV$ [142] and $W = 4.6eV$ [143] are valid for the model of Brandt et al. [142] as given by (R.1.12).
19	Lead Germanium Telluride	The values $\overline{E_{g_0}} = 0.21eV$, $g_v = 4$ [144] and $\phi_w \approx 6eV$ [145] are valid for the model of Vassilev [144] as given by (R.1.11).
20	Cadmium Antimonide	The values $a_1 = -32.3\times10^{-20}eVm^2, b_1 = -60.7\times10^{-20}eVm^2$, $a_2 = -16.3\times10^{-20}eVm^2$, $b_2 = -24.4\times10^{-20}eVm^2$, $a_3 = -91.9\times10^{-20}eVm^2, b_3 = -105\times10^{-20}eVm^2$, $A = 2.92\times10^{-10}eVm, B = -3.47\times10^{-10}eVm$, $G_3 = 1.3\times10^{-10}eVm, \Delta_3 = 0.070eV$ [119] and $\phi_w \approx 2eV$ [146]

Table 1.1. (Continued)

	Materials	Numerical Values of the Energy Band Constants
21	Cadmium Diphosphide	The values $\beta_1 = 8.6 \times 10^{-21} eVm^2$, $\beta_2 = 1.8 \times 10^{-21} (eVm)^2$, $\beta_4 = 0.0825 eV$, $\beta_5 = -1.9 \times 10^{-19} eVm^2$ [147] and $\phi_w \approx 5eV$ [148] are valid for the model of Chuiko [147] and is given by (R.1.20).
22	Zinc Diphosphide	The values $\beta_1 = 8.7 \times 10^{-21} eVm^2$, $\beta_2 = 1.9 \times 10^{-21} (eVm)^2$, $\beta_4 = 0.0875 eV$, $\beta_5 = -1.9 \times 10^{-19} eVm^2$ [147] and $W \approx 3.9eV$ [148] are valid for the model of Chuiko [147] and is given by (R.1.20).
23	Bismuth Telluride	The values $E_{g_0} = 0.145eV$, $\bar{\alpha}_{11} = 4.9$, $\bar{\alpha}_{22} = 5.92$, $\bar{\alpha}_{33} = 9.5$, $\bar{\alpha}_{23} = 4.22$, $g_s = 2$, $g_v = 6$ [115] and $\phi_w = 5.3eV$ [149] are valid for the model of Stordeur et al. [115] using (1.97).
24	Carbon Nanotube	The values $a_c = 0.144nm$ [150], $t_c = 2.7eV$ [151], $\bar{r}_0 = 0.7nm$ [152] and $W = 3.2eV$ [153] are valid for graphene band structure realization of carbon nanotube [152] and are given by (1.123) and (1.124).
25	Antimony	The values $\alpha_{11} = 16.7$, $\alpha_{22} = 5.98$, $\alpha_{33} = 11.61$, $\alpha_{23} = 7.54$ [154] and $W = 4.63eV$ [116] are valid for the model of Ketterson [154] and are given by (R.1.13) and (R.1.14) respectively.
26	Zinc Selenide	$m_{c2} = 0.16m_0$, $\Delta_2 = 0.42eV$, $E_{g_{02}} = 2.82eV$ [124] and $W = 3.2eV$ [155].
27	Lead Selenide	$m_t^- = 0.23m_0$, $m_l^- = 0.32m_0$, $m_t^+ = 0.115m_0$, $m_l^+ = 0.303m_0$, $P_{\parallel} \approx 138 meVnm$, $P_{\perp} = 471 meVnm$, $E_{g_0} = 0.28eV$ [156], $\varepsilon_{sc} = 21.0\varepsilon_0$ [124] and $W = 4.2eV$ [157-180].

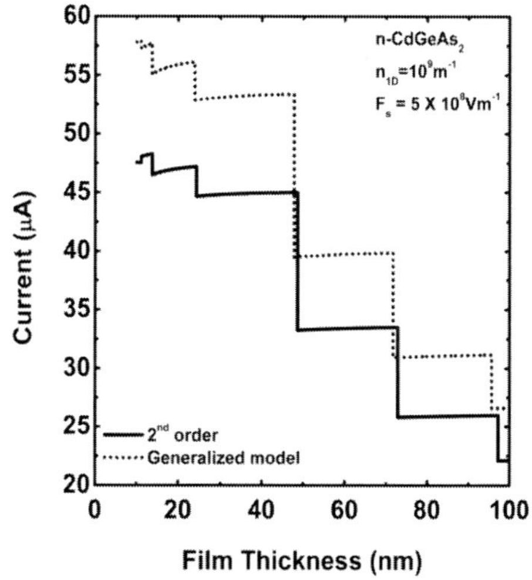

Figure 1.1. Plot of the FN field emission current as function of film thickness d_y for QWs of n-$CdGeAs_2$. The dotted and solid curves correspond to the generalized and the two band models of Kane respectively, where d_z = 30 nm.

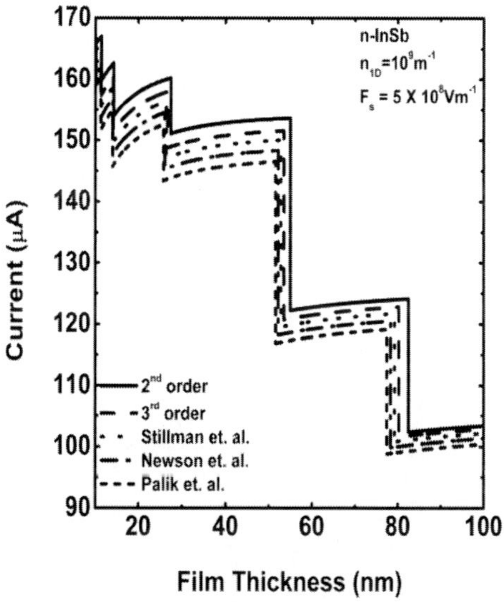

Figure 1.2. Plot of the FN field emission current as function of film thickness d_y for QWs of n-InSb in accordance with the three and two band models of Kane together with the models of Stillman, et al., Newson, et al. and Palik et al. respectively, where d_z = 30 nm.

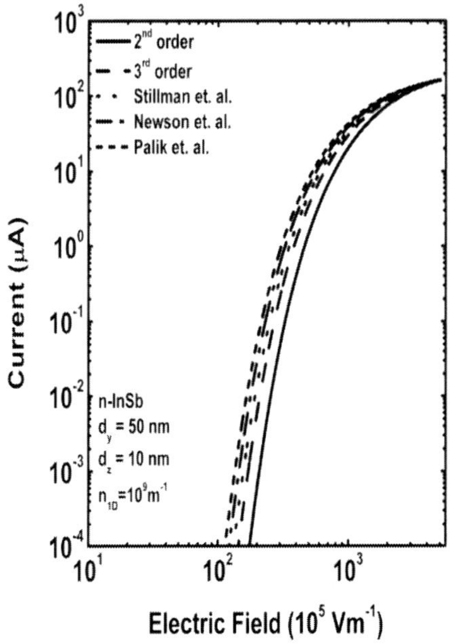

Figure 1.3. Plot of the FN field emission current as function of electric field for QWs of n-InSb in accordance with all the cases of Figure 1.2.

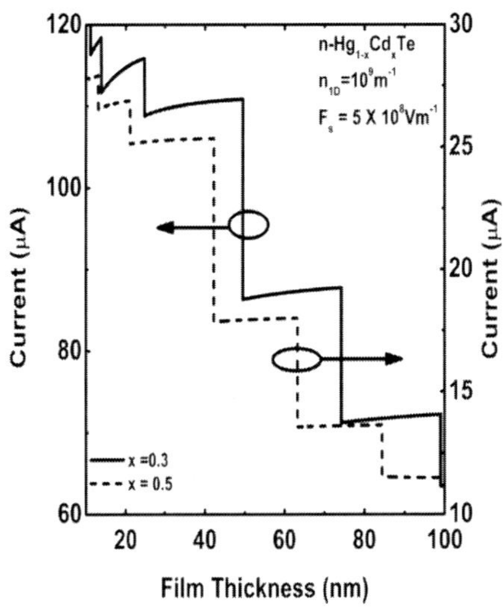

Figure 1.4. Plot of the FN field emission current as function of film thickness d_y for QWs of n-$Hg_{1-x}Cd_xTe$ in accordance with the two band model of Kane for two different values of alloy composition.

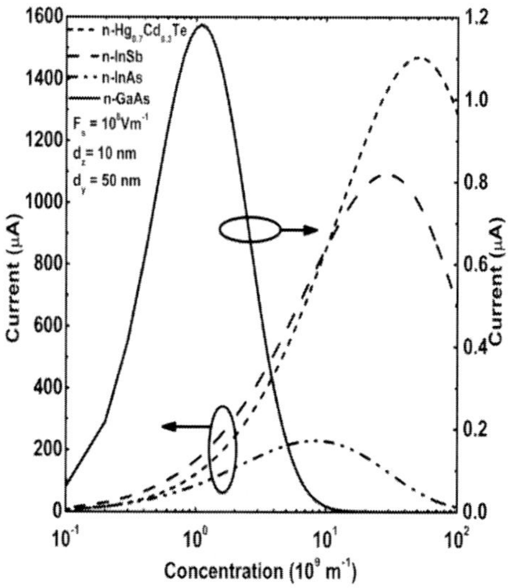

Figure 1.5. Plot of the FN field emission current as function of carrier concentration for QWs of n-$Hg_{0.7}Cd_{0.3}Te$, n-InSb, n-InAs and n-GaAs in accordance with the two band model of Kane at the lowest subband.

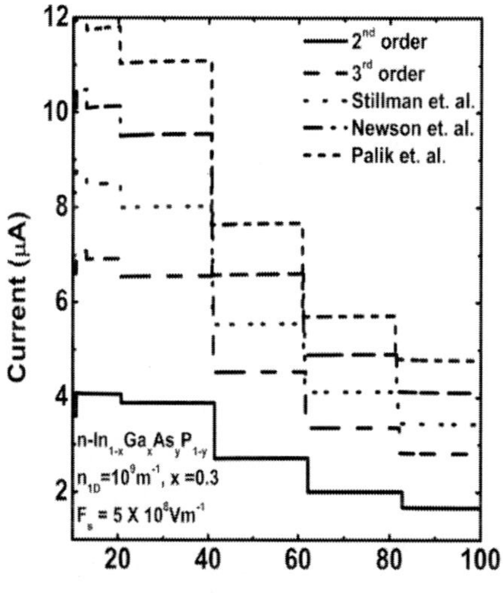

Figure 1.6. Plot of the field current as function of film thickness d_y for quantum wires of n-$In_{1-x}Ga_xAs_yP_{1-y}$ in accordance with the three and two band models of Kane together with the models of Stillman, et al., Newson, et al. and Palik et al. respectively, where $d_z = 30$ nm.

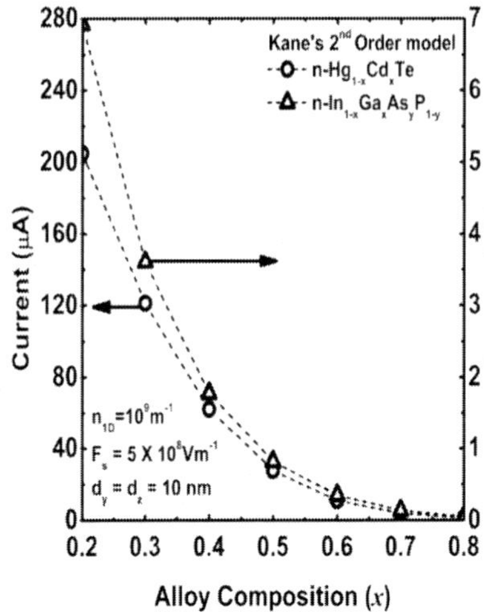

Figure 1.7. Plot of the field emitted current as function of alloy composition for QWs of n-$Hg_{1-x}Cd_xTe$ and $In_{1-x}Ga_xAs_yP_{1-y}$ in accordance with the two band model of Kane.

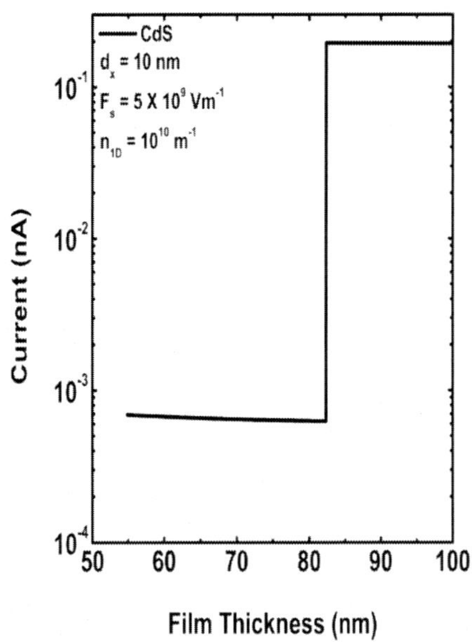

Figure 1.8. Plot of the field emitted current as function of film thickness d_y for QWs of CdS.

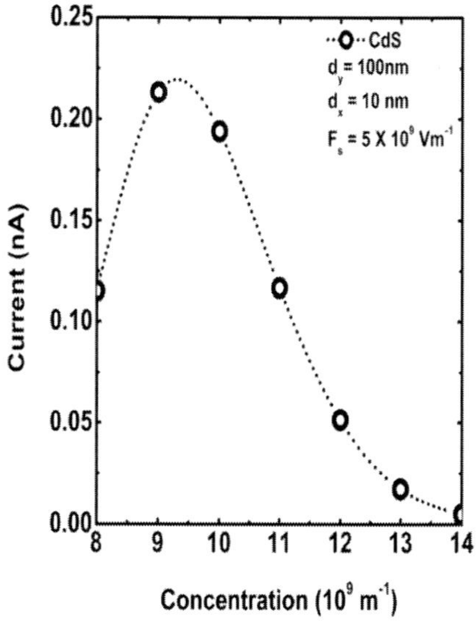

Figure 1.9. Plot of the field emitted current as function of concentration for QWs of CdS.

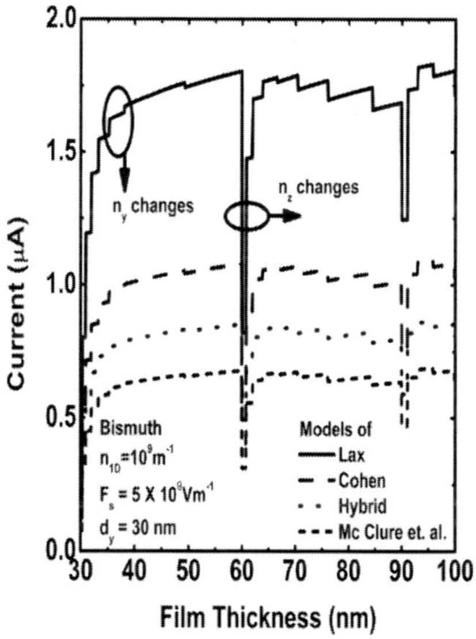

Figure 1.10. Plot of the field emitted current as function of film thickness for QWs of Bi in accordance with the models of Mc Clure and Choi, Hybrid, Cohen and Lax respectively.

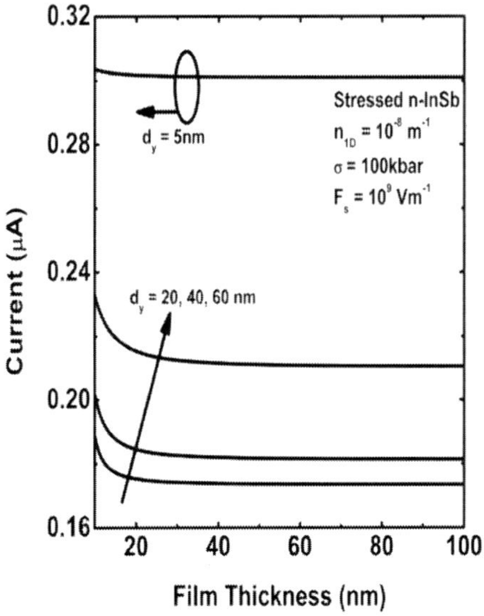

Figure 1.11. Plot of the field emitted current as function of film thickness for QWs of stressed n-InSb for different values of d_y.

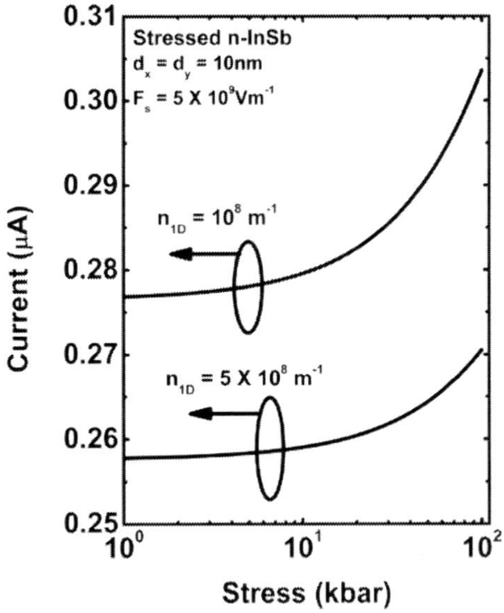

Figure 1.12. Plot of the field emitted current as function of stress for QWs of stressed n-InSb for two different values of carrier concentration as shown.

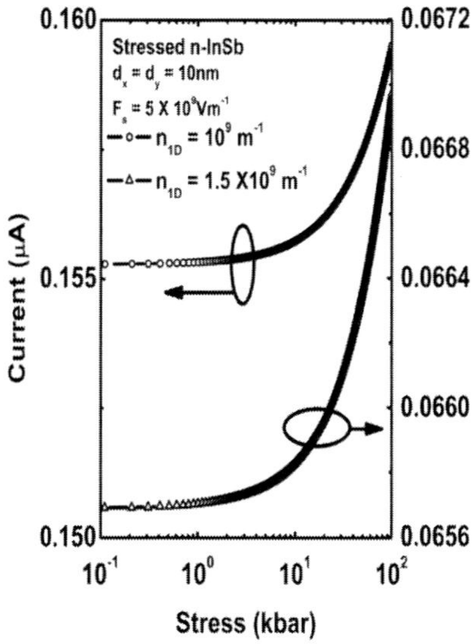

Figure 1.13. Plot of the field emitted current as function of stress for QWs of stressed n-InSb for two different values of carrier concentration as shown.

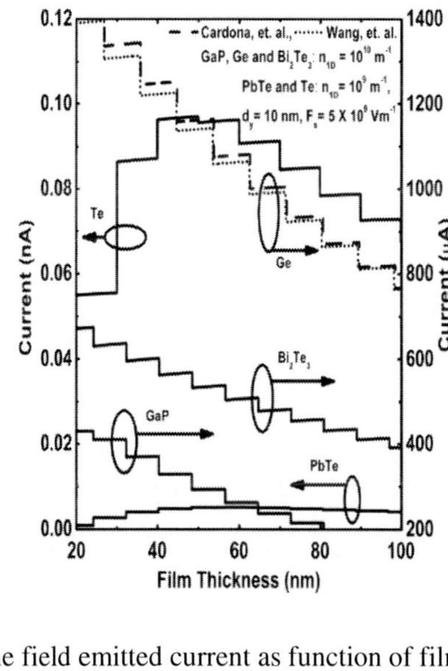

Figure 1.14. Plot of the field emitted current as function of film thickness for QWs of Ge (in accordance with both types of band models), GaP, Bi_2Te_3, PbTe and Te respectively.

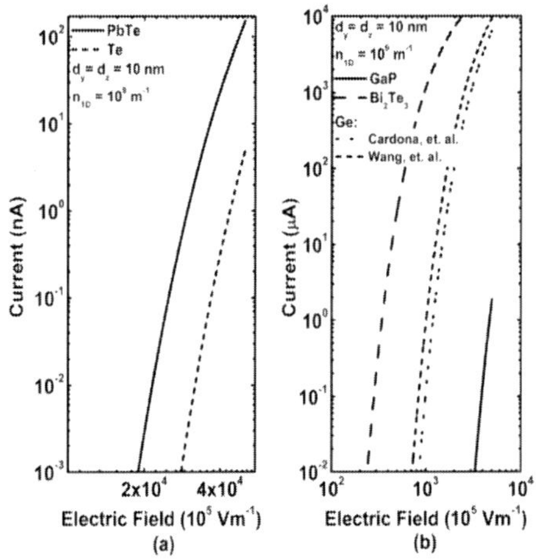

Figure 1.15. Plot of the field current as function of electric field for quantum wires of (a) PbTe and Te and (b) Gap, Bi_2Te_3 and Ge.

of Mc Clure and Choi, Hybrid, Cohen and Lax respectively. The Figure 1.11, exhibits the variation of the FNFE current as a function of film thickness for QWs of stressed materials taking stressed n-InSb as an example. The Figures 1.12 and 1.13 explore the stress dependence of the FNFE current from QWs of stressed n-InSb for different values of doping for the purpose of assessing the influence of carrier concentration on the field emitted current in this case. In Figure 1.14, the field emitted current as a function of film thickness has been plotted for QWs of n-Ge (in accordance with both types of band models of n-Ge), n-GaP, n-PbTe and p-Bi_2Te_3 respectively. The plot (a) of Figure 1.15 shows the variation of FNFE current on the electric field for QWs of PbTe and Te, while (b) exhibits the same for QWs of Ge, GaP and Bi_2Te_3 respectively.

The salient features of the above figures are described as follows:

From Figure 1.1, we observe that the field emission current exhibits a step-functional decreasing dependence with increase in film thickness for QWs of n-$CdGeAs_2$. The combined influence of the anisotropies of the energy band constants and the crystal field splitting is to enhance the field emitted current as compared with the corresponding which is based on two band model of Kane in the whole range of thicknesses as considered in Figure 1.1. The periodicity with respect to the film thickness is same in both the cases and is invariant of the energy band constants. It should be noted from the field emitted current in general, is a product of two quantities inside the summation signs. One of them is $F_0\ (\eta_{ij})$, ($i = 1, 2, 3...$ *and* $j = 1, 2, 3,...$) and the other one is $\exp(-\beta_{ij})$ which is a function of Fermi energy, electric field, effective mass and various parameters of the energy bands in a complex way. Thus, we observe that field emitted current depends totally on the product of these two functions within the summation signs and a pre-factor outside the summation sign. The product of these two functions ultimately determines the behavior of the field emitted current. Although we know [137] that the Fermi energy of low-dimensional systems decreases with increasing size, one cannot be always certain that I will decrease with increasing film thickness due to the particular form of field emitted current in case of QWs. If the rate of increase of $F_0\ (\eta_{ij})$ overcomes the rate of change of $\exp(-\beta_{ij})$, I will increase, whereas, for the opposite case, I will decrease. This important physical fact determines the magnitude of the field emitted current and its dependence with respect to any other physical variable.

From Figure 1.2, we observe that I decrease with increasing film thickness for the three and two band models of Kane together with the models of Stillman, et al., Newson, et al. and Palik, et al. respectively. The I exhibits a different magnitude which is the direct signature of the dispersion relation on the field emitted current. Numerical computations reflects the fact that the field emitted current for a 30 X 10 nm^2 size quantum wire can reach nearly 160 μA for n-InSb at low temperatures and at a field strength of 5 X 10^8 Vm^{-1} with a linear doping density of 10^9 m^{-1}. From Figure 1.3, it can be stated that field strength of nearly 10^7 Vm^{-1} is sufficient to produce a tenths of microamperes.

Incidentally, due to the velocity saturation phenomena, we observe that beyond 10^8 Vm^{-1}, I saturate converging to a unique value and become invariant of dispersion relations. The field emitted current from QWs of n-$Hg_{1-x}Cd_xTe$ as function of film thickness has been exhibited in Figure 1.4 and can be compared with the corresponding cases of earlier figures. It appears that for lower values of film thickness, I increase for a particular subband. As thickness increases, generation of different subbands occurs which leads to the overall decrease in *I*. The variation of *I* over a large range of carrier concentration has been plotted in Figure 1.5 for the quantum limit case. It appears that *I* initially increase due to low value of Fermi energy and as the concentration increases, the magnitude of the current decreases sharply exhibiting a peak. The amount of broadening is highly sensitive to the spectrum parameters of a particular semiconductor. It appears that for n-GaAs, the broadening is more as compared with others as shown in the figure. The Figure 1.6 exhibits the variation of *I* as function of film thickness for QWs of n-$In_{1-x}Ga_xAs_yP_{1-y}$ which shows almost a constant step functional dependence of the current. This implies that the difference in the Fermi energy and the quantized subband energy is almost invariant of film thickness. From Figure 1.7, one observes that as alloy composition increases, the field emitted current decreases for QWs of both ternary and quaternary compounds. For lower values of alloy composition, *I* from QWs of quaternary semiconductor is more than that of the corresponding ternary one, whereas for higher value of alloy composition, it converges. Step functional dependence of field emitted current with film thickness for QWs of CdS has been observed in Figure 1.8. We see a nearly constant field emitted current per subband, until a new subband is being occupied. Comparing with the earlier figures on the thickness dependence, we observe in this case an opposite trend because of the fact of overriding of the term $F_0(\eta_{ij})$ by $\exp(-\beta_{ij})$. The Figure 1.9 exhibits the fact that for p-CdS, the field current is of the order of a few nanoamperes even at field strength of 5×10^9 Vm^{-1}.

Composite oscillations in field emitted current from QWs of Bi with film thickness has been exhibited in Figure 1.10 for all types of energy band models as considered in this chapter. We observe that for $n_z = 1$, electrons will be populated in the various other subbands corresponding to n_y. After a certain value of film thickness, when $n_z = 2$, the redistribution of the electrons in the quantized energy levels is being repeated leading to a composite oscillations. It should be noted that this behavior is a general feature in thickness dependence of field emission from quantum wires. The energy band parameters determine the composite periodicity.

The effect of film thickness on the field current in stressed InSb has been plotted in Figure 1.11 in the quantum limit case, where the current decreases with increase in d_z for The effect of stress on the field emitted current has been exhibited in Figures 1.12 and 1.13 for different values of carrier concentration. From Figure 1.12, it appears that with the increase in stress, the current increases having a magnitude of few tenths of microamperes. With the increase in carrier degeneracy, the current almost reduces to about 10 times.

From Figure 1.14, we observe that the field current from QWs of Ge (in accordance with the models of Cardona, et al. and Wang et al. respectively), Bi_2Te_3, GaP decreases as film thickness increases because of the fact that the term $\exp(-\beta_{ij})$ dominates over the term $F_0(\eta_{ij})$ in the whole range of thickness as considered here. For QWs of PbTe and Te, the current initially increases and then decreases exhibiting the fact that of the dominancy of the two terms $F_0(\eta_{ij})$ and $\exp(-\beta_{ij})$. For QWs of Ge and Bi_2Te_3, the field current has been observed to be in the order of hundreds of microampere, where as for other 1D systems as considered in this figures the currents are in the order of few nanoamperes exhibiting the influence of the carrier dispersion relation of a particular semiconductor. In Figure 1.15 (a), the cut-in electric field for the field currents are in the order of 10^9 Vm^{-1} for QWs of PbTe and Te, while about 10^7 – 10^8 Vm^{-1} field strength is sufficient for field current to reach few microamperes.

The influence of quantum confinement is immediately apparent from Figures (1.1), (1.2), (1.5), (1.7), (1.9), (1.11) and (1.15) since the field emitted current depends strongly on the thickness of the quantum-confined materials in contrast with the corresponding bulk specimens. The current changes with increasing film thickness in an oscillatory way with different numerical magnitudes. It appears from the aforementioned figures that the FNFE exhibits spikes for particular values of film thickness which, in turn, depends on the particular band structure of the specific semiconductor. Moreover, the FNFE from QWs of different compounds can be smaller than of bulk specimens of the same materials, which is also a direct signature of quantum confinement. This oscillatory dependence will be less and less prominent with increasing film thickness. For bulk specimens of the same material, the FNFE will be found to increase continuously with increasing electron degeneracy in a non-oscillatory manner. The appearance of the discrete jumps in the respective figures are due to the redistribution of the electrons among the quantized energy levels when the size quantum number corresponding to the highest occupied level changes from one fixed value to the others.

With varying electron degeneracy, a change is reflected in the FNFE through the redistribution of the electrons among the size-quantized levels. It may be noted that at the transition zone from one sub band to another, the height of the peaks between any two sub-bands decreases with the increasing in the degree of quantum confinement and is clearly shown in all the curves. It should be noted that although, the FNFE changes in various manners with all the variables as evident from all the figures, the rates of variations are totally band-structure dependent.

It is imperative to state that the present investigation excludes the many-body, hot electron, broadening and the allied effects in the simplified theoretical formalism due to the absence of proper analytical techniques for including them for generalized systems as considered here. We have also approximated the variation of value of the work function from its bulk value in the present system. Our simplified approach will be appropriate for the purpose of comparisons when the methods of tackling the formidable problems after inclusion of the said effects for the generalized systems emerge. The results of this simplified approach get transformed to the well-known formulation of the FNFE for wide gap materials having parabolic energy bands. This indirect test not only exhibits the mathematical compatibility of the formulation but also shows the fact that this simple analysis is more generalized one, since one can obtain the corresponding results for materials having parabolic energy bands under certain limiting conditions from the present derivation. For the purpose of computer simulations for obtaining the plots of FNFE versus various external variables, we have taken very low temperatures since the quantization effects are basically low temperature phenomena together with the fact that the temperature dependence of all the energy band constants of all the semiconductors and their nanostructures as considered in this chapter are not available in the literature. Our results as formulated in this chapter are valid for finite temperatures and are useful in comparing the results for temperature variations of FNFE after the availability of the temperature dependences of such constants of various dispersion relations in this context. The experimental results for the verification of theoretical formulations of FNFE are still not available in the literature. It is worth noting that the nature of the curves of field emitted current with various physical variables based on our simplified formulations as presented here would be useful to analyze the experimental results when they materialize. The inclusion of the said effects would certainly increase the accuracy of the results although the qualitative features of FNFE would not change in the presence of the aforementioned effects.

It can be noted that on the basis of the dispersion relations of the various quantized structures as discussed above the effective electron mass, the Debye screening length, the plasma frequency, the activity coefficient, the carrier contribution to the elastic constants, the diffusion coefficient of minority carrers, the third-order non-linear optical susceptibility, the heat capacity, the dia and para-magnetic susceptibilities and the various important dc/ac transport coefficients can be probed for all types of QWs as considered here .Thus, our theoretical formulation comprises the dispersion relation dependent properties of various technologically important quantum-confined semiconductors having different band-structures. We have not considered other types of compounds in order to keep the presentation concise and succinct. With different sets of energy band parameters, we shall get different numerical values of the FNFE. The nature of variations of the FNFE as shown here would be similar for the other types of materials and the simplified analysis of this chapter exhibits the basic qualitative features of the FNFE. It may be noted that the basic aim of this chapter is not solely to demonstrate the influence of quantum confinement on the FNFE for a wide class of quantized materials but also to formulate the appropriate carrier statistics in the most generalized form, since the transport and other phenomena in modern nano-structured devices having different band structures and the derivation of the expressions of many important carrier properties are based on the temperature-dependent carrier statistics in such systems.

REFERENCES

[1] Fowler R. H., L. Nordheim, Proc. of Royal Society of London, Series-A, 119,173 (1928); A. Van Der Ziel, *Solid State Physical Electronics* (Prentice-Hall, Inc., Englewood Cliffs U.S.A. 1957 (p.176).

[2] Mitra B., K. P. Ghatak, *Phys. Lett. A,* 357, 146(1990); 142A, 401(1989); K. P. Ghatak, M. Mondal, *Jour. Mag. Mag. Mat.,* 74,203(1988); K. P. Ghatak, B. Mitra, *Phys. Lett.,* 156A. 233, (1991); K. P. Ghatak, A. Ghosal, S. N. Biswas, M. Mondal, *Proc. of SPIE,* USA, 1308, 356 (1990)

[3] Binh V. T., Ch. Adessi, *Phys. Rev. Lett.* 85,864(2000); R. G. Forbes, Ultramicroscopy, 79, 11(1999); J. W. Gadzuk, E. W. Plummer, Rev. Mod. Phys, 45,487(1973); J. M. Beebe, B. Kim, J. W. Gadzuk, C. D. Frisbie, J. G. Kushmerick, *Phys. Rev. Lett,* 97,026801(1999); Y. Feng, J. P. Verboncoeur, *Phys. Plasmas,* 12,103301(2005); W. S. Koh, L. K. Ang, *Nanotechnology*, 19,235402(2008); M. Razavy, *Quantum Theory of Tunneling* (World Scientific Publishing Co. Pte. Ltd, Singapore 2003).

[4] Baranchuk S. I., N. V. Mileshkina, *Soviet Phys. Solid State,* 23,1715(1981); P. G. Borzyak, A. A. dadykin, Soviet Phys. Docklady, 27, 335(1982); S. Bono, R. H. Good, Jr., Surface Sci., 134,272(1983); S. M. Lyth, S. R. P. Silva, *Appl. Phys. Letts.,* 90,173124(2007); C. Xu, X. Sun, *Int. Jour. of Nanotech.,* 1, 452 (2004); S. D. Liang, L. Chen, *Phys. Rev. Letts.*, 101,027602(2008).

[5] Heeres E. C., E. P. A. M. Bakkers, A. L. Roest, M. Kaiser, T. H, Oosterkamp, N. de Jonge, *Nano Lett.,* 7,536 (2007); L. Dong, J. Jiao, D. W. Tuggle, J. M. Petty, S. A. Elliff, M. Coulter, *Appl. Phys. Lett.,* 82, 1096(2003); S. Y. Li, P. Lin, C. Y. Lee, T. Y. Tseng, *Jour. of Appl. Phys*, 95, 3711(2004); N. N. Kulkarni, J. Bae, C. K. Shih, S.

K. Stanley, S. S. Coffee, J. G. Ekerdt, *Appl. Phys. Lett.*, 87, 213115 (2005).

[6] Senthil K., K. Yong, *Mat. Chem. and Phys.,* 112, 88, (2008); R. Zhou, H.C. Chang, V. Protasenko, M. Kuno, A. K. Singh, D. Jena, H. Xing, *Jour.of Appl. Phys.,* 101, 073704 (2007); K. S. Yeong, J. T. L. Thong, *Jour.of Appl. Phys.,* 100, 114325 (2006); C. H. Oon, S. H. Khong, C. B. Boothroyd, J. T. L. Thong, *Jour.of Appl. Phys.*, 99, 064309 (2006).

[7] Kim B. H., M. S. Kim, K. T. Park, J. K. Lee, D. H. Park, J. Joo, S. G. Yu, S. H. Lee, *Appl. Phys. Lett.,* 83, 539 (2003); Z. S. Wu, S. Z. Deng, N. S. Xu, J. Chen, J. Zhou, J. Chen, *Appl. Phys. Lett.,* 80, 3829 (2002); Y. W. Zhu, T. Yu, F. C. Cheong, X. J. Xu, C. T. Lim, V. B. C. Tan, J. T. L. Thong, C. H. Sow, *Nanotechnology,* 16, 88 (2005); Y. W. Zhu, H. Z. Zhang, X. C. Sun, S. Q. Feng, J. Xu, Q. Zhao, B. Xiang, R. M. Wang, D. P. Yu, *Appl. Phys. Lett.,* 83,144(2003); S. Bhattacharjee, T. Chowdhury, *Appl. Phys. Lett.,* 95, 061501 (2009); S. Kher, A. Dixit, D. N. Rawat, M. S. Sodha, *Appl. Phys. Lett.* 96, 044101 (2010)

[8] Shigeo I., W.Teruo, O. Kazuyoshi, T. Masateru, U.Satoshi, N.Norio, Jour. of Vac. Sc. and Tech. B:, 13, 487(2009); C. A. Spindt, I. Brodie, L. Humphrey, E. R. Westerberg, *Jour. of Appl. Phys*, 47, 5248 (2009); Q. Fu, A. V. Nurmikko, L. A. Kolodziejski, R. L. Gunshor, J. W. Wu, *Appl. Phys. Lett.*, 51, 578 (2009).

[9] Petroff P. M., A. C. Gossard, W. Wiegmann, *Appl. Phys. Letts.* 45, 620 (1984); J. M. Gaines, P. M. Petroff, H. Kroemar, R. J. Simes, R. S. Geels, J. H. English, *J. Vac. Sci. Tech.* B 6, 1378 (1988).

[10] Cilbert J., P. M. Petroff, G. J. Dolan, S. J. Pearton, A. C. Gossard, J. H. English, *Appl. Phys. Letts*. 49, 1275 (1986).

[11] Fujui, H. Saito, *Appl. Phys. Letts.* 50, 824 (1987).

[12] Sasaki H., *Jpn. J. Appl. Phys.* 19, 94 (1980).

[13] Petroff P. M., A. C. Gossard, R. A. Logan, W. Weigmann, *Appl. Phys. Lett.* 41 635 (1982).

[14] Temkin H., G. J. Dolan, M. B. Panish, S. N. G. Chu, *Appl. Phys. Lett.* 50, 413 (1988); I. Miller, A. Miller, A. Shahar, U. Koren, P. J. Corvini, *Appl. Phys. Lett.* 54, 188 (1989).

[15] Chang L. L., H. Esaki, C. A. Chang, L. Esaki, *Phys. Rev. Letts.* 38, 1489 (1977); K. Less, M. S. Shur, J. J. Drunnond, H. Morkoc, *IEEE Trans. Electron. Devices* ED-30, 07 (1983); G. Bastard, Wave mechanics applied to semiconductor heterostructures, Halsted; Les Ulis, Les Editions de Physique, New York (1988); M. J. Kelly, Low dimensional semiconductors: materials, physics, technology, devices, Oxford University Press, Oxford (1995); C. Weisbuch, B. Vinter, Quantum semiconductor structures, Boston Academic Press, Boston, (1991).

[16] Linch N. T., *Festkorperprobleme,* 23, 27 (1985).

[17] Sciferes D. R., C. Lindstrom, R. D. Burnham, W. Streifer, T. L. Paoli, *Electron. Letts.* 19, 170 (1983).

[18] Solomon P. M., *Proc. IEEE,* 70, 489 (1982); T. E. Schlesinger, T. Kuech, *Appl. Phys. Lett.* 49, 519 (1986).

[19] Kasemet D., C. S. Hong, N. B. Patel, P. D. Dapkus, *Appl. Phys. Letts.*, 41, 912 (1982); K. Woodbridge, P. Blood, E. D. Pletcher, P. J. Hulyer, *Appl. Phys. Lett.*, 45, 16 (1984); S. Tarucha, H. O. Okamoto, *Appl. Phys. Letts.* 45, 16 (1984); H. Heiblum, D. C. Thomas, C. M. Knoedler, M. I. Nathan, *Appl. Phys. Letts.* 47, 1105 (1985).

[20] Aina O., M. Mattingly, F. Y. Juan, P. K. Bhattacharyya, *Appl. Phys. Letts.* 50, 43 (1987).

[21] Suemune I., L. A. Coldren, *IEEE, J. Quant. Electronic.* 24, 1178 (1988).

[22] Miller D. A. B., D. S. Chemla, T. C. Damen, J. H. Wood, A. C. Burrus, A. C. Gossard, W. Weigmann, *IEEE, J. Quant. Electron.* 21, 1462 (1985).

[23] Harrison P., Quantum Wells, Wires and Dots, John Wiley and Sons, Ltd, (2002); B. K. Ridley, *Electrons and Phonons in semiconductors multilayers,* Cambridge University Press, Cambridge (1997); V. V.

Martin, A. A. Kochelap, M. A. Stroscio, *Quantum Heterostructures,* Cambridge University Press, Cambridge (1999).

[24] Lent C. S., D. J. Kirkner, *J. Appl. Phys.* 67, 6353 (1990); F. Sols, M. Macucci, U. Ravaioli, K. Hess, *Appl. Phys. Lett.* 54, 350 (1980).

[25] Kim C. S., A. M. Satanin, Y. S. Joe, R. M. Cosby, *Phys. Rev. B*, 60, 10962 (1999).

[26] Midgley S., J. B. Wang, *Phys. Rev. B* 64, 153304 (2001).

[27] Sugaya T., J. P. Bird, M. Ogura, Y. Sugiyama, D. K. Ferry, K. Y. Jang, *Appl. Phys. Lett.* 80, 434 (2002).

[28] Kane B., G. Facer, A. Dzurak, N. Lumpkin, R. Clark, L. PfeiKer, K. West, *Appl. Phys. Lett.* 72, 3506 (1998).

[29] Dekker C., *Physics Today*, 52, 22 (1999).

[30] Yacoby A., H. L. Stormer, N. S. Wingreen, L. N. Pfeiffer, K. W. Baldwin, K. W. West, *Phys. Rev. Lett.* 77, 4612 (1996).

[31] Hayamizu Y., M. Yoshita, S. Watanabe, H. A. L. PfeiKer, K. West, *Appl. Phys. Lett.* 81, 4937 (2002).

[32] Frank S., P. Poncharal, Z. L. Wang, W. A. Heer, *Science* 280, 1744 (1998).

[33] Kamiya I., I. I. Tanaka, K. Tanaka, F. Yamada, Y. Shinozuka, H. Sakaki, *Physica E* 13, 131 (2002).

[34] Geim A. K., P. C. Main, N. LaScala, L. Eaves, T. J. Foster, P. H. Beton, J. W. Sakai, F. W. Sheard, M. Henini, G. Hill et al., *Phys. Rev. Lett.* 72, 2061 (1994).

[35] Melinkov A. S., V. M. Vinokur, *Nature* 415, 60 (2002).

[36] Schwab K., E. A. Henriksen, J. M. Worlock, M. L. Roukes, *Nature* 404, 974 (2000).

[37] Kouwenhoven L., *Nature* 403, 374 (2000).

[38] Komiyama S., O. Astafiev, V. Antonov, H. Hirai, *Nature* 403, 405 (2000).

[39] Paspalakis E., Z. Kis, E. Voutsinas, A. F. Terziz, *Phys. Rev. B* 69, 155316 (2004).

[40] Jefferson J. H., M. Fearn, D. L. J. Tipton, T. P. Spiller, *Phys. Rev. A* 66, 042328 (2002).

[41] Appenzeller J., C. Schroer, T. Schapers, A. Hart, A. Froster, B. Lengeler, H. Luth, *Phys. Rev. B* 53, 9959 (1996).
[42] Appenzeller J., C. Schroer, *J. Appl. Phys.* 87, 31659 (2002).
[43] Debray P., O. E. Raichev, M. Rahman, R. Akis, W. C. Mitchel, *Appl. Phys. Lett.* 74, 768 (1999).
[44] Kasemet D., C. S. Hong, N. B. Patel, P. D. Dapkus, *Appl. Phys. Letts.* 41, 912 (1982); K. Woodbridge, P. Blood, E. D. Pletcher, P. J. Hulyer, *Appl. Phys. Lett.* 45, 16 (1984); S. Tarucha, H. O. Okamoto, *Appl. Phys. Letts.* 45, 16 (1984);
[45] Nathan M. I., *Appl. Phys. Letts.* 47, 1105 (1985).
[46] Suemune, L. A. Coldren, IEEE, *J. Quant. Electronic.* 24, 1178 (1988).
[47] Shay J. L., J. W. Wernick, *Ternary Chalcopyrite Semiconductors-Growth, Electronic Properties and Applications*, (Pergamon Press, London 1975).
[48] Rowe J. W., J. L. Shay, *Phys. Rev.* B, 3, 451 (1973).
[49] Kildal H., *Phys. Rev.* B, 10, 5082 (1974).
[50] Bodnar J., in Proc. Int. Conf. of the Physics of Narrow-gap Semiconductors (Polish Science Publishers, Warsaw, 1978); G. P. Chuiko, N. N. Chuiko, *Sov. Phys. Semicond.* 15, 739 (1981); K. P. Ghatak, S. N. Biswas, *Proc. of SPIE,* 1484, 149 (1991).
[51] Rogalski A., *Jour. of Alloys and Comp.,* 371, 53, (2004)
[52] Baumgartner A., A. Chaggar, A. Patanè, L. Eaves, M. Henini, *Appl. Phys. Lett.* 92, 091121 (2008).
[53] Devenson J., R. Teissier, O. Cathabard, A. N. Baranov, *Proc. SPIE* 6909, 69090U (2008).
[54] Passmore B. S., J. Wu, M. O. Manasreh, G. J. Salamo, *Appl. Phys. Lett.* 91, 233508 (2007).
[55] Mikhailova M., N. Stoyanov, I. Andreev, B. Zhurtanov, S. Kizhaev, E. Kunitsyna, K. Salikhov, Y. Yakovlev, *Proc. SPIE* 6585, 658526 (2007).
[56] Kruppa W., J. B. Boos, B. R. Bennett, N. A. Papanicolaou, D. Park, R. Bass, *Electron. Lett.* 42, 688 (2006).

[57] Nag B. R., *Electron Transport in Compound Semiconductors*, (Springer Verlag, 1980); E. O. Kane, in *Semiconductors and Semimetals,* Vol. 1, Ed. By R. K. Willardson, A. C. Beer (Academic Press, New York 1966) p. 75.

[58] Stillman G. E., C. M. Wolfe and J. O. *Dimmock in semiconductors and semimetals,* Ed. R. K. Willardon and A. C. Beer 12, p 169 (1977) Academic Press, IV,V, USA.

[59] Newson D. J., and A. Karobe, Semiconductor. *Sci. and Tech.* 3, 786 (1988).

[60] Palik E. D., G. S. Picus, S. Teither and R. E. Wallis, *Phys. Rev*. 122, 475 (1961).

[61] Lu P. Y., C. H. Wung, C. M. Williams, S. N. G. Chu, C. M. Stiles, *Appl. Phys. Letts.* 49 (1986) 1372.

[62] Taskar N. R., I. B. Bhat, K. K. Prat, D. Terry, H. Ehasani, S. K. Ghandhi, *J. Vac. Sci. Tech.* 7A (1989) 281.

[63] Koch F., *Springer Series in Solid States Sciences* (Vol. 53 pp. 20, Springer-Verlag, Germany, 1984).

[64] Tomasetta L. R., H. D. Law, R. C. Eden, I. Reyhimy, K. Nakano, *IEEE, J. Quant. Electron.* 14 (1978) 800.

[65] Yamato T., K. Sakai, S. Akiba, Y. Suematsu, *IEEE J. Quantum Electronics* 14 (1978) 95.

[66] Pearsall T. P., B. I. Miller, R. J. Capik, *Appl. Phys. Letts.* 28 (1976) 499.

[67] Washington M. A., R. E. Nahory, M. A. Pollack, E. D. Beeke, *Appl. Phys. Letts.* 33 (1978) 854.

[68] Timmons M. I., S. M. Bedair, R. J. Markunas, J. A. Hutchby, *Proceedings of the 16th IEEE Photovoltaic Specialist Conference* (IEEE, San Diego, California 666 (1982)).

[69] Zapien J. A., Y. K. Liu, Y. Y. Shan, H. Tang, C. S. Lee, S. T. Lee, *Appl. Phys. Lett*. 90, 213114 (2007).

[70] Park M., *Proc. SPIE*, 2524, 142, (1995).

[71] Hur S. G., E. T. Kim, J. H. Lee, G. H. Kim, S. G. Yoon, *Electrochem. Solid-State Lett.* 11, H176 (2008); H. Kroemer, *Rev. of Mod. Phys.*, 73, 783, (2001); T. Nguyen Duy, J. Meslage, G.

Pichard, *Jour. of Crys. Growth,* 72, 490 (1985); T. Aramoto, F. Adurodija, Y. Nishiyama, T. Arita, A. Hanafusa, K. Omura, A. Morita: *Solar Energy Mat. & Solar Cells,* 75, 211, (2003); H. B. Barber, *Jour. of Elect. Mat.,* 25, 1232, (1996); S. Taniguchi, T. Hino, S. Itoh, K. Nakano, N. Nakayama, A. Ishibashi, M.Ikeda, *Elect. Lett.,* 32, 552 (1996).

[72] Hopfield J. J., *J. Appl. Phys.*, 32, 2277 (1961).

[73] Belosludov R. V., A. A. Farajian, H. Mizuseki, K. Miki, Y. Kawazoe, *Phys. Rev*. B, 75, 113411, (2007); J. Heremans, C. M. Thrush, Y. -M Lin, S. Cronin, Z. Zhang, M. S. Dresselhaus, J. F. Mansfield, *Phys. Rev*. B. ¸61, 2921. (2000)].

[74] Shoenberg D., *Proc. Roy. Soc.* (London) 170, 341 (1939); B. Abeles, S. Meiboom, *Phys. Rev.* 101, 544 (1956).

[75] Lax B., J. G. Mavroides, H. J. Zieger, R. J. Keyes, *Phys. Rev*. Letts. 5, 241 (1960).

[76] Kao Y. H., *Phys. Rev*. 129, 1122 (1963).

[77] Dinger R. J., A. W. Lawson, *Phys. Rev*. B, 3, 253 (1971).

[78] Koch J. F., J. D. Jensen, *Phys. Rev*. 184, 643 (1969).

[79] Cohen M. H., *Phys. Rev*. 121, 387 (1961).

[80] Takaoka S., H. Kawamura, K. Murase, S. Takano, *Phys. Rev*. B 13, 1428 (1976).

[81] McClure J. W., K. H. Choi, *Solid State Comm.* 21, 1015 (1977).

[82] Agrawal G. P., and N. K. Dutta, *Semiconductor Lasers,* Van Nostrand Reinhold, New York, (1993), p. 547; S. Chatterjee and U. Pal, *Opt. Eng.,* Bellingham, 32, 2923, (1993); T. K. Chaudhuri, *Int. J. Energy Res.* 16, 481, (1992); J. H. Dughaish, Physica B 322, 205, (2002); C. Wood, *Rep. Prog. Phys.* 51, 459, (1988).

[83] Hsu K. F., S. Loo, F. Guo, W. Chen, J. S. Dyck, C. Uher, T. Hogan, E. K. Polychroniadis, and M. G. Kanatzidis, *Science* 303, 818 (2004); J. Androulakis, K. F. Hsu, R. Pcionek, H. Kong, C. Uher, J. J. D'Angelo, A. Downey, T. Hogan, and M. G. Kanatzidis, *Adv. Mater., Weinheim, Ger.* 18, 1170 (2006); P. F. P. Poudeu, J. D'Angelo, A. D. Downey, J. L. Short, T. P. Hogan, and M. G. Kanatzidis, *Angew. Chem., Int. Ed.* 45, 3835, (2006); P. F. Poudeu,

J. D'Angelo, H. Kong, A. Downey, J. L. Short, R. Pcionek, T. P. Hogan, C. Uher, and M. G. Kanatzidis, *J. Am. Chem. Soc.* 128, 14347, (2006); J. R. Sootsman, R. J. Pcionek, H. Kong, C. Uher, and M. G. Kanatzidis, *Chem. Mater.* 18, 4993 (2006).

[84] Mountvala A. J., and G. Abowitz, *J. Am. Ceram. Soc.* 48, 651 (1965); E. I. Rogacheva, I. M. Krivulkin, O. N. Nashchekina, A. Yu. Sipatov, V. A. Volobuev, and M. S. Dresselhaus, *Appl. Phys. Lett.* 78, 3238, (2001); H. S. Lee, B. Cheong, T. S. Lee, K. S. Lee, W. M. Kim, J. W. Lee, S. H. Cho, and J. Y. Huh, *Appl. Phys. Lett.* 85, 2782, (2004); K. Kishimoto, M. Tsukamoto, and T. Koyanagi, *J. Appl. Phys.* 92, 5331, (2002).

[85] Rogacheva E. I., O. N. Nashchekina, S. N. Grigorov, M. A. Us, M. S. Dresselhaus, and S. B. Cronin, *Nanotechnology* 14, 53 (2003).

[86] Rogacheva E. I., O. N. Nashchekina, A. V. Meriuts, S. G. Lyubchenko, M. S. Dresselhaus, and G. Dresselhaus, *Appl. Phys. Lett.* 86, 063103 (2005); E. I. Rogacheva, S. N. Grigorov, O. N. Nashchekina, T. V. Tavrina, S. G. Lyubchenko, A. Yu. Sipatov, V. V. Volobuev, A. G. Fedorov, and M. S. Dresselhaus, *Thin Solid Films* 493, 41 (2005).

[87] Qiu X., Y. Lou, A. C. S. Samia, A. Devadoss, J. D. Burgess, S. Dayal, and C. Burda, *Angew. Chem., Int.* Ed. 44, 5855 (2005).

[88] Wang C., G. Zhang, S. Fan, and Y. Li, J. *Phys. Chem. Solids* 62, 1957 (2001); B. Poudel, W. Z. Wang, D. Z. Wang, J. Y. Huang, and Z. F. Ren, *J. Nanosci. Nanotechnol.* 6, 1050 (2006); B. Zhang, J. He, and T. M. Tritt, *Appl. Phys. Lett.* 88, 043119 (2006); W. Heiss, H. Groiss, E. Kaufmann, G. Hesser, M. Böberl, G. Springholz, F. Schäffler, K. Koike, H. Harada, and M. Yano, *Appl. Phys. Lett.* 88, 192109 (2006).

[89] Akimov B. A., V. A. Bogoyavlenskiy, L. I. Ryabova, and V. N. Vasil'kov, *Phys. Rev.* B 61, 16045 (2000); Ya. A. Ugai, A. M. Samoilov, M. K. Sharov, O. B. Yatsenko, and B. A. Akimov, *Inorg. Mater.* 38, 12 (2002); Ya. A. Ugai, A. M. Samoilov, S. A. Buchnev, Yu. V. Synorov, and M. K. Sharov, *Inorg. Mater.* 38, 450 (2002); A. M. Samoilov, S. A. Buchnev, Yu. V. Synorov, B. L. Agapov, and A.

M. Khoviv, *Inorg. Mater.* 39, 1132 (2003); A. M. Samoilov, S. A. Buchnev, E. A. Dolgopolova, Yu. V. Synorov, and A. M. Khoviv, Inorg. Mater. 40, 349 (2004); H. Murakami, W. Hattori, and R. Aoki, Physica C 269, 83 (1996); H. Murakami, W. Hattori, Y. Mizomata, and R. Aoki, *Physica C* 273, 41 (1996); H. Murakami, R. Aoki, and K. Sakai, *Thin Solid Films*, 27, 343 (1999).

[90] Volkov B. A., L. I. Ryabova, and D. R. Khokhlov, *Phys. Usp.* 45, 819 (2002), and references therein.

[91] Hüe F., M. Hÿtch, H. Bender, F. Houdellier, A. Claverie, *Phys. Rev.* Lett. 100, 156602 (2008); S. Banerjee, K. A. Shore, C. J. Mitchell, J. L. Sly, M. Missous, *IEE Proc., Circuits Devices Syst.* 152, 497 (2005); M. Razeghi, A. Evans, S. Slivken, J. S. Yu, J. G. Zheng, V. P. Dravid, Proc. SPIE, 5840, 54 (2005); R. A. Stradling, *Semicond. Sci. Technol.,* 6, *C52*, (1991).

[92] Weimer P. K., *Proc. IEEE,* 52, 608 (1964).

[93] Ribakovs G. and A. A. Gundjian, *IEEE, J. Quant. Electron.* QE-14, 42, (1978).

[94] Dey S. K., *J. Vac. Sc. Tech.,* 10, 227 (1973); S. J. Lynch, *Thin Solid Films,* 102, 47, (1983).

[95] Kudzin V. V., V. S. Kulakov, D. R. Pape', S. V. Kulakov, and V. V. Molotok, *IEEE, Ultrasonics Symposium,* 1, 749, (1997).

[96] Hatami F., V. Lordi, J. S. Harris, H. Kostial, W. T. Masselink, *J. Appl. Phys.* 97, 096106 (2005).

[97] Wessels B. W., *J. Electrochem. Soc.* 122, 402 (1975); D. W. L. Tolfree, *J. Sci. Instrum.,* 41, 788, 1964; P. B. Hart, *Proc. the IEEE,* 61, 880, 1973.

[98] Hines M. A., G. D. Scholes, *Adv. Mater.* 15, 1844, (2003); C. A. Wang, R. K. Huang, D. A. Shiau, M. K. Connors, P. G. Murphy, P. W. O'Brien, A. C. Anderson, D. M. DePoy, G. Nichols, M. N. Palmisiano, *Appl. Phys. Lett,* 83, 1286, (2003); C. W. Hitchcock, R. J. Gutmann, J. M. Borrego, I. B. Bhat, G. W. Charache, *IEEE Trans. Electron. Devices,* 46, 2154, (1999).

[99] Goldsmid H. J., R. W. Douglas, *Brit. J. Appl. Phys.* 5, 386 (1954).

[100] Rosi F. D., B. Abeles, R. V. Jensen, *J. Phys. Chem. Sol.* 10, 191 (1959); T. M. Tritt (ed.), *Semiconductors and Semimetals, Vols. 69, 70 and 71: Recent Trends in Thermoelectric Materials Research I, II and III* (Academic Press, USA, 2000); D. M. Rowe (Ed.), *CRC Handbook of Thermoelectrics* (CRC Press, USA, 1995); D. M. Rowe, C. M. Bhandari, *Modern Thermoelectrics* (Reston Publishing Company, Virginia, 1983); D. M. Rowe (ed.), *Thermoelectrics Handbook: Macro to Nano* (CRC Press, USA, 2006).

[101] Choi H., M. Chang, M. Jo, S. J. Jung, H. Hwang, *Electrochem. Solid-State Lett.* 11, H154 (2008). S. Cova, M. Ghioni, A. Lacaita, C. Samori, F. Zappa, *Appl. Opt.*, 35, 1956, (1996); H. W. H. Lee, B. R. Taylor, S. M. Kauzlarich, *Nonlinear Optics: Materials, Fundamentals, and Applications*, (Technical Digest, 12, 2000); E. Brundermann, U. Heugen, A. Bergner, R. Schiwon, G. W. Schwaab, S. Ebbinghaus, D. R. Chamberlin, E. E. Haller, M. Havenith, *29th International Conference on Infrared and Millimeter Waves and 12th International Conference on Terahertz Electronics*, 283, (2004).

[102] Baranov A. N., T. I. Voronina, N. S. Zimogorova, L. M. Kauskaya, Y. P. Yakoviev, *Sov. Phys. Semicond.* 19, 1676 (1985); M. Yano, Y. Suzuki, T. Ishii, Y. Matsushima, M. Kimata, *Jpn. J. Appl. Phys.* 17, 2091 (1978); F. S. Yuang, Y. K. Su, N. Y. Li, *Jpn. J. Appl. Phys.* 30, 207 (1991); F. S. Yuang, Y. K. Su, N. Y. Li, K. J. Gan, *J. Appl. Phys.* 68, 6383 (1990); Y. K. Su, S. M. Chen, *J. Appl. Phys.* 73, 8349 (1993); S. K. Haywood, A. B. Henriques, N. J. Mason, R. J. Nicholas, P. J. Walker, *Semicond. Sci. Technol.* 3, 315 (1988).

[103] Young C., W. W. Anderson, L. B. Anderson, *Trans. Electron Dev., IEEE,* 24, 492, (1977).

[104] Gordon R. L., V. I. Neeley, H. R. Curtin, *Proc. IEEE,* 54, 2014, (1966).

[105] Weimer P. K., *Proc. IRE,* 50, 1462, (1962).

[106] Lee M. J., S. W. Wright, C. P. Judge, P. Y. Cheung, Display Research Conference, International Conference Record, 211, (1991).

[107] Abramowitz M., I. A. Stegun, Handbook of Mathematical Functions (Dover Publications, New York, U.S.A., 1965).

[108] Haung A., *Theoretical Solid State Physics* (vol. 1, Pergamon Press, Oxford).

[109] Flugge S., Practical quantum mechanics (Springer Verlag, Germany, 1974); M. Razavy, *Quantum Theory of Tunneling* (World Scientific publishing Co. Pte. Ltd, Singapore 2003).

[110] Dimmock J. O., in "*The Physics of semimetals and narrowgap semiconductors*" Ed. by D. L. Carter, R. T. Bates (Pergamon Press, Oxford, 1971).

[111] Seiler D. G., B. D. Bajaj and A. E. Stephens, *Phys. Rev*. B 16, 2822 (1977); A. V. Germaneko and G. M. Minkov, *Phys. Stat. Sol*. (b) 184, 9 (1994); G. L. Bir and G. E. Pikus, *Symmetry and Strain – Induced effects in Semiconductors Nauka,* Russia (1972). (in Russian); M. Mondal and K. P. Ghatak, *Phys. Stat. Sol*. (b) 135, K21 (1986).

[112] Bouat J., and J. C. Thuillier, *Surface Science,* 73,528 (1978).

[113] Rees G. J., *Phys. of Compounds*, Proc. of the 13th Inter. Nat. Conf. Ed. F. G. Fumi, pp. 1166, North Holland Company, (1976).

[114] Emtage P. R., *Phys. Rev*. 138, A246 (1965).

[115] Stordeur M., W. Kuhnberger, *Phys. Stat. sol.* (b), 69, 377(1975); D. R. Lovett, *Semimetals and Narrow-Bandgap Semiconductor*,(Pion Limited, UK, 1977); H. Kohler, *Phys. Stat. Sol.* (b), 74,591(1976).

[116] Cardona M., W. Paul, H. Brooks Helv, *Acta Phys.* 33, 329 (1960); A. F. Gibson in *"Proceeding of International school of physics "ENRICO FERMI"* course XIII, Ed. R. A Smith, Academic Press, 171 (1963).

[117] Wang C. C., N. W. Ressler, *Phys. Rev.* 2, 1827 (1970).

[118] Mathur P. C., S. Jain, *Phys. Rev.*19, 1359(1979).

[119] Singh M., P. R. Wallace, S. D. Jog and E. Arushanov, *J. Phys. Chem.* Solids, 45, 409, (1984); Y. Yamada, *J. Phys. Soc. Japan,* 35,1600(1973),37,606(1974); J. Cisowski, J. C Portal, E. K. Erushanov, J. M. Broto, S. Huant and L. C. Brumel, *Phys. Stat. Sol* (b), 121, 289 (1984).

[120] Arushanov E. A., A. A. Kaynzev, A. N. Natepov and S. I. Radautsan, *Sov. Phys. Semicond.* 15, 828 (1981).

[121] Hong K. S., R. F. Speyer and R. A. Condrate, *J. Phys. Chem. Solids,* 51, 969 (1990).

[122] Mondal M., S. Banik and K. P. Ghatak, *J. Low. Temp. Phys.* 74, 423 (1989). M. J. Gelten, C. V. M. VanEs, F. A. P. Blom, J. W. F. Jongencelen, *Solid Stste Commun.,* 33, 833 (1980); A. A. El-Shazly; H. S. Soliman, H. E. A. El-Sayed and D. A. A. El-Hady, *J. Vac.,* 47, 53, (1996).

[123] Adachi S., *Properties of Group-IV, III-V and II-VI Semiconductors,* John Wiley and Sons, 2005.

[124] Madelung O., *Semiconductors: Data Handbook*, 3rd ed. (Springer-Verlag, Germany, (2003).

[125] Adachi S., *J. Appl. Phys.,* 58, R1, (1985).

[126] Adachi S., "*GaAs and Related Materials: Bulk Semiconductors and Superlattice Properties,"* (World Scientific, USA, 1994).

[127] Hansen G. L., J. L. Schmit and T. N. Casselman, *J. Appl. Phys.* 63, 7079 (1982).

[128] Wenus J., J. Rutkowski, and A. Rogalski, *IEEE Trans. Elect. Dev.,* 48, 1326, (2001).

[129] Adachi S., *J. Appl. Phys.* 53, 8775 (1982); S. K. Sutradhar, D. Chattopadhyay and B. R. Nag, *Phys. Rev.* (b) 25, 4069 (1982).

[130] Seiler D. G., W. M. Beeker and L. M. Roth, *Phys. Rev.* 1, 764 (1970).

[131] Skryabinskii I. V., Yu. I. Ukhanov, *Sov. Phys. Solid State,* 14, 2838, (1973).

[132] Tiwari S., and S. Tiwari, *Cryst. Res. Technol.,* 41, 78, (2006).

[133] Lowney J. R., and S. D. Senturia, *J. Appl. Phys.,* 47, 1771 (1976).

[134] Spicer W. E., and G. J. Lapeyre *Phys. Rev.* 139, A565 (1965); G. M. T. Foley and P. N. Langenberg, *Phys. Rev.* B, 15, 4850 (1977).

[135] Lovett D. R., In: *Semimetals and Narrow Band Gap Semiconductors* (Pion Limited, London, 1977).

[136] Wu C. C., and C. J. Lin, *J. Low. Temp. Phys.* 57, 469 (1984); *J.Rose* and R. Schuchardt, *Phys. Stat. Sol.* (b),117, 213 (1983).

[137] Ghatak K. P., S. Bhattachaya, D. De, *Einstein Relation in Compound Semiconductors and Nanostructures*, (Springer Series in Material Sciences, 116, 2009.
[138] Ivanov-Omskii V. I., A. Sh. Mekhtisev, S. A. Rustambekova and E. N. Ukraintsev, *Phys. Stat. Sol. (b)* 119, 159 (1983).
[139] Kim H., K. Cho, H. Song, B. Min, J. Lee, G. Kim, S. Kim, S. H. Kim and T. Noh, *Appl. Phys. Lett.* 83, 4619 (2003).
[140] Reynolds R. A., M. J. Brau, R. A. Chapman, *J. Phys. Chem. Solids* 29, 755, (1968).
[141] O'Shaughnessy J., and C. Smith, *Solid State Communications,* 8, 481, (1970).
[142] Brandt N. B., V. N. Davydov, V. A. Kulbachinskii and O. M. Nikitina, *Sov. Phys. Sol. Stat.* 29, 1014, (1987).
[143] Viculis L. M., J. J. Mack, O. M. Mayer, H. T. Hahn and R. B. Kaner, *J. Mater. Chem.*, 15, 974 (2005).
[144] Vassilev L. A., *Phys. Stat. Sol.* (b), 121, 203 (1984).
[145] Partin D. L., *Superlattices and Microstructures,* 1, 131, (1985).
[146] Turner W. J., A. S. Fischler, and W. E. Reese, *Phys. Rev.* 121, 759 (1961).
[147] Chuiko G. P., *Sov. Phys. Semi.*, 19 (12), 1381, (1985).
[148] W. E. Swank, P. G.Le Comber, *Phys. Rev.* 153, 844, (1967).
[149] D. Haneman, *J. Phys. Chem. Solids,* 11, 205, (1959).
[150] Wei N., G. Wu and J. Dong, *Phys. Lett. A* 325, 403 (2004).
[151] Reich S., J. Maultzsch, C. Thomsen and P. Ordejo´n, *Phys. Rev.* B, 66, 035412, (2006).
[152] Lundstrom M. S., J. Guo, "*Nanoscale Transistors: Device Physics, Modeling and Simulation,*" (Springer, USA, 2006); J. W. Mintmire and C. T. White, *Phys. Rev. Lett.,* 81, 2506 (1998).
[153] Buonocore F., F. Trani, D. Ninno, A. Di Matteo, G. Cantele and G. Iadonisi, *Nanotechnology,* 19, 025711, (2008).
[154] Ketterson J. B., *Phys. Rev.* 129, 18 (1963).
[155] Vilão R. C., J. M. Gil, A. Weidinger, H. V. Alberto, J. Piroto Duarte, N. A. de Campos, R. L. Lichti, K. H. Chow, S. P. Cottrell and S. F. J. Cox, *Phys. Rev.* B, 77, 235212, (2008).

[156] Kang I. and F. W. Wise, *Phys. Rev.* B, *J. Opt. Soc. Am. B,* 14, 1632, (1997).

[157] Cui D., J. Xu, S. Y. Xu, G. Paradee, B. A. Lewis and M. D. Gerhold, *IEEE, Trans. Elect. Dev.* 5, 362 (2006).

[158] Ivchenko E. L., G. E. Pikus *Sov. Phys. Semicond.* 13, 579 (1979).

[159] Bangert E., P. Kastner, *Phys. Stat. Sol* (b) 61, 503(1974).

[160] Foley G. M. T., P. N. Langenberg, *Phys. Rev.* B,15B,4850 (1977).

[161] Ivanov-Omskii V. I., A. Sh. Mekhtisev, S. A Rustambekova, E. N Ukraintsev, *Phys. Stat. Sol.* (b) 119,159 (1983).

[162] Zhang H. I., *Phys. Rev.* B, 1, 3450 (1970).

[163] Rossler U., *Solid State Commun.* 49, 943 (1984).

[164] Johnson J., D. H. Dickey, *Phys. Rev.* 1,2676(1970).

[165] Agafonov V. G., P. M. Valov, B. S. Ryvkin, I. D. Yarashetskin, *Sov. Phys. Semiconduct.* 12,1182 (1978).

[166] Averkiev N. S., V. M. Asnin, A. A. Bakun, A. M. Danishevskii, E. L. Ivchenko, G. E. Pikus, A. A. Rogachev, *Sov. Phys. Semicond.* 18, 379 and 402 (1984).

[167] Cunningham R. W., *Phys. Rev.* 167,761(1968).

[168] Yekimov A. I., A. A. Onushchenko, A. G. Plyukhin, A. I. L. Efros, *J. Expt. Theor. Phys.* 88, 1490 (1985).

[169] Roman B. J., A. W. Ewald, *Phys. Rev.* B5, 3914 (1972).

[170] Ghatak K. P., S. Chakrabarti, B. Chatterjee, M. Debbarma, N. Debbarma, *Journal of Advanced Science, Engineering and Medicine*, 11,914, (2019); S. K. Biswas, A. K. Pradhan, K. Bhattacharya, S. Maiti, M. Mitra, S. Chakrabarti, B. Chatterjee, K. P. Ghatak, *Journal of Advanced Science, Engineering and Medicine*, 11,901, (2019); K. P. Ghatak, S. Chakrabarti, B. Chatterjee, *Journal of Advanced Science, Engineering and Medicine*, 11,901, (2019).

[171] Paul R., P. K. Das, M. Mitra, K. P. Ghatak, *Journal of Advanced Science, Engineering and Medicine*, 11,903, (2019); PK Das, KP Ghatak, *Journal of nanoscience and nanotechnology,* 19, 2909, (2019).

[172] Ghatak K. P., S. Chakrabarti, B. Chatterjee, *Journal of Materials Focus*, 7, 361 (2018); K. P. Ghatak, S. Chakrabarti, B. Chatterjee, P.

K. Das, P. Dutta, A. Halder, *Journal of Materials Focus*, 7, 390 (2018); B. Chatterjee, N. Debbarma, M. Mitra, T. Datta, K. P. Ghatak, *Journal of Nanoscience and Nanotechnology*, 17, 3352 (2017).

[173] Kamakhya Prasad Ghatak, Debashis De, *Journal of Materials Focus*, 6, 114 (2017); P. K. Das, P. Dutta, A. Halder, J. Pal, N. Debbarma, S. Debbarma, K. P. Ghatak, *Journal of Materials Focus,* 6, 167 (2017); P. K. Das, P. Dutta, A. Halder, R. Bhattacharjee, K. P. Ghatak, *Journal of Materials Focus*, 6, 133 (2017).

[174] Bhattacharjee R., K. P. Ghatak, *Journal of Nanoscience and Nanotechnology*, 17, 640 (2017); M. Mitra, T. N. Sen, T. Datta, R. Bhattacharjee, LS Singh, KP Ghatak, *Journal of Nanoscience and Nanotechnology*, 17, 256 (2017).

[175] Paitya N., and K. P. Ghatak, *Quantum Matter*, 5,191 (2016); B. Chatterjee, S. Chakrabarti, S. K. Sen, M. Mitra and K. P. Ghatak, *Quantum Matter*, 5, 85, (2016); M. Mitra, M. Chakraborty, S. Debbarma, S. Chakraborty, S. K. Sen, B. Chatterjee, K. P. Ghatak, *Quantum Matter*, 5, 58, (2016).

[176] Sen T. N., K. P. Ghatak, J*our. of Nanosci. and Nanotechno.,* 16, 1229, (2016); S. Debbarma, K. P. Ghatak, *Journal of Nanosci. and Nanotechno.,* 16, 1095, (2016).

[177] Chatterjee B., S. Chakrabarti, M. Chakraborty, K. P. Ghatak, Reviews in Theoretical Science, 3, 428, (2015); S. M. Adhikari, K. P. Ghatak, *Quantum Matter*, 4, 599, (2015); S. Chakrabarti, B. Chatterjee, S Debbarma, K. P. Ghatak, *Journal of Nanoscience and Nanotechnology*, 9, 6460, (2015); S. M. Adhikari, A. Karmakar, K. P. Ghatak, *Journal of Reviews in Theo. Sci*. 3, 273, (2015).

[178] Mitra M., B. Chatterjee, K. P. Ghatak, *Jour. of Comput. and Theo. Nanosci.*, 12,1527, (2015); M. Mitra, S. Chakrabarti, M. Chakraborty, S. Debbarma, K. P. Ghatak, *Jour. of Compu. and Theo. Nanosci.*, 12, 1898, (2015); K. P. Ghatak, L. S. Singh, K. Sarkar, N. Debbarma, M. Debbarma, *Journal of Materials Focus*, 4, 85, (2015).

[179] Chakraborty M., K. P. Ghatak, *Quantum Matter*, 4, 104, (2015); K. Sarkar, M. Chakraborty, S Chakravarti, B. Chatterjee, K. P. Ghatak, *Jour. of Nanoeng. and Nanoman.*, 5,43, (2015); S Debbarma, K. P. Ghatak, *Journal of Reviews in Theo. Sci.*, 3,16, (2015).

[180] Chatterjee B., N. Debbarma, S. Debbarma, S. Chakrabarti, K. P. Ghatak, *Adv. Sci., Eng. and Med.*, 6, 1177, (2014); S Debbarma, S. Chakravarti, N. Debbarma, M. Mitra, K. P. Ghatak, *Jour. of Adv. Phys.*, 3, 213, (2014); S. Debbarma, N. Debbarma, B. Chatterjee, S. M. Adhikari, K. P. Ghatak, *Adv. Sci., Eng. and Med.*, 6, 1024, (2014); S. Chakrabarti, S. K. Sen, S. Chakraborty, L. S. Singh, K. P. Ghatak, *Adv. Sci., Eng. and Med.*, 6, 1042, (2014).

In: Quantum Wires: An Overview
Editor: Kamakhya Prasad Ghatak
ISBN: 978-1-53617-676-6

Chapter 2

THE EFFECTIVE MASS IN HEAVILY DOPED QUANTUM WIRES

M. Debbarma[1], R. Paul[2], B. Cahtterjee[3], S. Chakrabarti[3], D. De[4], S. Das[5] and P. K. Das[6,*]

[1]Department of Physics, Women's College, Tripura, India
[2]Department of Computer Science and Engineering, University of Engineering and Management, Kolkata, India
[3]Department of Computer Science and Engineering, University of Engineering and Management and Institute of Engineering and Management, Kolkata and Jaipur, Salt Lake Electronics Complex, Kolkata, and Gurukul, Jaipur, Rajasthan, India
[4]Department of Computer Science and Engineering, Maulana Abul Kalam Azad University of Technology, Bidhannagar, Kolkata, West Bengal
[5]Department of Applied Physics, University of Calcutta, Kolkata, India
[6]Department of Basic Science Institute of Engineering and Management, Management House, Salt Lake, Kolkata, West Bengal, India

[*] Corresponding Author's Email: prabir.das@iemcal.com.

ABSTRACT

In this chapter we study the effective mass (EM) of the carriers in heavily doped quantum Wires of (HDQWS) in this chapter in Sections 2.1–2.14, we have investigated the EMs in NWs of HD non-linear optical, III-V, II-VI, stressed Kane type materials by formulating the electron statistics including all types of anisotropies of the energy band constants within the frame work of Kane's theory. It appears that the EM in HDNWs of Cd3As2and CdGeAs2 at the lowest subband in both the cases are strong cross-sectional functions of the dimensions which converge to their corresponding bulk values at larger dimensions. The effect of crystal field splitting in case of dispersion relation of Cd_3As_2 lets the asymptotic fall to be closer to the bulk value. It appears from these figures that the difference in the energy band models in predicting the EM is almost in significant. Hence for all practical purposes for determination of EM, the second order model of Kane can fit well. Under certain limiting conditions all the results for all the models get simplified into the well-known's expressions of the parabolic energy bands.

INTRODUCTION

It is well-known that the concept of the EM in quantum materials is one of the basic pillars in the realm of modern low dimensional quantum science and technology [1]. Among the various definitions of the effective mass (e.g., effective acceleration mass, density-of-state effective mass, concentration effective mass, conductivity effective mass, Faraday rotation effective mass, etc.) [2], it is the effective momentum mass that should be regarded as the basic quantity [3], since it is this mass which appears in the description of transport phenomena and all other properties of the carriers in materials having arbitrary band structures [3] together with the fact that only the basic definition of mass generates the Einstein's $E = mc^2$ (where E, m and c are the energy, mass and velocity of light respectively) equation for photons and under the condition of stationary frame of reference [4]. It can be shown that it is the effective momentum mass which enters in various transport coefficients and plays the most dominant role in explaining the experimental results of different scattering mechanisms through Boltzmann's transport equation [5]. The carrier degeneracy in materials influences the effective mass when it is energy dependent. Under degenerate conditions, only the electrons at the Fermi surface of n-type materials participate in the conduction process and hence, the effective momentum mass of the electrons corresponding to the Fermi level would be of interest in electron transport under such conditions. The Fermi energy is again determined by the carrier energy spectrum and the electron statistics and therefore, these two features would determine the dependence of the EM in degenerate n-type materials under the degree of carrier degeneracy. In recent years, various energy wave vector dispersion relations have been proposed [6-15] which have created the interest in studying the EM in such materials under external conditions. Besides the physical properties of various quantum materials have been studied intensively [16]. The nature of variations of the EM has widely been studied in the literature [17-33]. Some of the significant features, which have emerged from these investigations, are:

1. The EM increases monotonically with increasing carrier concentration.
2. The EM increases with doping in heavily doped materials in the presence of band tails.
3. The nature of variations is significantly influenced by the energy band constants of various materials having different band structures.
4. The EM oscillates with inverse quantizing magnetic field due to SdH effect. The EM in Bismuth under magnetic quantization depends both on the Fermi energy and the magnetic quantum number due to the presence of band nonparabolicity only.
5. The EM increases with the magnitude of the quantizing electric field in n-channel inversion layers of III-V materials and depends on the sub-band index for both weak and strong electric field limits.
6. The EM in ultrathin films of nonlinear optical materials depends on the Fermi energy and size quantum numbers due to the specific dispersion relations.

7. The EM has significantly different values in superlattices and also in the presence of quantum confined superlattices of small gap materials with graded interfaces together with effective mass of lattices.

It is well known that heavy doping and carrier degeneracy are the keys to unlock the important properties of semiconductors and they are especially instrumental in dictating the characteristics of Ohomic contacts and Schottky contacts respectively [34-44]. It is an amazing fact that although the heavily doped semiconductors (HDS) have been investigated in the literature but the study of the corresponding EMs of HDS is still one of the open research problems. It is well known that the band tails are being formed in the forbidden zone of the HDS and can be explained by the overlapping of the impurity band with the conduction and valence bands [45]. Kane [46] and Bonch Bruevich [47] have independently derived the theory of band tailing for semiconductors having unperturbed parabolic energy bands. Kane's model [46] was used to explain the experimental results on tunneling [48] and the optical absorption edges [49, 50] in this context. Halperin and Lax [51] developed a model for band tailing applicable only to the deep tailing states. Although Kane's concept is often used in the literature for the investigation of band tailing [56, 57], it may be noted that this model [48, 56] suffers from serious assumptions in the sense that the local impurity potential is assumed to be small and slowly varying in space coordinates [57]. In this respect, the local impurity potential may be assumed to be a constant. In order to avoid these approximations, we have developed in this book, the electron energy spectra for HDS for studying the EM based on the concept of the variation of the kinetic energy [58, 67] of the electron with the local point in space coordinates. This kinetic energy is then averaged over the entire region of variation using a Gaussian type potential energy. On the basis of the E–k dispersion relation, we have obtained the electron statistics for different HDS for the purpose of numerical computation of the respective EMs. It may be noted that, a more general treatment of many-body theory for the DOS of HDS merges with one-electron theory under macroscopic conditions [45]. Also, the experimental results for the Fermi energy and others are the average effect of this macroscopic case. So, the present treatment of the one-electron system is more applicable to the experimental point of view and it is also easy to understand the overall effect in such a case [44]. In a HDS, each impurity atom is surrounded by the electrons, assuming a regular distribution of atoms, and it is screened independently [86, 89, 90]. The interaction energy between electrons and impurities is known as the impurity screening potential. This energy is determined by the inter-impurity distance and the screening radius (popularly known as the Debye screening length). The screening length changes with the band structure. Furthermore, these entities are important for HDS in characterizing the semiconductor properties [91-94] and the modern electronic devices [52, 59].The works on Fermi energy and the screening length in an n-type GaAs have already been initiated in the literature [60], based on Kane's model. Incidentally, the limitations of Kane's model [45, 63], as mentioned above, are also present in their studies.

In this chapter in sections 2.1,2.2, 2.3,2.4,2.5,2.6,2.7,2.8,2.9,2.10, 2.11, 2.12, 2.13 and 2.14, we have investigated the EMs in NWs of HD non-linear optical, III-V, II-VI, stressed Kane type, Te, GaP, $PtSb_2$, Bi_2Te_3, Ge, GaAs, II-V, lead germanium telluride and zinc and cadmium phosphides respectively. The section 4.3 contains the summary and conclusion pertaining to this chapter. The section 4 presents 19 open research problems.

THEORETICAL BACKGROUND

The EM in NWs of HD Nonlinear Optical Materials

Thus the generalized unperturbed electron energy spectrum for the bulk specimens of the nonlinear optical materials in the absence of band tails can be expressed as

$$\frac{\hbar^2 k_z^2}{2m_\parallel^*}+(\frac{b_\parallel}{b_\perp}\frac{c_\perp}{c_\parallel})\frac{\hbar^2 k_s^2}{2m_\perp^*}=\{\frac{E(\alpha E+1)(b_\parallel E+1)}{(c_\parallel E+1)}+\frac{\alpha b_\parallel}{c_\parallel}[\delta E+\frac{2}{9}(\Delta_\parallel^2-\Delta_\perp^2)]-(\frac{2}{9})\frac{\alpha b_\parallel}{c_\parallel}\frac{(\Delta_\parallel^2-\Delta_\perp^2)}{(c_\parallel E+1)}\}$$
$$-(\frac{\hbar^2 k_s^2}{2m_\perp^*})\{(\frac{b_\parallel}{b_\perp}\frac{c_\perp}{c_\parallel})[(\frac{\delta}{2}+\frac{\Delta_\parallel^2-\Delta_\perp^2}{6\Delta_\parallel})\frac{\alpha_\parallel}{\alpha_\parallel E+1}+(\frac{\delta}{2}-\{\frac{\Delta_\parallel^2-\Delta_\perp^2}{6\Delta_\parallel}\})\frac{c_\parallel}{c_\parallel E+1}]\}$$
$$where,\ b_\parallel\equiv 1/(E_g+\Delta_\parallel), c_\perp\equiv 1/(E_g+\frac{2}{3}\Delta_\perp), b_\perp\equiv 1/(E_g+\Delta_\perp), c_\parallel= 1/(E_g+\frac{2}{3}\Delta_\parallel)\ and\ \alpha=1/E_g \tag{1}$$

The Gaussian distribution $F(V)$ of the impurity potential is given by [122]

$$F(V)=(\pi\eta_g^2)^{1/2}\exp(-V^2/\eta_g^2) \tag{2}$$

where, η_g is the impurity screening potential. It appears from (1.4) that the variance parameter η_g is not equal to zero, but the mean value is zero. Further, the impurities are assumed to be uncorrelated and the band mixing effect has been neglected in this simplified theoretical formalism.

We have to average the kinetic energy in the order to obtain the EM in nonlinear optical materials in the presence of band tails. Using the (1) and (2), we get

$$[\frac{\hbar^2 k_z^2}{2m_\parallel^*}\int_{-\infty}^{E}F(V)dV]+[(\frac{b_\parallel}{b_\perp}\frac{c_1}{c_\parallel})\frac{\hbar^2 k_s^2}{2m_\perp^*}\int_{-\infty}^{E}F(V)dV]=\{\int_{-\infty}^{E}\frac{(E-V)[\alpha(E-V)+1][b_\parallel(E-V)+1}{[c_\parallel(E-V)+1]}F(V)dV$$
$$+\frac{\alpha b_\parallel}{c_\parallel}[\delta\int_{-\infty}^{E}(E-V)F(V)dV+\frac{2}{9}(\Delta_\parallel^2-\Delta_\perp^2)\int_{-\infty}^{E}F(V)dV]-(\frac{2}{9})\frac{\alpha b_\parallel}{c_\parallel}(\Delta_\parallel^2-\Delta_\perp^2)\int_{-\infty}^{E}\frac{F(V)dV}{[c_\parallel(E-V)+1]}\}-$$
$$(\frac{\hbar^2 k_s^2}{2m_\perp^*})\{(\frac{b_\parallel}{b_\perp}\frac{c_\perp}{c_\parallel})[(\frac{\delta}{2}+\frac{\Delta_\parallel^2-\Delta_\perp^2}{6\Delta_\parallel})\alpha\int_{-\infty}^{E}\frac{F(V)dV}{[\alpha(E-V)+1]}+(\frac{\delta}{2}-\frac{\Delta_\parallel^2-\Delta_\perp^2}{6\Delta_\parallel})c_\parallel\int_{-\infty}^{E}\frac{F(V)dV}{[c_\parallel(E-V)+1]}]\} \tag{3}$$

The (3) can be rewritten as

$$\frac{\hbar^2 k_z^2}{2m_\parallel^*}I(1)+(\frac{b_\parallel}{b_\perp}\frac{c_1}{c_\parallel})\frac{\hbar^2 k_s^2}{2m_\perp^*}I(1)==\{I_3(c_\parallel)+\frac{\alpha b_\parallel}{c_\parallel}[\delta I(4)+\frac{2}{9}(\Delta_\parallel^2-\Delta_\perp^2)I(1)]-(\frac{2}{9})\frac{\alpha b_\parallel}{c_\parallel}(\Delta_\parallel^2-\Delta_\perp^2)I_6(c_\parallel)\}-$$
$$(\frac{\hbar^2 k_s^2}{2m_\perp^*})\{(\frac{b_\parallel}{b_\perp}\frac{c_1}{c_\parallel})[(\frac{\delta}{2}+\frac{\Delta_\parallel^2-\Delta_\perp^2}{6\Delta_\parallel})\alpha I(\alpha)+(\frac{\delta}{2}\{\frac{\Delta_\parallel^2-\Delta_\perp^2}{6\Delta_\parallel}\})c_\parallel I(c_\parallel)]\} \tag{4}$$

where, $I(1) = \int_{-\infty}^{E} F(V)dV$ (5)

$$I_3(c_{\|}) = \int_{-\infty}^{E} \frac{(E-V)[\alpha(E-V)+1][b_{\|}(E-V)+1]}{[c_{\|}(E-V)+1]} F(V)dV$$

$$I_3(c_{\|}) = \int_{-\infty}^{E} \frac{(E-V)[\alpha(E-V)+1][b_{\|}(E-V)+1]}{[c_{\|}(E-V)+1]} F(V)dV$$

$$I_3(c_{\|}) = \int_{-\infty}^{E} \frac{(E-V)[\alpha(E-V)+1][b_{\|}(E-V)+1]}{[c_{\|}(E-V)+1]} F(V)dV \quad (6)$$

$$I(4) = \int_{-\infty}^{E} (E-V)F(V)dV \quad (7)$$

$$I(\alpha) = \int_{-\infty}^{E} \frac{F(V)dV}{[\alpha(E-V)+1]} \quad (8)$$

Substituting $E - V \equiv x$ and, $\frac{x}{\eta_g} \equiv t_0$ we get from (1.7)

$$I(1) = (\exp(-\frac{E^2}{\eta_g^2}) \Big/ \sqrt{\pi}) \int_0^{\infty} \exp[-t_0^2 + (\frac{2Et_0}{\eta_g})] dt_0$$

Thus

$$I(1) = [\frac{1 + Erf(E/\eta_g)}{2}] \quad (9)$$

where, $Erf(E/\eta_g)$ *is the error function of* $((E/\eta_g))$

From (7), one can write

$$I(4) = (1/\eta_g\sqrt{\pi}) \int_{-\infty}^{E} (E-V)\exp(-V^2 \big/ \eta_g^2) dV = \frac{E}{2}[1 + Erf(E \big/ \eta_g)] - \{\frac{1}{\sqrt{\pi\eta_g^2}} \int_{-\infty}^{E} V\exp(-V^2/\eta_g^2)dV\}$$

(10)

After computing this simple integration, one obtains
Thus,

$$I(4) = \eta_g \exp(-E^2/\eta_g^2)(2\sqrt{\pi}) + \frac{E}{2}(1 + Erf(E/\eta_g)) = \gamma_0(E, \eta_g) \quad (11)$$

From (8), we can write

$$I(\alpha) = \frac{1}{\sqrt{\pi\eta_g^2}} \int_{-\infty}^{E} \frac{\exp(-V^2/\eta_g^2)dV}{[\alpha(E-V)+1]} \quad (12)$$

when, $V \to \pm\infty, \frac{1}{[\alpha(E-V)+1]} \to 0$ and; $\exp(-V^2/\eta_g^2) \to 0$

Thus (12) can be expressed as

$$I(\alpha) = (1/\alpha\eta_g\sqrt{\pi})\int_{-\infty}^{\infty} \exp(-t^2)(u-t)^{-1} dt \quad (13)$$

where, $\frac{V}{\eta_g} = t$ and $u \equiv (\frac{1+\alpha E}{\alpha\eta})$

It is well known that

$$W(Z) = (i/\pi)\int_{-\infty}^{\infty} (Z-t)^{-1} \exp(-t^2)dt \quad (14)$$

In which $i = \sqrt{-1}$ and Z, in general, is a complex number.
We also know

$$W(Z) = \exp(-Z^2)Erfc(-iZ) \quad (15)$$

where, $Erfc(Z) \equiv 1 - Erf(Z)$.

Thus, $Erfc(-iu) = 1 - Erf(-iu)$

Since, $Erf(-iu) = -Erf(iu)$

Thus,

$$I(\alpha) = [-i\sqrt{\pi}/\alpha\eta_g]\exp(-u^2)[1 + Erf(\mathrm{iu})] \tag{16}$$

We also know that

$$Erf(x+iy) = Erf(x) + (\frac{e^{-x^2}}{2\pi x})[(1-\cos(2xy)) + i\sin(2xy)$$

$$+ i\sin(2xy) + \frac{2}{\pi}e^{-x^2}\sum_{p=1}^{\infty}\frac{\exp(-p^2/4)}{(p^2+4x^2)}][f_p(x,y) + ig_p(x,y) + \varepsilon(x,\mathrm{y})] \tag{17}$$

where, $f_p(x,y) = [2x - 2x\cosh(py)\cos(2xy) + p\sinh(py)\sin(2xy)]$, $g_p(x,y) \equiv [2x\cosh(py)\sin(2xy) + p\sinh(py)\cos(2xy)] |\varepsilon(x,y)| \approx 10^{-15} |Erf(x+iy)|$

Substituting $x = 0$ and $y = u$ in (2.19), one obtains,

$$Erf(iu) = (\frac{2i}{\pi})\sum_{p=1}^{\infty}\{\frac{\exp(-p^2/4)}{p}\sinh(pu)\} \tag{18}$$

Therefore, one can write

$$I(\alpha) = C_{21}(\alpha, E, \eta_g) - iD_{21}(\alpha, E, \eta_g) \tag{19}$$

where

$C_{21}(\alpha,E,\eta_g) = [\frac{2}{\alpha\eta_g\sqrt{\pi}}]\exp(-u^2)[\sum_{p=1}^{\infty}\{\frac{\exp(-p^2/4)}{p}\sinh(pu)\}]$ and $D_{21}(\alpha,\mathrm{E},\eta_g) = [\frac{\sqrt{\pi}}{\alpha\eta_g}\exp(-u^2)]$

The (19) consists of both real and imaginary parts and therefore, $I(\alpha)$ is complex, which can also be proved by using the method of analytic continuation of Complex Analysis.

The integral $I_3(c_{\|})$ in (6) can be written as

$$I_3(c_{\|}) = (\frac{\alpha b_{\|}}{c_{\|}})I(5) + (\frac{\alpha c_{\|} + b_{\|}c_{\|} - \alpha b_{\|}}{c_{\|}^2})I(4) + \frac{1}{c_{\|}}(1-\frac{\alpha}{c_{\|}})(1-\frac{b_{\|}}{c_{\|}})I(1) - \{\frac{1}{c_{\|}}(1-\frac{\alpha}{c_{\|}})(1-\frac{b_{\|}}{c_{\|}})I(c_{\|})\} \tag{20}$$

where

$$I(5) = \int_{-\infty}^{E} (E-V)^2 F(V)dV \tag{21}$$

From (21) one can write

$$I(5) = \frac{1}{\sqrt{\pi\eta_g^2}}[E^2\int_{-\infty}^{E} \exp(\frac{-V^2}{\eta_g^2})dV - 2E\int_{-\infty}^{E} V\exp(\frac{-V^2}{\eta_g^2})dV + \int_{-\infty}^{E} V^2\exp(\frac{-V^2}{\eta_g^2})dV]$$

The evaluations of the component integrals lead us to write

$$I(5) = \frac{\eta_g E}{2\sqrt{\pi}}\exp(\frac{-E^2}{\eta_g^2}) + \frac{1}{4}(\eta_g^2 + 2E^2)[1 + Erf(\frac{E}{\eta_g})] = \theta_0(E,\eta_g) \tag{22}$$

Thus combining the aforementioned equations, $I_3(c_{\|})$ can be expressed as

$$I_3(c_{\|}) = A_{21}(E,\eta_g) + iB_{21}(E,\eta_g) \tag{23}$$

$$where,\ A_{21}(E,\eta)=[\frac{\alpha b_{\|}}{c_{\|}}[\frac{\eta_g E}{2\sqrt{\pi}}\exp(\frac{-E^2}{\eta_g^2})-\frac{1}{4}(\eta_g^2+2E^2)\{1+Erf(\frac{E}{\eta_g})\}]$$

$$+[\frac{\alpha c_{\|}+b_{\|}c_{\|}-\alpha b_{\|}}{c_{\|}^2}]\{\frac{E}{2}[1+Erf(E/\eta)]+\frac{\eta\exp(-E^2/\eta_g^2)}{2\sqrt{\pi}}\}$$

$$+\frac{1}{c_{\|}}(1-\frac{\alpha}{c_{\|}})(1-\frac{b_{\|}}{c_{\|}})\frac{1}{2}[1+Erf(E/\eta)]-\{\frac{2}{c_{\|}^2\eta_g\sqrt{\pi}}(1-\frac{\alpha}{c_{\|}})(1-\frac{b_{\|}}{c_{\|}})\exp(-u_1^2)\}$$

$$[\sum_{p=1}^{\infty}\{\frac{\exp(-p^2/4)}{p}\sinh(pu_1)\}]$$

$$u_1\equiv[\frac{1+c_{\|}E}{c_{\|}\eta_g}]\ and\ B_{21}(E,\eta_g)\equiv\frac{\sqrt{\pi}}{c_{\|}^2\eta_g}(1-\frac{\alpha}{c_{\|}})(1-\frac{b_{\|}}{c_{\|}})\exp(-u_1^2)$$

Therefore, the combination of all the appropriate integrals together with algebraic manipulations leads to the expression of the dispersion relation of the conduction electrons of HD nonlinear optical materials forming Gaussian band tails as

$$\frac{h^2k_z^2}{2m_{\|}^{*}T_{21}(E,h_g)}+\frac{h^2k_s^2}{2m_{\wedge}^{*}T_{22}(E,h_g)}=1 \tag{24}$$

where, $T_{21}(E,h_g)$ and $T_{22}(E,h_g)$ have both real and complex parts and are given by

$$T_{21}(E,\eta_g)=[T_{27}(E,\eta_g)+iT_{28}(E,\eta_g)],\ T_{27}(E,\eta_g)=[\frac{T_{23}(E,\eta_g)}{T_5(E,\eta_g)}]$$

$$T_{23}(E,\eta_g)\equiv[A_{21}(E,\eta_g)+\frac{\alpha b_{\|}}{c_{\|}}[\delta\gamma_0(E,\eta_g)+\frac{1}{9}(\Delta_{\|}^2-\Delta_{\perp}^2)[1+Erf(E/\eta_g)]]$$

$$-\{\frac{2}{9}(\frac{\alpha b_{\|}}{c_{\|}})(\Delta_{\|}^2-\Delta_{\perp}^2)G_{21}(c_{\|},E,\eta_g)\}],$$

$$G_{21}(E,\eta_g)\equiv\frac{2}{c_{\|}\eta_g\sqrt{\pi}}\exp(-u_1^2)\sum_{p=1}^{\infty}\{\frac{\exp(-p^2/4)}{p}\sinh(pu_1)\},T_5(E,\eta_g)\equiv\frac{1}{2}[1+Erf(E/\eta_g)],$$

$$T_{28}(E,\eta_g) \equiv [\frac{T_{24}(E,\eta_g)}{T_5(E,\eta_g)}], T_{24}(E,\eta_g) \equiv [B_{21}(E,\eta_g) + \frac{2}{9}\frac{\alpha b_{\|}}{c_{\|}}(\Delta_{\|}^2 - \Delta_{\perp}^2) H_{21}(c_{\|}, E, \eta_g)]$$

$$H_{21}(c_{\|}, E, \eta_g) \equiv [\frac{\sqrt{\pi}}{\eta_g c_{\|}} \exp(-u_1^2)], T_{22}(E,\eta_g) \equiv [T_{29}(E,\eta_g) + iT_{30}(E,\eta_g)],$$

$$T_{29}(E,\eta_g) \equiv \frac{T_3(E,\eta_g)\mathrm{T}_{25}(E,\eta_g) - T_{24}(E,\eta_g)T_{26}(E,\eta_g)}{[(T_{25}(E,\eta_g))^2 + (T_{26}(E,\eta_g))^2]},$$

$$T_{25}(E,\eta_g) \equiv [(\frac{b_{\|}}{b_{\perp}}\frac{c_{\perp}}{c_{\|}})\frac{1}{2}[1 + Erf(\frac{E}{\eta_g})] + (\frac{b_{\|}}{b_{\perp}}\frac{c_{\|}}{c_{\perp}})(\frac{\delta}{2} + [\frac{\Delta_{\|}^2 - \Delta_{\perp}^2}{6\Delta_{\|}}])\alpha_{\|} C_{21}(\alpha_{\|}, E, \eta_g)$$

$$+(\frac{b_{\|}c_{\perp}}{b_{\perp}})(\frac{\delta}{2} - [\frac{\Delta_{\|}^2 - \Delta_{\perp}^2}{6\Delta_{\|}}])G_{21}(\alpha_{\|}, E, \eta_g)]$$

$$C_{21}(\alpha, E, \eta_g) \equiv [\frac{2}{\alpha\sqrt{\pi}\eta_g}\exp(-u^2)[\sum_{p=1}^{\infty}\frac{\exp(-p^2/4)}{p}\sinh(pu)]]$$

$$T_{26}(E,\eta_g) \equiv (\frac{b_{\|}}{b_{\perp}}\frac{c_{\perp}}{c_{\|}})(\frac{\delta}{2} - \frac{\Delta_{\|}^2 - \Delta_{\perp}^2}{6\Delta_{\|}})\alpha D_{21}(\alpha, E, \eta_g) + \frac{b_{\|}c_{\perp}}{b_{\perp}}(\frac{\delta}{2} - \frac{\Delta_{\|}^2 - \Delta_{\perp}^2}{6\Delta_{\|}})H_{21}(c_{\|}, E, \eta_g)$$

and

$$T_{30}(E,\eta_g) \equiv \frac{T_{24}(E,\eta_g)T_{25}(E,\eta_g) + T_{23}(E,\eta_g)T_{26}(E,\eta_g)}{[(T_{25}(E,\eta_g))^2 + (T_{26}(E,\eta_g))^2]}$$

From (26), it appears that the energy spectrum in HD nonlinear optical materials is complex. The complex nature of the electron dispersion law in HD materials occurs from the existence of the essential poles in the corresponding electron energy spectrum in the absence of band tails. It may be noted that the complex band structures have already been studied for bulk materials and super lattices without heavy doping and bears no relationship with the complex electron dispersion law as indicated by (26). The physical picture behind the formulation of the complex energy spectrum in HDS is the interaction of the impurity atoms in the tails with the splitting constants of the valance bands. More is the interaction; more is the prominence of the complex part than the other case. In the absence of band tails, h_g ® 0, and there is no interaction of the impurity atoms in the tails with the spin orbit constants. As a result, there exist no complex energy spectrum and (26) gets converted into (2) when h_g ® 0. Besides, the complex spectra are not related to same evanescent modes in the band tails and the conduction bands.

The DR of the 1D electrons in this case can be written as

$$\frac{\hbar^2(n_z\pi/d_z)^2}{2m_{\|}^{*}T_{21}(E,\eta_g)}+\frac{\hbar^2(n_y\pi/d_y)^2}{2m_{\|}^{*}T_{22}(E,\eta_g)}+\frac{\hbar^2k_x^2}{2m_{\|}^{*}T_{21}(E,\eta_g)}=1 \tag{25}$$

where, n_z(=1, 2, 3, …), d_z are the size quantized number and the nano thickness along the z-direction respectively, n_y(=1, 2, 3, …), d_y are the size quantized number and the nano thickness along the z-direction respectively.

The 1D DOS function per sub-band is given by

$$N_{1D}(E)=\frac{2g_v}{\pi}\frac{\partial k_x}{\partial E} \tag{26}$$

Thus by using (4.1) and (4.2) the total DOS function in this case can be written as

$$N_{1DHD\Gamma}(E)=\frac{2g_v}{\pi}\sum_{n_y=1}^{n_{y\max}}\sum_{n_z=1}^{n_{z\max}}T_1'(E,n_y,n_z,\eta_g)H(E-E'_{1HDNW}) \tag{27}$$

where, E'_{1HDNW} is the complex sub-band energy which can be expressed in this case as

$$\frac{\hbar^2(n_z\pi/d_z)^2}{2m_{\|}^{*}T_{21}(E'_{1HDNW},\eta_g)}+\frac{\hbar^2(n_y\pi/d_y)^2}{2m_{\|}^{*}T_{22}(E'_{1HDNW},\eta_g)}=1 \tag{28}$$

and

$$T_{1HDNW}(E,n_y,n_z,\eta_g)=[[1-\frac{\hbar^2(n_z\pi/d_z)^2}{2m_{\|}^{*}T_{21}(E,\eta_g)}+\frac{\hbar^2(n_y\pi/d_y)^2}{2m_{\|}^{*}T_{22}(E,\eta_g)}]\frac{2m_{\|}^{*}T_{21}(E,\eta_g)}{\hbar^2}]^{1/2} \tag{29}$$

The EM in this case in given by

$$m^*(E_{F1HDNW}, n_y, n_z, \eta_g) = \frac{\hbar^2}{2}[\text{Re}\,al\ Part\ of\ \frac{\partial}{\partial(E_{F1HDNW})}[T_{1HDNW}(E, n_y, n_z, \eta_g)]^2] \tag{30}$$

where $E_{F\ 1HDNW}$ in the Fermi energy in this case.

Thus, we observe that the EM is the function of size quantum numbers in both the directions and the Fermi energy due to the combined influence of the crystal filed splitting constant and the anisotropic spin-orbit splitting constants respectively. Besides it is a function of η_g due to which the EM exists in the band gap, which is otherwise impossible.

The electron concentration per unit length in this case can be written as

$$n_{01D} = \frac{2g_v}{\pi}\text{Re}\,al\ Part\ of \sum_{n_y=1}^{n_{y_{max}}}\sum_{n_z=1}^{n_{z_{max}}}[T_1(E_{F1HDNW}, n_y, n_z, \eta_g) + \sum_{r=1}^{s} L(r) T_1(E_{F1HDNW}, n_y, n_z, \eta_g)]] \tag{31}$$

The EM in NWs of HD III-V Materials

(i) Three Band Model of Kane

The dispersion relation of the 1D electrons in this case can be written as

$$\frac{\hbar^2 k^2}{2m_c} = T_{31}(E, \eta_g) + iT_{32}(E, \eta_g) \tag{32}$$

$$where,\ T_{31}(E,\eta_g) \equiv (\frac{2}{1+Erf(E/\eta_g)})[\frac{ab}{c}\theta_0(E,\eta_g) + [\frac{\alpha c + bc - \alpha b}{c^2}]\gamma_0(E,\eta_g) + \frac{1}{c}(1-\frac{\alpha}{c})(1-\frac{b}{c})\frac{1}{2}[1+Erf(\frac{E}{\eta_g})]$$
$$-\frac{1}{c}(1-\frac{\alpha}{c})(1-\frac{b}{c})\frac{2}{c\eta_g\sqrt{\pi}}\exp(-u_2^2)[\sum_{p=1}^{\infty}\frac{\exp(-p^2/4)}{p}\sinh(pu_2)]], b \equiv (E_g+\Delta)^{-1}, c \equiv ()^{-1}$$
$$u_2 \equiv \frac{1+cE}{c\eta_g}\ \text{and}\ T_{32}(E,\eta_g) \equiv (\frac{2}{1+Erf(E/\eta_g)})\frac{1}{c}(1-\frac{\alpha}{c})(1-\frac{b}{c})\frac{\sqrt{\pi}}{c\eta_g}\frac{\sqrt{\pi}}{c\eta_g}\exp(-u_2^2)$$

Thus, the complex energy spectrum occurs due to the term $T_{32}(E,\eta_g)$ and this imaginary band is quite different from the forbidden energy band.

The Dispersion Relation of the 1D electrons in this case can be written following (1.48) as

$$\frac{\hbar^2(n_z\pi/d_z)^2}{2m_c}+\frac{\hbar^2(n_y\pi/d_y)^2}{2m_c}+\frac{\hbar^2k_x^2}{2m_c}=T_{31}(E,\eta_g)+iT_{32}(E,\eta_g) \tag{33}$$

The DOS function in this case assume the form

$$N_{1DHD\Gamma}(E,\eta_g)=\frac{2g_v}{\pi}\text{Re}\,al\;part\,of\sum_{n_y=1}^{n_{y_{\max}}}\sum_{n_z=1}^{n_{z_{\max}}}\times T_2'(E,n_y,n_z,\eta_g)H(E-E_{2HDNW}') \tag{34}$$

where

$$T_2(E,n_y,n_z,\eta_g)=[[T_{31}(E,\eta_g)+iT_{32}(E,\eta_g)-[\frac{\hbar^2(n_z\pi/d_z)^2}{2m_c}+\frac{\hbar^2(n_y\pi/d_y)^2}{2m_c}]]\frac{2m_c}{\hbar^2}]^{1/2}$$

In (34) E_{2HDNW}' is the sub-band energy in this case which can be expressed as

$$\frac{\hbar^2(n_z\pi/d_z)^2}{2m_c}+\frac{\hbar^2(n_y\pi/d_y)^2}{2m_c}+\frac{\hbar^2k_z^2}{2m_c}=T_{31}(E_{2HDNW}',\eta_g)+iT_{32}(E_{2HDNW}',\eta_g) \tag{35}$$

The EM in this case is given by

$$m^*(E_{2HDNW}',\eta_g)=m_c[T_{31}(E_{2HDNW}'\eta_g)] \tag{36}$$

The electron concentration per unit length can be written as

$$n_{01D} = \frac{2g_v}{\pi} \operatorname{Re}al\ part\ of \sum_{n_y=1}^{n_{y\max}} \sum_{n_z=1}^{n_{z\max}} [T_2(E'_{F2HDNW}, n_y, n_z, \eta_g) + \sum_{r=1}^{s} L(r) T_2(E'_{F2HDNW}, n_y, n_z, \eta_g)] \tag{37}$$

(ii) Two band model of Kane

$$\frac{\hbar^2 k^2}{2m_c} = \gamma_2(E, \eta_g) \tag{38}$$

$$where,\ \gamma_2(E, \eta_g) \equiv [\frac{2}{1 + Erf(E/\eta_g)}][\gamma_0(E, \eta_g) + \alpha\theta_0(E, \eta_g)]$$

The DR of the 1D electrons in this case can be written as

$$\frac{\hbar^2 (n_z \pi / d_z)^2}{2m_c} + \frac{\hbar^2 (n_y \pi / d_y)^2}{2m_c} + \frac{\hbar^2 k_x^2}{2m_c} = \gamma_2(E, \eta_g) \tag{39}$$

The DOS function in this case assume the form

$$N_{1DHD\Gamma}(E, \eta_g) = \frac{2g_v}{\pi} \sum_{n_y=1}^{n_{y\max}} \sum_{n_z=1}^{n_{z\max}} T_3'(E, n_y, n_z, \eta_g) H(E - E'_{3HDNW}) \tag{40}$$

where

$$T_3(E, n_y, n_z, \eta_g) = [[\gamma_2(E, \eta_g) - [\frac{\hbar^2 (n_z \pi / d_z)^2}{2m_c} + \frac{\hbar^2 (n_y \pi / d_y)^2}{2m_c}]] \frac{2m_c}{\hbar^2}]^{1/2}$$

In (40) E'_{2HDNW} is the sub-band energy in this case which can be expressed as

$$\frac{\hbar^2 (n_z \pi / d_z)^2}{2m_c} + \frac{\hbar^2 (n_y \pi / d_y)^2}{2m_c} = \gamma_2(E'_{3HDNW}, \eta_g) \tag{41}$$

The EM in this case is given by

$$m^*(E'_{F1HDNW},\eta_g)=m_c[\gamma_2(E'_{F1HDNW}\eta_g)] \tag{42}$$

The electron concentration per unit length can be written as

$$n_{01D}=\frac{2g_v}{\pi}\sum_{n_y=1}^{n_{y_{\max}}}\sum_{n_z=1}^{n_{z_{\max}}}[T_3(E_{F1HDNW},n_y,n_z,\eta_g)+\sum_{r=1}^{s}L(r)[T_3(E_{F1HDNW},n_y,n_z,\eta_g)]] \tag{43}$$

(iii) Parabolic energy bands

The dispersion relation in HDS whose energy spectrum in the absence of band tails obeys the parabolic energy bands is given by

$$\frac{\hbar^2k^2}{2m_c}=\gamma_3(E,\eta_g) \tag{44}$$

where, $\gamma_3(E,\eta_g)\equiv[\frac{2}{(1+Erf(E/\eta_g))}]\gamma_0(E,\eta_g)$.

Since the dispersion relation in accordance with the said model is an all zero function with no pole in the finite complex plane, therefore the HD counterpart will be totally real, which is also apparent form the expression (78).

The EM in this case can be written as

$$m^*(E_{Fh},\eta_g)=m_c\{\gamma_3(E_{Fh},\eta_g)\}' \tag{45}$$

In the absence of band tails, $n_g\to 0$ and the EM assumes the form

$$m^*(E_F)=m_c \tag{46}$$

It is well-known that the EM in unperturbed parabolic energy bands is a constant quantity in general excluding cross-fields configuration. However, the same mass in the corresponding HD bulk counterpart becomes a complicated function of Fermi energy and the impurity potential together with the fact that the EM also exists in the band gap solely due to the presence of finite η_g. The DR of the 1D electron in this case can be written as

$$\frac{\hbar^2(n_z\pi/d_z)^2}{2m_c}+\frac{\hbar^2(n_y\pi/d_y)^2}{2m_c}+\frac{\hbar^2k_x^2}{2m_c}=\gamma_3(E,\eta_g) \tag{47}$$

The DOS function in this case assume the form

$$N_{1DHD\Gamma}(E,\eta_g)=\frac{2g_v}{\pi}\sum_{n_y=1}^{n_{y\max}}\sum_{n_z=1}^{n_{z\max}}T_4'(E,n_y,n_z,\eta_g)H(E-E_{4HDNW}') \tag{48}$$

$$where,\ T_4(E,n_y,n_z,\eta_g)=[[\gamma_3(E,\eta_g)-[\frac{\hbar^2(n_z\pi/d_z)^2}{2m_c}+\frac{\hbar^2(n_y\pi/d_y)^2}{2m_c}]]\frac{2m_c}{\hbar^2}]^{1/2}$$

In (48) E_{3HDNW}' is the sub-band energy in this case which can be expressed as

$$\frac{\hbar^2(n_z\pi/d_z)^2}{2m_c}+\frac{\hbar^2(n_y\pi/d_y)^2}{2m_c}=\gamma_3(E_{3HDNW}',\eta_g) \tag{49}$$

The EM in this case is given by

$$m^*(E_{F1HDNW}',\eta_g)=m_c[\gamma_3'(E_{F1HDNW},\eta_g)] \tag{50}$$

The electron concentration per unit length can be written as

$$n_{01D} = \frac{2g_v}{\pi} \sum_{n_y=1}^{n_{y\max}} \sum_{n_z=1}^{n_{z\max}} [T_4(E_{F1HDNW}, n_y, n_z, \eta_g) + \sum_{r=1}^{s} L(r)[T_4(E_{F1HDNW}, n_y, n_z, \eta_g)]] \tag{51}$$

(iv) The Model of Stillman et al.

$$\frac{\hbar^2 k^2}{2m_c} = I_{12}(E, \eta_g) \tag{52}$$

$where,\ I_{12}(E,\eta_g) = a_{11}[1-(1-a_{12}\gamma_3(E,\eta_g))^{1/2}], a_{11} \equiv (\frac{\hbar^2 \bar{t}_{11}}{4m_c \bar{t}_{12}}),\ and\ a_{12} = \frac{4\bar{t}_{12}}{\bar{t}_{11}^2}$

The EEM can be written as

$$m^*(E_{F_h}, \eta_g) = m_c\{I_{12}(E_{F_h}, \eta_g)\}' \tag{53}$$

The DR of the 1D electrons in this case can be written as

$$\frac{\hbar^2(n_z\pi/d_z)^2}{2m_c} + \frac{\hbar^2(n_y\pi/d_y)^2}{2m_c} + \frac{\hbar^2 k_x^2}{2m_c} = \theta_4(E, \eta_g) \tag{54}$$

$where,\ \theta_4(E,\eta_g) = I_{12}(E,\eta_g)$

The DOS function in this case assumes the form

$$N_{1DHD\Gamma}(E, \eta_g) = \frac{2g_v}{\pi} \sum_{n_y=1}^{n_{y\max}} \sum_{n_z=1}^{n_{z\max}} T_5'(E, n_y, n_z, \eta_g) H(E - E_{5HDNW}') \tag{55}$$

$where,\ T_5(E, n_y, n_z, \eta_g) = [[\theta_4(E,\eta_g) - [G_2(n_y, n_z)]]\frac{2m_c}{\hbar^2}]^{1/2}$

In (55) E'_{5HDNW} is the sub-band energy in this case which can be expressed as

$$G_2(n_y,n_z) = \theta_4(E'_{5HDNW},\eta_g) \tag{56}$$

The EM in this case is given by

$$m^*(E'_{F1HDNW},\eta_g) = m_c[\theta'_4(E_{F1HDNW},\eta_g)] \tag{57}$$

The electron concentration per unit length can be written as

$$n_{01D} = \frac{2g_v}{\pi} \sum_{n_y=1}^{n_{y_{\max}}} \sum_{n_z=1}^{n_{z_{\max}}} [T_5(E_{F1HDNW},n_y,n_z,\eta_g) + \sum_{r=1}^{s} L(r)[T_5(E_{F1HDNW},n_y,n_z,\eta_g)]] \tag{58}$$

(v) The Model of Palik et al.

Under the condition of heavy doping forming Gaussian band tails assumes the form

$$\frac{\hbar^2 k^2}{2m_c} = I_{13}(E,\eta_g) \tag{59}$$

$where,\ I_{13}(E,\eta_g) = \bar{b}_{12}[\bar{a}_{12} - ((\bar{a}_{12})^2 - 4\bar{B}_{11}\gamma_3(E,\eta_g))^{1/2}]$

The EM can be written as

$$m^*(E_{F_h},\eta_g) = m_c\{I_{13}(E_{F_h},\eta_g)\}' \tag{60}$$

The DR of the 1D electrons in this case can be written as

$$\frac{\hbar^2(n_z\pi/d_z)^2}{2m_c}+\frac{\hbar^2(n_y\pi/d_y)^2}{2m_c}+\frac{\hbar^2k_x^2}{2m_c}=\theta_5(E,\eta_g) \tag{61}$$

where, $\theta_5(E,\eta_g)=I_{13}(E,\eta_g)$

The DOS function in this case assumes the form

$$N_{1DHD\Gamma}(E,\eta_g)=\frac{2g_v}{\pi}\sum_{n_y=1}^{n_{y\max}}\sum_{n_z=1}^{n_{z\max}} T_6'(E,n_y,n_z,\eta_g)H(E-E_{6HDNW}') \tag{62}$$

where, $T_6(E,n_y,n_z,\eta_g)=[[\theta_5(E,\eta_g)-[G_2(n_y,n_z)]]\frac{2m_c}{\hbar^2}]^{1/2}$

In (62) E_{5HDNW}' is the sub-band energy in this case which can be expressed as

$$G_2(n_y,n_z)=\theta_5(E_{6HDNW}',\eta_g) \tag{63}$$

The EM in this case is given by

$$m^*(E_{F1HDNW}',\eta_g)=m_c[\theta_5'(E_{F1HDNW},\eta_g)] \tag{64}$$

The electron concentration per unit length can be written as

$$n_{01D}=\frac{2g_v}{\pi}\sum_{n_y=1}^{n_{y\max}}\sum_{n_z=1}^{n_{z\max}} [T_6(E_{F1HDNW},n_y,n_z,\eta_g)+\sum_{r=1}^{s} L(r)[T_6(E_{F1HDNW},n_y,n_z,\eta_g)]] \tag{65}$$

The EM in NWs of HD II-VI Materials

Therefore the dispersion relation of the carriers in HD II-VI materials in the presence of Gaussian band tails can be expressed as

$$\gamma_3(E,\eta_g) = a_0' k_s^2 + b_0' k_z^2 \pm \bar{\lambda}_0 k_s \tag{66}$$

Thus, the energy spectrum in this case is real since the corresponding E-k relation in the absence of band tails as given by (66) is a no pole function in the finite complex plane.

The 1D DR in NW of HD II-VI materials can be written as

$$\gamma_3(E,\eta_g) = a_0'[(\frac{n_x\pi}{d_x})^2 + (\frac{n_y\pi}{d_y})^2] \pm \bar{\lambda}_0[(\frac{n_x\pi}{d_x})^2 + (\frac{n_y\pi}{d_y})^2]^{1/2} + \frac{\hbar^2 k_x^2}{2m_{\|}^*} \tag{67}$$

The DOS function in this case assumes the form

$$N_{1DHD\Gamma}(E,\eta_g) = \frac{g_v}{\pi}\sum_{n_x=1}^{n_{x\max}}\sum_{n_y=1}^{n_{y\max}} T_7'(E,n_x,n_y,\eta_g)H(E-E_{13HDNW}') \tag{68}$$

$$where,\ T_7(E,n_y,n_z,\eta_g) = [[\gamma_3(E,\eta_g) - [G_3,\pm(n_x,n_y)]]\frac{2m_{\|}^*}{\hbar^2}]^{1/2}$$

$$and\ G_3,\pm(n_x,n_y) = a_0'[(\frac{n_x\pi}{d_x})^2 + (\frac{n_y\pi}{d_y})^2] \pm \bar{\lambda}_0[(\frac{n_x\pi}{d_x})^2 + (\frac{n_y\pi}{d_y})^2]^{1/2}$$

and E_{13HDNW}' is the sub-band energy in this case which can be expressed as

$$\gamma_3(E_{13HDNW}',\eta_g) = a_0'[(\frac{n_x\pi}{d_x})^2 + (\frac{n_y\pi}{d_y})^2] \pm \bar{\lambda}_0[(\frac{n_x\pi}{d_x})^2 + (\frac{n_y\pi}{d_y})^2]^{1/2} \tag{69}$$

The EM in this case is given by

$$m^*(E_{F1HDNW},\eta_g)=m_{\|}^*\gamma_3'[\theta_5'(E_{F1HDNW},\eta_g)] \tag{70}$$

The electron concentration per unit length can be written as

$$n_{01D}=\frac{2g_v}{\pi}\sum_{n_x=1}^{n_{x\max}}\sum_{n_y=1}^{n_{y\max}}[T_7(E_{F1HDNW},n_x,n_y,\eta_g)+\sum_{r=1}^{s}L(r)[T_7(E_{F1HDNW},n_x,n_y,\eta_g)]] \tag{71}$$

The EM in NWs of HD IV-VI Materials

1. Dimmock Model

The 1D electron dispersion law in NW of HD IV-VI materials in accordance with the Dimmock model can be expressed as

$$\gamma_2(E,\eta_g)+\alpha\gamma_3(E,\eta_g)=(\frac{\hbar^2}{2x_4}(\frac{n_x\pi}{d_x})^2+\frac{\hbar^2}{2x_5}(\frac{n_y\pi}{d_y})^2)+\alpha\gamma_3(E,\eta_g)\frac{\hbar^2}{2x_6}k_z^2-(1+\alpha\gamma_3(E,\eta_g))$$
$$\times(\frac{\hbar^2}{2x_1}(\frac{n_x\pi}{d_x})^2+\frac{\hbar^2}{2x_2}(\frac{n_y\pi}{d_y})^2)-\alpha(\frac{\hbar^2}{2x_1}(\frac{n_x\pi}{d_x})^2+\frac{\hbar^2}{2x_2}(\frac{n_y\pi}{d_y})^2)\times(\frac{\hbar^2}{2x_4}(\frac{n_x\pi}{d_x})^2+\frac{\hbar^2}{2x_5}(\frac{n_y\pi}{d_y})^2)$$
$$-\alpha(\frac{\hbar^2}{2x_1}(\frac{n_x\pi}{d_x})^2+\frac{\hbar^2}{2x_2}(\frac{n_y\pi}{d_y})^2)\frac{\hbar^2}{2x_6}k_z^2-(1+\alpha\gamma_3(E,\eta_g))\frac{\hbar^2}{2x_3}k_z^2$$
$$-\alpha\frac{\hbar^2}{2x_3}k_z^2(\frac{\hbar^2}{2x_4}(\frac{n_x\pi}{d_x})^2+\frac{\hbar^2}{2x_5}(\frac{n_y\pi}{d_y})^2)-\alpha\frac{\hbar^4k_z^4}{2x_3x_6}$$

$$=\frac{\hbar^2}{2m_1}(\frac{n_x\pi}{d_x})^2+\frac{\hbar^2}{2m_2}(\frac{n_y\pi}{d_y})^2\frac{\hbar^2}{2m_3}k_z^2 \tag{72}$$

Equation (72) can be written as

$$k_z = T_{36}(E,\eta_g,n_x,n_y)$$
$$\text{where, } T_{36}(E,\eta_g,n_x,n_y) = [(2C_{22})^{-1}[-B_{HD}(E,\eta_g,n_x,n_y) + \sqrt{B_{HD}^2(E,\eta_g,n_x,n_y) + 4C_{22}A_{HD}(E,\eta_g,n_x,n_y)}]]$$
$$C_{22} = (\frac{\hbar^4}{4x_3x_6}), B_{HD}(E,\eta_g,n_x,n_y) = [\alpha(\frac{\hbar^2}{2x_1}(\frac{n_x\pi}{d_x})^2 + \frac{\hbar^2}{2x_2}(\frac{n_y\pi}{d_y})^2)\frac{\hbar^2}{2x_6} + (1+\alpha\gamma_3(E,\eta_g))\frac{\hbar^2}{2x_3} + \frac{\hbar^2}{2m_3}$$
$$+\alpha\frac{\hbar^2}{2x_3}(\frac{\hbar^2}{2x_4}(\frac{n_x\pi}{d_x})^2 + \frac{\hbar^2}{2x_5}(\frac{n_y\pi}{d_y})^2)] \tag{73}$$

and

$$A_{HD}(E,\eta_g,n_x,n_y) = [-[(\frac{\hbar^2}{2m_1}(\frac{n_x\pi}{d_x})^2 + \frac{\hbar^2}{2m_2}(\frac{n_y\pi}{d_y})^2)\gamma_2(E,\eta_g) + \alpha\gamma_3(E,\eta_g)] \times (\frac{\hbar^2}{2x_4}(\frac{n_x\pi}{d_x})^2 + \frac{\hbar^2}{2x_5}(\frac{n_y\pi}{d_y})^2)$$
$$-\alpha(\frac{\hbar^2}{2x_1}(\frac{n_x\pi}{d_x})^2 + \frac{\hbar^2}{2x_2}(\frac{n_y\pi}{d_y})^2) \times (\frac{\hbar^2}{2x_4}(\frac{n_x\pi}{d_x})^2 + \frac{\hbar^2}{2x_5}(\frac{n_y\pi}{d_y})^2) - (1+\alpha\gamma_3(E,\eta_g))\frac{\hbar^2}{2x_1}(\frac{n_x\pi}{d_x})^2 + \frac{\hbar^2}{2x_2}(\frac{n_y\pi}{d_y})^2)$$

The DOS function in this case assumes the form

$$N_{1DHD\Gamma}(E,\eta_g) = \frac{g_v}{\pi}\sum_{n_x=1}^{n_{x_{\max}}}\sum_{n_y=1}^{n_{y_{\max}}} T'_{36}(E,n_x,n_y,\eta_g)H(E-E'_{14HDNW}) \tag{74}$$

In (74) E'_{14HDNW} is the sub-band energy in this case which can be expressed as

$$0 = \gamma_3(E'_{14HDNW},\eta_g,n_x,n_y) \tag{75}$$

The EM in this case is given by

$$m^*(E_{F1HDNW},\eta_g,n_x,n_y) = \frac{\hbar^2}{2}\frac{\partial}{\partial E}[T_{36}^2(E_{F1HDNW},n_x,n_y,\eta_g)] \tag{76}$$

The electron concentration per unit length can be written as

$$n_{01D} = \frac{2g_v}{\pi}\sum_{n_x=1}^{n_{x_{\max}}}\sum_{n_y=1}^{n_{y_{\max}}} [T_{36}(E_{F1HDNW},n_x,n_y,\eta_g) + \sum_{r=1}^{s} L(r)[T_{36}(E_{F1HDNW},n_x,n_y,\eta_g)]] \tag{77}$$

The 1D DR in NW of HD IV-VI materials in the absence of band tails can be expressed as

$$E(1+\alpha E)+\alpha E(\frac{\hbar^2}{2x_4}(\frac{n_x\pi}{d_x})^2+\frac{\hbar^2}{2x_5}(\frac{n_y\pi}{d_y})^2)+\alpha E\frac{\hbar^2}{2x_6}k_z^2-(1+\alpha E)$$
$$(\frac{\hbar^2}{2x_1}(\frac{n_x\pi}{d_x})^2+\frac{\hbar^2}{2x_2}(\frac{n_y\pi}{d_y})^2)-\alpha(\frac{\hbar^2}{2x_1}(\frac{n_x\pi}{d_x})^2+\frac{\hbar^2}{2x_2}(\frac{n_y\pi}{d_y})^2)\times(\frac{\hbar^2}{2x_4}(\frac{n_x\pi}{d_x})^2+\frac{\hbar^2}{2x_5}(\frac{n_y\pi}{d_y})^2)$$
$$-\alpha(\frac{\hbar^2}{2x_1}(\frac{n_x\pi}{d_x})^2+\frac{\hbar^2}{2x_2}(\frac{n_y\pi}{d_y})^2)\times(\frac{\hbar^2}{2x_4}(\frac{n_x\pi}{d_x})^2+\frac{\hbar^2}{2x_5}(\frac{n_y\pi}{d_y})^2)-\alpha(\frac{\hbar^2}{2x_1}(\frac{n_x\pi}{d_x})^2+\frac{\hbar^2}{2x_2}(\frac{n_y\pi}{d_y})^2)\frac{\hbar^2}{2x_6}k_z^2$$
$$-(1+\alpha E)\frac{\hbar^2}{2x_3}k_z^2-\alpha\frac{\hbar^2}{2x_3}k_z^2(\frac{\hbar^2}{2x_4}(\frac{n_x\pi}{d_x})^2+\frac{\hbar^2}{2x_5}(\frac{n_y\pi}{d_y})^2)$$
$$-\alpha\frac{\hbar^4k_z^4}{4x_3x_6}=\frac{\hbar^2}{2m_1}(\frac{n_x\pi}{d_x})^2+\frac{\hbar^2}{2m_2}(\frac{n_y\pi}{d_y})^2+\frac{\hbar^2}{2m_3}k_z^2 \tag{78}$$

Equation (78) can be written as

$$k_z=T_{40}(E,\eta_g,n_x,n_y) \tag{79}$$

where, $T_{40}(E,\eta_g,n_x,n_y)=[(2C_{22})^{-1}[-B_0(E,n_x,n_y)+\sqrt{B_0^2(E,n_x,n_y)+4C_{22}A_0(E,n_x,n_y)}]]$

where,

$$B_0(E,n_x,n_y)=[\alpha(\frac{\hbar^2}{2x_1}(\frac{n_x\pi}{d_x})^2+\frac{\hbar^2}{2x_2}(\frac{n_y\pi}{d_y})^2)\frac{\hbar^2}{2x_6}+(1+\alpha E)\frac{\hbar^2}{2x_3}-\frac{\hbar^2}{2x_6}$$
$$+\frac{\hbar^2}{2m_3}+\alpha\frac{\hbar^2}{2x_3}(\frac{\hbar^2}{2x_4}(\frac{n_x\pi}{d_x})^2+\frac{\hbar^2}{2x_5}(\frac{n_y\pi}{d_y})^2)]$$

and

$$A_0(E,n_x,n_y)=[-[(\frac{\hbar^2}{2m_1}(\frac{n_x\pi}{d_x})^2+\frac{\hbar^2}{2m_2}(\frac{n_y\pi}{d_y})^2)E(1+\alpha E)+\alpha E(\frac{\hbar^2}{2x_4}(\frac{n_x\pi}{d_x})^2+\frac{\hbar^2}{2x_5}(\frac{n_y\pi}{d_y})^2)$$
$$-\alpha(\frac{\hbar^2}{2x_1}(\frac{n_x\pi}{d_x})^2+\frac{\hbar^2}{2x_2}(\frac{n_y\pi}{d_y})^2)\times(\frac{\hbar^2}{2x_4}(\frac{n_x\pi}{d_x})^2+\frac{\hbar^2}{2x_5}(\frac{n_y\pi}{d_y})^2)-(1+\alpha E)(\frac{\hbar^2}{2x_1}(\frac{n_x\pi}{d_x})^2+\frac{\hbar^2}{2x_2}(\frac{n_y\pi}{d_y})^2)]$$

The DOS function in this case assumes the form

$$N_{1D\Gamma}(E)=\frac{g_v}{\pi}\sum_{n_x=1}^{n_{x_{\max}}}\sum_{n_y=1}^{n_{y_{\max}}}T_{40}'(E,n_x,n_y)H(E-E_{22}') \tag{80}$$

In (80) E'_{22} is the sub-band energy in this case which can be expressed as

$$0 = T_{40}(E'_{22}, n_x, n_y) \tag{81}$$

The EM in this case is given by

$$m^*(E_{F1D}, n_x, n_y) = \frac{\hbar^2}{2}\frac{\partial}{\partial E}[T_{40}^2(E_{F1D}, n_x, n_y)] \tag{82}$$

The electron concentration per unit length can be written as

$$n_{01D} = \frac{g_v}{\pi}\sum_{n_x=1}^{n_{x_{\max}}}\sum_{n_y=1}^{n_{y_{\max}}}[T_{40}(E_{F1D}, n_x, n_y) + \sum_{r=1}^{s} L(r)[T_{40}(E_{F1D}, n_x, n_y)]] \tag{83}$$

2. Bangert and Kastner Model

The dispersion relation of the conduction electrons in bulk specimens of IV-VI semiconductors in accordance with the model of Bangert and Kastner [147] is given by

$$\omega_1(\mathrm{E})k_s^2 + \omega_2(\mathrm{E})k_z^2 = 1 \tag{84}$$

$$where, \omega_1(E) = (2E)^{-1}[\frac{(\bar{R})^2}{E_{g_0}(1+\alpha_1 E)} + \frac{(\bar{S})^2}{\Delta'_c(1+\alpha_3 E)} + \frac{(\bar{Q})^2}{\Delta''_c(1+\alpha_3 E)}]$$

and $\omega_2(\mathrm{E}) = (2E)^{-1}[\frac{(\bar{\mathrm{A}})^2}{\mathrm{E}_{g0}(1+\alpha_1\,\mathrm{E})} + \frac{(\bar{\mathrm{S}}+\bar{\mathrm{Q}})^2}{\Delta_c''(1+\alpha_3\,\mathrm{E})}]$, $(\bar{R})^2 = 2.3\times 10^{-10}(eVm)^2$,

$(\bar{S})^2 = 4.6(\bar{R})^2, \alpha_1 = \frac{1}{\mathrm{E}_{g0}}, \alpha_2 = \frac{1}{\Delta_c'}, \alpha_3 = \frac{1}{\Delta_c''}, \frac{D}{\mu} = \frac{n_0}{|e|}$ Re*al* part of$[\frac{\partial n_0}{\partial(E_{Fi} - e_{ii})}]^{-1}$

The electron energy spectrum in heavily doped IV-VI materials in this case can be expressed as

$$2I(4)=k_s^2[\{c_1(\alpha_1 E,E_g)-iD_1(\alpha_1,E,E_g)\}\frac{(\bar{R})^2}{E_{g_0}}+\{c_2(\alpha_2 E,E_g)-iD_2(\alpha_2,E,E_g)\}\frac{(\bar{S})^2}{\Delta_c'}$$

$$+\{c_3(\alpha_3 E,E_g)-iD_3(\alpha_3 E,E_g)\}\frac{(\bar{Q})^2}{\Delta_c''}]+k_z^2[\frac{2(\bar{A})^2}{E_{g_0}}\{c_1(\alpha_1 E,E_g)-iD_1(\alpha_1 E,E_g)\}$$

$$+\frac{(\bar{S}+\bar{Q})^2}{\Delta_c''}\{c_3(\alpha_3 E,E_g)-iD_3(\alpha_3 E,E_g)\}] \tag{85}$$

where, $\alpha_1=\frac{1}{E_g},\alpha_2=\frac{1}{\Delta_c'},\alpha_3=\frac{1}{\Delta_c''},G_t=\frac{1+\alpha_i E}{\eta_g\alpha_i}$

$c_i(\alpha_i E,\eta_g)=[\frac{2}{\alpha_i\eta_g\sqrt{\pi}}]\exp(-u_i^2)\times[\sum_{p=1}^{\infty}\{\exp(-p^2/4)(\sinh(pu_i))\}p^{-1}]$,

$i=1,2,3,4,...$ and $D_i(\alpha_i,E,\eta_g)=[\frac{\sqrt{\pi}}{\alpha_i\eta_g}]\exp(-u_i^2)$

Therefore (85) can be written as,

$$F_1(E,\eta_g)k_s^2+F_2(E,\eta_g)k_z^2=1 \tag{86}$$

where, $F_1(E,\eta_g)=[2\gamma_0(E,\eta_g)]^{-1}[\frac{(\bar{R})^2}{E_g}\{C_1(\alpha_1 E,E_g)-iD_1(\alpha_1,E,E_g)\}+\frac{(\bar{S})^2}{\Delta_c'}\{C_2(\alpha_2,E,E_g)-iD_2(\alpha_2,E,E_g)\}$

$+\frac{(\bar{Q})^2}{\Delta_c''}\{C_3(\alpha_3,E,E_g)-iD_3(\alpha_3,E,E_g)\}]$

and

$F_2(E,\eta_g)=[2\gamma_0(E,\eta_g)]^{-1}[\frac{2(\bar{A})^2}{E_g}\{C_1(\alpha_1 E,E_g)-iD_1(\alpha_1,E,E_g)\}$

$+\frac{(\bar{S}+\bar{Q})^2}{\Delta_c''}\{C_3(\alpha_3,E,E_g)-iD_3(\alpha_3,E,E_g)\}]$

Since $F_1(E,\eta_g)$ and $F_2(E,\eta_g)$ are complex, the energy spectrum is also complex in the presence of Gaussian band tails.

The 1D DR in NW of HD II-VI materials can be written as

$$F_1(E,\eta_g)=[(\frac{n_x\pi}{d_x})^2+(\frac{n_y\pi}{d_y})^2]F_2(E,\eta_g)k_z^2=1 \tag{87}$$

The (87) can be written as

$$k_z=T_{60}(E,\eta_g,n_x,n_y) \tag{88}$$

where, $T_{60}(E,\eta_g,n_x,n_y)=[[1-$

$F_1(E,\eta_g)][(\frac{n_x\pi}{d_x})^2+(\frac{n_y\pi}{d_y})^2]][F_2(E,\eta_g)]^{-1}]^{1/2}$

The DOS function in this case assumes the form

$$N_{1DHD\Gamma}(E,\eta_g)=\frac{2g_v}{\pi}\sum_{n_x=1}^{n_{x\max}}\sum_{n_y=1}^{n_{y\max}} T'_{40}(E,n_x,n_y,\eta_g)H(E-E'_{15HDNW}) \tag{89}$$

In (89) E'_{15HDNW} is the sub-band energy in this case which can be expressed as

$$0=T_{60}(E'_{15HDNW},\eta_g,n_x,n_y) \tag{90}$$

The EM in this case is given by

$$m^*(E_{F1HDNW},\eta_g,n_x,n_y)=\frac{\hbar^2}{2}\frac{\partial}{\partial E}[T_{40}^2(E_{F1HDNW},\eta_g,n_x,n_y)] \tag{91}$$

The electron concentration per unit length can be written as

$$n_{01D}=\frac{2g_v}{\pi}\sum_{n_x=1}^{n_{x\max}}\sum_{n_y=1}^{n_{y\max}} [T_{40}(E_{F1HDNW},\eta_g,n_x,n_y)+\sum_{r=1}^{s} L(r)[T_{40}(E_{F1HDNW},\eta_g,n_x,n_y)]] \tag{92}$$

The EM in NWs of HD Stressed Kane Type Materials

The use of (92) can be written as

$$(E-\alpha_1)k_x^2+(E-\alpha_2)K_Y^2+(E-\alpha_3)k_z^2=t_1E^3-t_2E^2+t_3E+t_4 \quad (93)$$

where, $\alpha_1\equiv[E_g-C_1\varepsilon-(\bar{a}_0+C_1)\varepsilon+\frac{3}{2}\bar{b}_0\varepsilon_{xx}-\frac{\bar{b}_0}{2}\varepsilon+(\frac{\sqrt{3}}{2})\varepsilon_{xy}\bar{d}_0]$,

$\alpha_2\equiv[E_g-C_1\varepsilon-(\bar{a}_0+C_1)\varepsilon+\frac{3}{2}\bar{b}_0\varepsilon_{xx}-\frac{\bar{b}_0}{2}\varepsilon-(\frac{\sqrt{3}}{2})\varepsilon_{xy}\bar{d}_0]$,

$\alpha_3\equiv[E_g-C_1\varepsilon-(\bar{a}_0+C_1)\varepsilon+\frac{3}{2}\bar{b}_0\varepsilon_{zz}-\frac{\bar{b}_0}{2}\varepsilon],t_1\equiv(\frac{3}{2B_2^2}),t_2\equiv(\frac{1}{2B_2^2})[6(E_g-C_1\varepsilon)+3C_1\varepsilon]$,

$t_3\equiv(\frac{1}{2B_2^2})[3(E_g-C_1\varepsilon)^2+6C_1\varepsilon(E_g-C_1\varepsilon)-2C_2^2\varepsilon_{xy}^2]$

and $t_4\equiv(\frac{1}{2B_2^2})[3-C_1^{\varepsilon}(E_g-C_1\varepsilon)^2+2C_2^2\varepsilon_{xy}^2]$.

The (93) can be written as

$$Ek^2-T_{17}k_x^2-T_{27}k_y^2-T_{37}k_z^2=[q_{67}E^3-R_{67}E^2+V_{67}E+\rho_{67}] \quad (94)$$

where, $T_{17}=\alpha_1$, $T_{27}=\alpha_2$, $T_{37}=\alpha_3$, $t_1=q_{67}$, $t_2=R_{67}$, $t_3=V_{67}$ and $t_4=\rho_{67}$.

Under the condition of heavy doping, (1.179b) can be written as

$$I(4)k^2-T_{17}I(1)k_x^2-T_{27}I(1)k_y^2-T_{37}k_z^2I(1)=\left[q_{67}I(6)-R_{67}I(5)+V_{67}I(4)+\rho_{67}I(1)\right]$$

where,

$$I(6)=\int_{-\infty}^{E}(E-V)^3F(V)dV \quad (95)$$

The (95) can be written as

$$I(6) = E^3 I(1) - 3E^2 I(7) + 3EI(8) - I(9) \tag{96}$$

In which,

$$I(7) = \int_{-\infty}^{E} VF(V)dV \tag{97}$$

$$I(8) = \int_{-\infty}^{E} V^2 F(V)dV \tag{98}$$

$$I(9) = \int_{-\infty}^{E} V^3 F(V)dV \tag{99}$$

Using (1.4), together with simple algebraic manipulations, one obtains

$$I(7) = \frac{-\eta_g}{2\sqrt{\pi}} \exp(\frac{-E^2}{\eta_g^2}) \tag{100}$$

$$I(8) = \frac{\eta_g^2}{4}[1 + Erf(\frac{E}{\eta_g})] \tag{101}$$

and

$$I(9) = \frac{-\eta_g^3}{2\sqrt{\pi}} \exp(\frac{-E^2}{\eta_g^2})[1 + \frac{E^2}{\eta_g^2}] \tag{102}$$

Thus (98) can be written as

$$I(6) = [\frac{E}{2}[1 + Erf(\frac{E}{\eta_g})][E^2 + \frac{3}{2}\eta_g^2] + \frac{\eta_g}{2\sqrt{\pi}} \exp(\frac{-E^2}{\eta_g^2})[4E^2 + \eta_g^2]] \tag{103}$$

Thus, combining the appropriate equations, the dispersion relations of the conduction electrons in HD stressed materials can be expressed as

$$P_{11}(E,\eta_g)k_x^2 + Q_{11}(E,\eta_g)k_y^2 + S_{11}(E,\eta_g)k_z^2 = 1 \tag{104}$$

where, $P_{11}(E,\eta_g) \equiv [\frac{\gamma_0(E,\eta_g) - (T_{17}/2)[1+Erf(E/\eta_g)]}{\Delta_{14}(E,\eta_g)}]$

$$\Delta_{14}(E,\eta_g) \equiv [q_{67}\{\frac{E}{2}[1+Erf(\frac{E}{\eta_g})][E^2+\frac{3}{2}\eta_g^2] + \frac{\eta_g}{2\sqrt{\pi}}\exp(\frac{-E^2}{\eta_g^2})[4E^2+\eta_g^2]\},$$

$$-R_{67}\theta_0(E,\eta_g) + V_{67}\gamma_0(E,\eta_g) + \frac{\rho_{67}}{2}[1+Erf(E/\eta_g)]],$$

$$Q_{11}(E,\eta_g) \equiv [\frac{\gamma_0(E,\eta_g) - (T_{27}/2)[1+Erf(E/\eta_g)]}{\Delta_{14}(E,\eta_g)}]$$

and

$$S_{11}(E,\eta_g) \equiv [\frac{\gamma_0(E,\eta_g) - (T_{37}/2)[1+Erf(E/\eta_g)]}{\Delta_{14}(E,\eta_g)}]$$

Thus, the energy spectrum in this case is real since the dispersion relation of the corresponding materials in the absence of band tails as given by (1.184) has no poles in the finite complex plane.

The 1D DR in NW of HD II-VI materials can be written as

$$P_{11}(E,\eta_g)(\frac{\pi n_x}{d_x})^2 + Q_{11}(E,\eta_g)(\frac{\pi n_y}{d_y})^2 + S_{11}(E,\eta_g)k_z^2 = 1 \tag{105}$$

can be written as

$$k_z = T_{70}(E,\eta_g,n_x,n_y) \tag{106}$$

where, $T_{70}(E,\eta_g,n_x,n_y) = [[1 - P_{11}(E,\eta_g)(\frac{\pi n_x}{d_x})^2 + Q_{11}(E,\eta_g)(\frac{\pi n_y}{d_y})^2][S_{11}(E,\eta_g)]^{-1}]^{1/2}$

The DOS function in this case assumes the form

$$N_{1DHD\Gamma}(E,\eta_g)=\frac{2g_v}{\pi}\sum_{n_x=1}^{n_{x_{\max}}}\sum_{n_y=1}^{n_{y_{\max}}} T'_{70}(E,n_x,n_y,\eta_g)H(E-E'_{30HDNW}) \tag{107}$$

In (107) E'_{30HDNW} is the sub-band energy in this case which can be expressed as

$$0=T_{70}(E'_{30HDNW},\eta_g,n_x,n_y) \tag{108}$$

The EM in this case is given by

$$m^*(E_{F1HDNW},\eta_g,n_x,n_y)=\frac{\hbar^2}{2}\frac{\partial}{\partial E}[T_{70}^2(E_{F1HDNW},\eta_g,n_x,n_y)] \tag{109}$$

The electron concentration per unit length can be written as

$$n_{01D}=\frac{2g_v}{\pi}\sum_{n_x=1}^{n_{x_{\max}}}\sum_{n_y=1}^{n_{y_{\max}}} [T_{70}(E_{F1HDNW},n_x,n_y,\eta_g)+\sum_{r=1}^{s} L(r)[T_{70}(E_{F1HDNW},n_x,n_y,\eta_g)]] \tag{110}$$

RESULTS AND DISCUSSION

The variation of the EM in HD NWs of different materials along the transport direction has been exhibited in the figures below at various conditions. Throughout our formalism, we have assumed the HD NWs to be of rectangular cross-sectional dimensions so that the usual "particle-in-a-box" concept can be applied along the quantized directions. The HD NWs are assumed to be degenerately doped with a carrier density starting from 10^8 m^{-1}. While deriving the closed form analytical solutions of EM in all the materials, we have also assumed that the constants of the energy band structures of the materials are independent of thickness in the range beyond 5nm. Keeping this trend in view, we have assumed the invariant property of the material energy spectrum constants and evaluated the EM in HD NWs of Cd_3As_2 and $CdGeAs_2$ as function of wire thickness along their respective transport directions in Figures 2.1 and 2.2. It appears from Figures 2.1 and 2.2 that the EM at the lowest subband in both the cases are strong cross-sectional functions of the dimensions which converge to their corresponding bulk values at larger dimensions. The effect of crystal field splitting in case of dispersion relation of Cd_3As_2 lets the asymptotic fall to be closer to the bulk value. It should be noted that all the curves have been evaluated at T = 4 K where the average thermal energy i.e., $E_F + k_BT$ is very less than that of the difference of the adjacent subband energies. This leads the carrier to reside in the lowest subbands only.

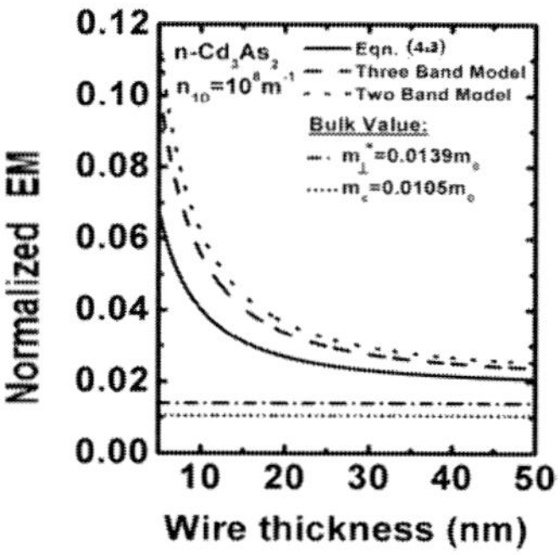

Figure 2.1. Plot of the normalizedEMas function of wire thickness for NWs of HD Cd_3As_2. The plots for three and two band models of Kane have also been exhibited with their corresponding anisotropic bulk values.

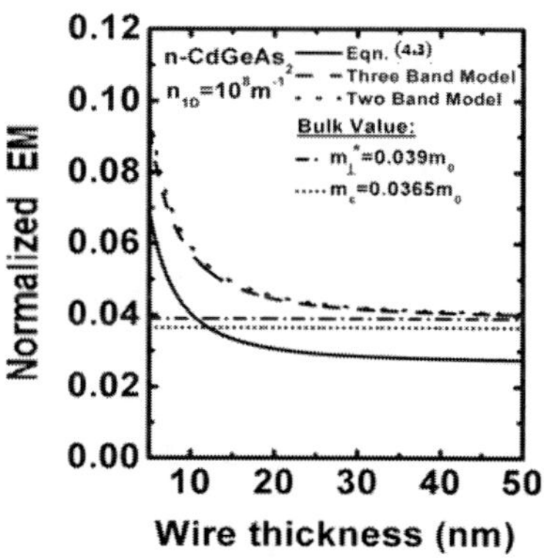

Figure 2.2. Plot of the normalizedEMas a function of wire thickness for NWs of HD $CdGeAs_2$ for all the cases of Figure 2.1.

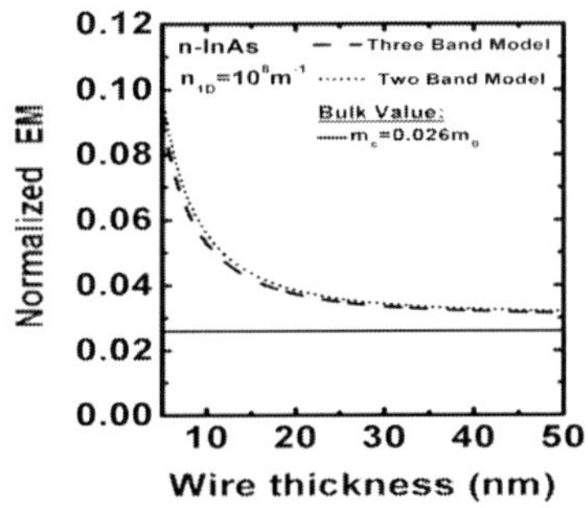

Figure 2.3. Plot of the normalizedEMas function of wire thickness for NWs of HD InAs considering the three and two band models of Kane with the corresponding isotropic bulk value.

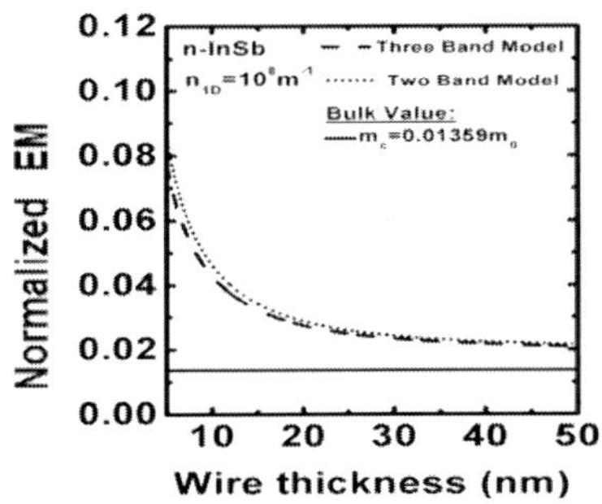

Figure 2.4. Plot of the normalizedEMas function of wire thickness for NWs of HD InSb considering the three and two band models of Kane with the corresponding isotropic bulk value.

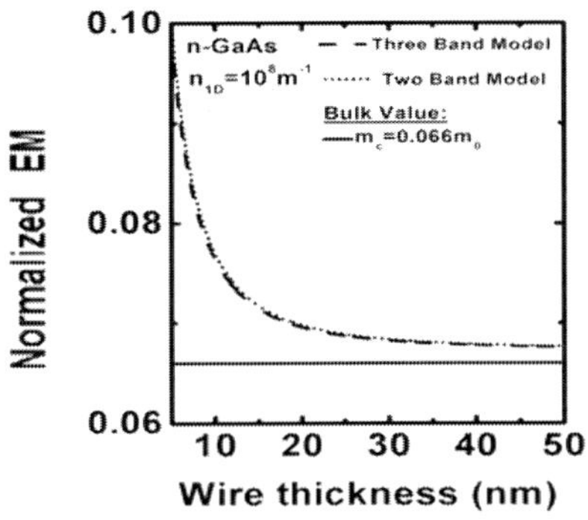

Figure 2.5. Plot of the normalizedEMas function of wire thickness for NWs of HD GaAs considering the three and two band models of Kane with the corresponding isotropic bulk value.

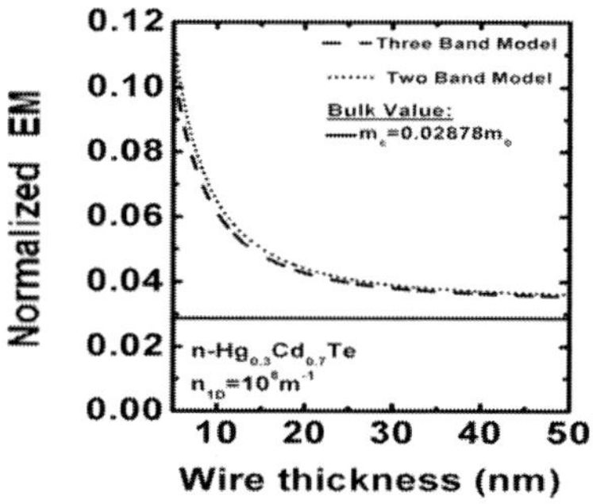

Figure 2.6. Plot of the normalized EM as function of wire thickness for NWs of HD $Hg_{1-x}Cd_xTe$ considering the three and two band models of Kane with the corresponding isotropic bulk value at x = 0.3.

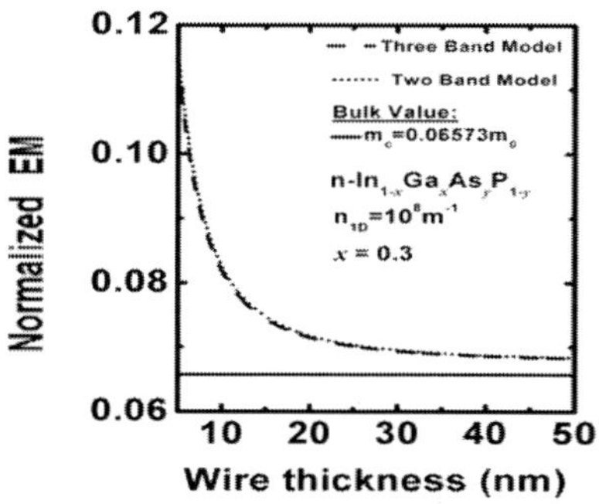

Figure 2.7. Plot of the normalizedEMas function of wire thickness for NWs of HD $In_{1-x}Ga_xAs_{1-y}P_y$ considering the three and two band models of Kane with the corresponding isotropic bulk value at x = 0.3.

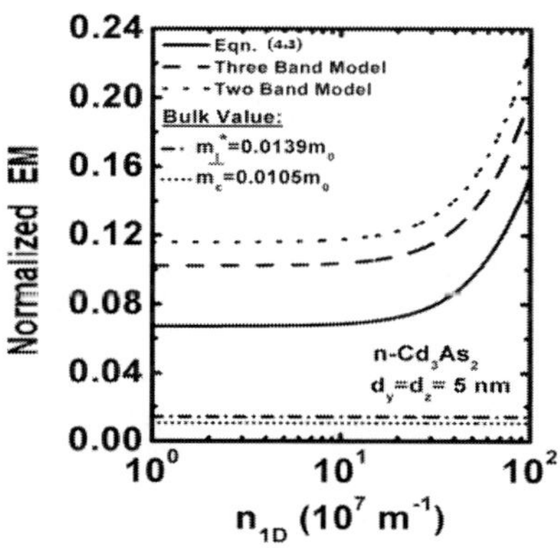

Figure 2.8. Plot of the normalizedEMas function of carrier degeneracy for HD n-Cd_3As_2 NW for all cases of Figure 2.1.

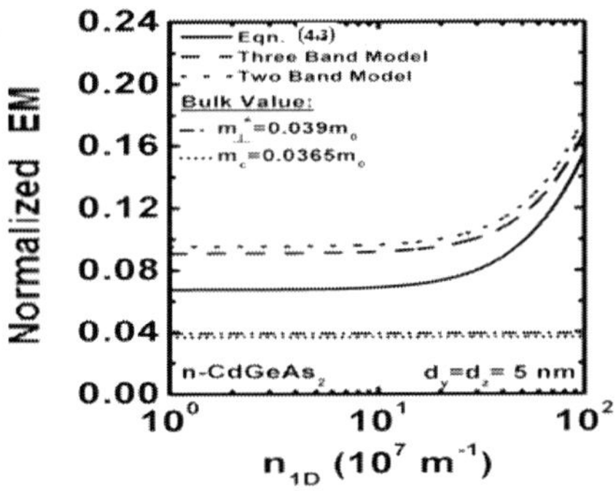

Figure 2.9. Plot of the normalizedEMas function of carrier degeneracy for NWs of HD $CdGeAs_2$ for all cases of Figure 2.2.

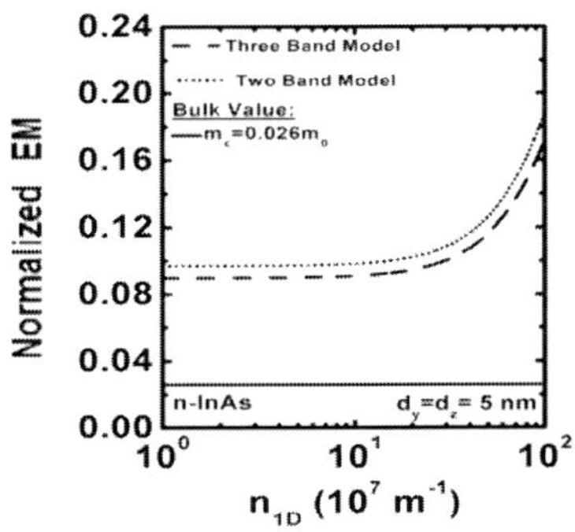

Figure 2.10. Plot of the normalizedEMas function of carrier degeneracy for n-InAs NW for all cases of Figure 2.3.

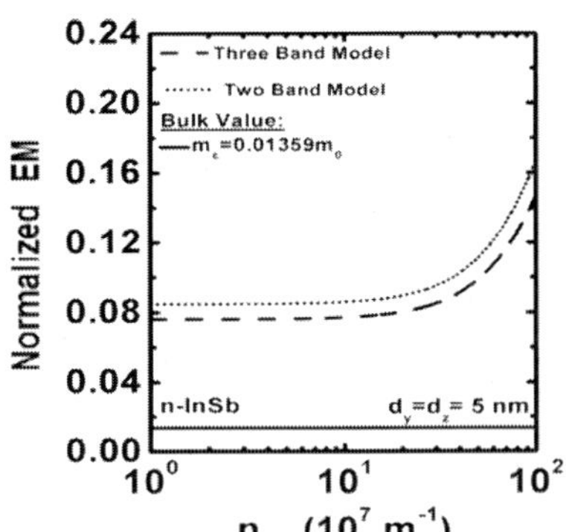

Figure 2.11. Plot of the normalizedEMas function of carrier degeneracy for HD n-InSb NW for all cases of Figure 2.4.

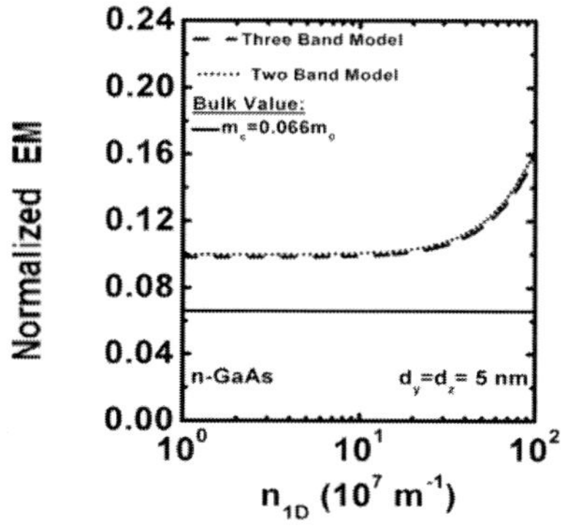

Figure 2.12. Plot of the normalized EM as function of carrier degeneracy for HD n-GaAs NW for all cases of Figure 2.5.

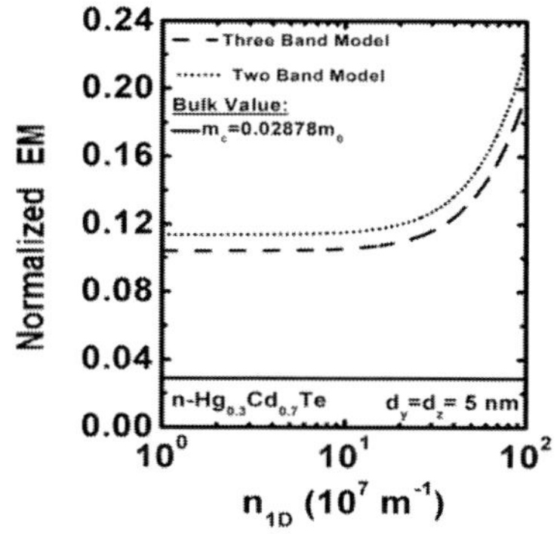

Figure 2.13. Plot of the normalizedEMas function of carrier degeneracy for HD n-$Hg_{1-x}Cd_xTe$ NW for all cases of Figure 2.6.

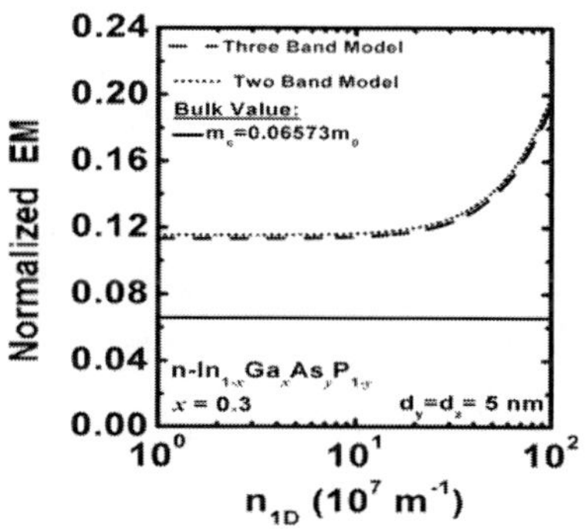

Figure 2.14. Plot of the normalizedEMas function of carrier degeneracy for HD n-$In_{1-x}Ga_xAs_{1-y}P_y$ NW for all cases of Figure 2.7.

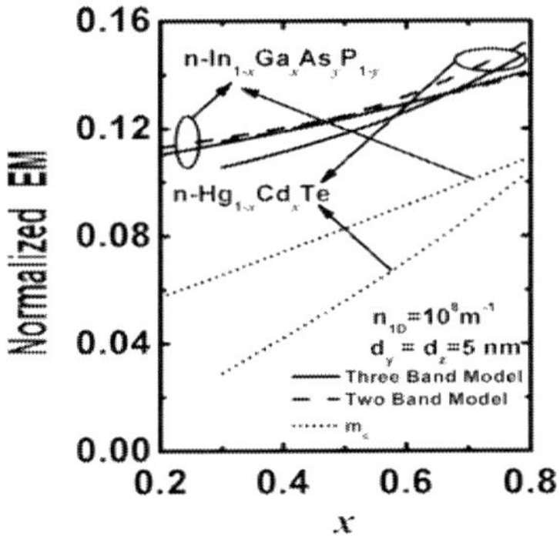

Figure 2.15. Plot of the normalizedEMas function of alloy composition for HD n-$Hg_{1-x}Cd_xTe$ and n-$In_{1-x}Ga_xAs_{1-y}P_y$ NWs.

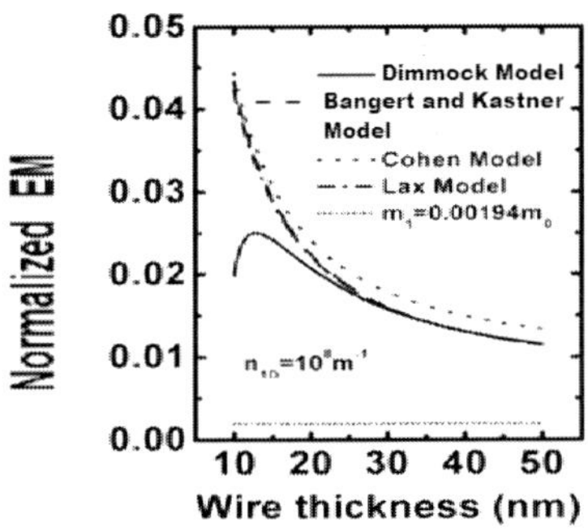

Figure 2.16. Plot of the normalizedEMas function of wire thickness for HD Bismuth NW.

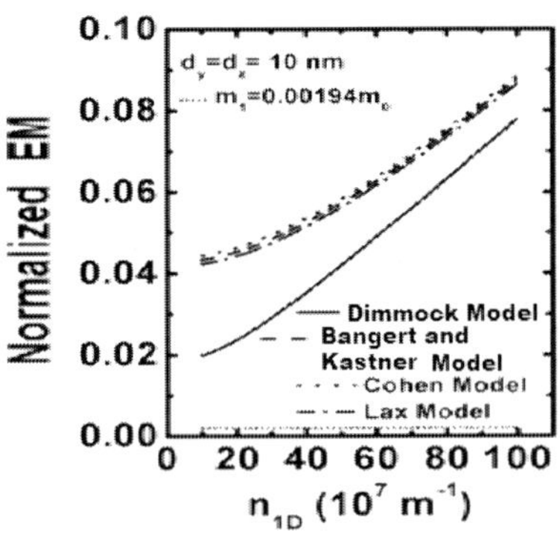

Figure 2.17. Plot of the normalizedEMas function of carrier degeneracy for HD Bismuth NW.

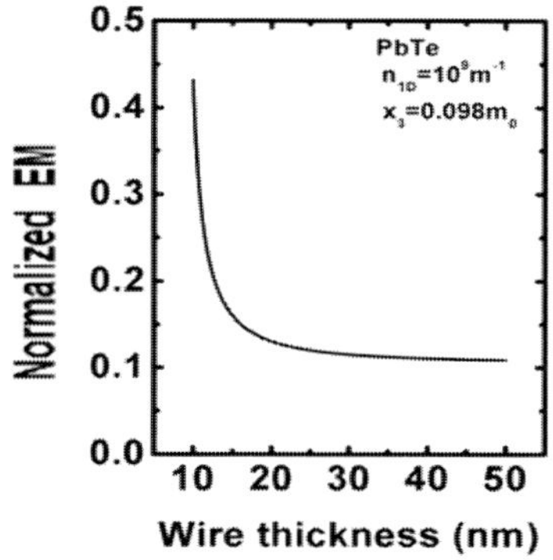

Figure 2.18. Plot of the normalizedEMas function of carrier degeneracy for HD PbTe NW.

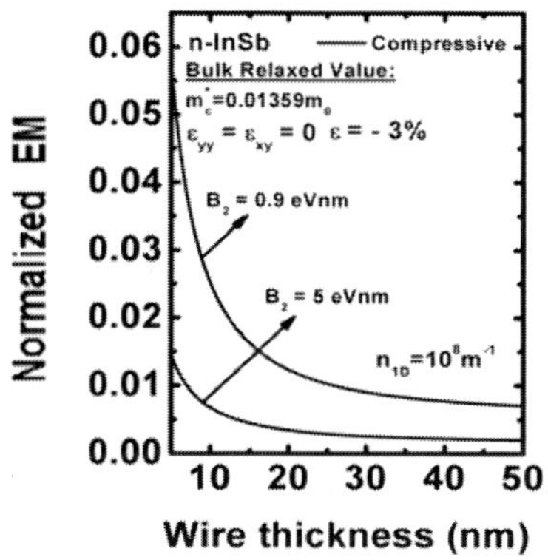

Figure 2.19. Plot of the normalized EM as function of wire thickness for stressed HD InSb NW.

Figures 2.4-2.5 exhibit the variation of the EM with the wire thickness for III-V materials namely InAs, InSb and GaAs in accordance with the well-known standard non-parabolic dispersion relation of Kane. The reader is expected to evaluate the EM for other models as derived in this chapter for III-V materials. It appears from these figures that the difference in the energy band models in predicting the EM is almost insignificant. Hence for all practical purposes for determination of EM, the second order model of Kane can fit well. It should be noted that the EM for these materials as presented here can be compared with that of the EM in 2D systems as given in chapter 1. A quick view can lead to interpret that the EM for both the case are same. However, it should also be kept in mind about the difference in the wire thickness and carrier concentration. All the curves in this chapter have been evaluated at those concentrations for which the EM stand close to that of their corresponding 2D systems. Further in deriving the results, we have assumed that the conduction band valley does not splits along the channel transport direction, which is a usual case with Silicon nanowire along [110] and [111] valleys (chapter 8). Figures 2.6 and 2.7 exhibit the variation of the EM for $Hg_{1-x}Cd_xTe$ and $In_{1-x}Ga_xAs_{1-y}P_y$ considering all the aforementioned cases at x = 0.3. In Figures 2.8-2.14, we have exhibited the variation of the EM as function of carrier degeneracy. It appears from the said figures that the EM increases with the increase in the degeneracy. The EM rises sharply above 10^7 m^{-1} for all the materials in an exponential way due to the presence of the Fermi-Dirac probability factor in the respective carrier concentration equation. Figure 2.15 exhibits the variation of the EM as function of the alloy composition in ternary and quaternary systems. It appears that as in the previous cases of quantum confinements, the EM in this case also exhibits an increasing variation with x. We have also plotted the variation of the bulk effective mass as x varies to present a comparative view. The influence of band non-parabolicity on the EM in these two materials can easily be seen. In both the cases, we see that the EM is a slow variation function of x due to the change in band gap. The effect of dimensionality on nano-wire of HD PbS has been exhibited in Figures 2.16 and 2.17 for the energy band models of Dimock, Bangert and Kastner, Cohen and Lax. It appears that as the dimension reduces in Figure 2.16, the EM increases which is generally accepted. However, using Dimock model the EM tends to decrease in the sub-5 nm regime there by unfolding the validity of the model in this zone. However at large thicknesses, the entire model tends to their corresponding bulk value $0.00194m_0$. The variation of the EM as function of carrier degeneracy has further been plotted using the aforementioned band structure models. It appears from the two figures that the effect of different band structure models has significantly less deviation from one another except for model of Dimock. The EM is found to increase almost linear with degeneracy 2 X 10^8 m^{-1} and beyond using the all the models. In case of PbTe, the EM rises sharply with decreasing wire thickness below sub-15 nm from the bulk value $0.098m_0$ at carrier degeneracy of 10^9 m^{-1}, which can affect the carrier mobility strongly.

The effect of strain on stressed InSb nano-wires has been exhibited in Figure 2.19 for two different momentum matrix elements to signify its importance as dimension reduces. A compressive strain of 3% has been applied along x and z directions to predict the variation of the EM along the transport direction. At this point, we could not provide the influence of strain on the energy band structure of InSb nano-wire due to the lack in both experimental and simulation investigations.

REFERENCES

[1] Ono, T., Y. Fujimoto and S. Tsukamoto, Quan. Matt., 1, 4 (2012); H. De Raedt, F. Jin and K. Michielsen, Quan. Matt., 1, 20 (2012); R. Hofmann and D. Kaviani, Quan. Matt., 1, 41 (2012); M. Narayanan and A. J. Peter, Quan. Matt., 1, 53 (2012); S. Olszewski, Quan. Matt., 1, 59 (2012); A. Heidari, N. Heidari and M. Ghorbani, Quan. Matt., 1, 86 (2012); S. Olszewski, Quan. Matt., 1, 127 (2012); V. Sajfert, P. Mali, N Bednar, N. Pop, D. Popov and B. Tosic, Quan. Matt., 1, 134 (2012); B. Tuzun and S. Erkoc, Quan. Matt., 1, 136 (2012), A. Heidari, N. Heidari and M. Ghorbani, Quan. Matt., 1, 149 (2012); C. Falquez, R. Hofmann and T. Baumbach, Quan. Matt., 1, 153 (2012); I. S Amiri, A. Nikoukar, J. Ali and P. P Yupapin, Quan. Matt., 1, 159 (2012); N. Paitya, S. Bhattacharya, D. De and K. P Ghatak, Quan. Matt., 1, 63 (2012); S. Bhattacharya, D. De, S. Ghosh and K. P Ghatak, *J. of Comp. and theo. Nano Sc.,* 10, 1 (2013); S. Bhattacharya, D. De, N. Paitya, S. M Adhikari and K. P Ghatak, Adv. Sc. Lett.16, 348 (2012); D. De, S. Bhattacharya, S. Ghosh and K. P Ghatak, *Adv. Sc. Engg. and Medicine*, 4, 211 (2012); N. Paitya, S. Bhattacharya, D. De, S. Ghosh and K. P Ghatak, J. of Nanoengg. and Nanomanufac., 2, 211 (2012); N. Paitya, S. Bhattacharya, D. De and K. P Ghatak, *Adv. Sc. Engg. and Medi.,* 4, 96 (2012); N. Paitya

and K. P Ghatak, *J. of Adv. Phys.*, 1, 1 (2012); S. Bhattacharya, D. De and K. P Ghatak, *J. of Sc. Engg. and Medi.*, 6, 5 (2012); N. Paitya and K. P Ghatak, *J. of Nanoengg. and Nanomanufac.*, 2, 1 (2012).

[2] Adachi, S., *J. Appl. Phys.* 58, R11 (1985); R. Dornhaus, and G. Nimtz, *Springer Tracts in Modern Physics*, Berlin-Heidelberg, N., Springer (1976), Vol. 78, pp-1.

[3] Zawadzki, W., *Handbook of Semiconductor Physics*, (Paul, W., ed.), Amsterdam, North Holland, (1982), Vol. 1, pp-719; I. M. Tsidilkovski, Cand. *Thesis Leningrad University SSR* (1955).

[4] Sen, T. N. and K. P. Ghatak, *Quantum Matter* 5, 1(2016).

[5] Bass, F. G. and I. M. Tsidilkovski, Ivz. *Acad. Nauk Azerb SSR* 10, 3 (1966).

[6] B. Mitra and K. P. Ghatak, Phys. Scrip. 40, 776 (1989); S. K. Biswas, A. R. Ghatak, A. Neogi, A. Sharma, S. Bhattacharya and K. P. Ghatak, *Phys.* E 36, 163 (2007).

[7] Mondal, M., S. Banik and K. P. Ghatak, *J. Low Temp. Phys.* 74, 423 (1989).

[8] Ghatak, K. P., S. Bhattacharya, H.Saikia and A. Sinha, *J. Comp. Theor. Nanoscience*, 3, 1 (2006).

[9] Ghatak, K. P. and S. N. Biswas, J. Vac. Sc. Tech. 7B, 104 (1989); P. K. Chakraborty, S. Choudhury and K. P. Ghatak, *Phys. B* 387, 333 (2007).

[10] Bhattacharya, S S., Pahari, D. K. Basu and K. P. Ghatak, J. Comp. Theo. *Nanoscience*, 3, 280 (2006); A. Sinha, A. K. Sharma, R. Barui, A. R. Ghatak, S. Bhattacharya, and K. P. Ghatak, *Phys. B* 391, 141 (2007).

[11] Choudhury, S., L. J. Singh and K. P. Ghatak, *Nanotechnology*, 15, 180 (2004).

[12] Chowdhary, S., L. J. Singh and K. P. Ghatak, *Phys. B* B365, 5 (2005); P. K. Chakraborty, A. Sinha, S. Bhattacharya and K. P. Ghatak, *Phys. B* 390, 325 (2007); K. P. Ghatak, J. P. Banerjee and D. Bhattacharyya, *Nanotechnology*, 7, 110 (1996).

[13] Chakraborty, P. K., B. Nag and K. P. Ghatak. *J. Phys. Chem. Solids*, 64, 2191 (2003).

[14] Ghatak, K. P., J. P. Banerjee and B. Nag, *J. Appl. Phys.* 83, 1420 (1998).

[15] Nag, B. and K. P. Ghatak, *J. Phys. Chem. Solids* 59, 713 (1998); B. Nag and K. P. Ghatak, *J. Phys. Chem. Sol.* 58, 427 (1997).

[16] Ghatak, K. P., D. K. Basu and B. Nag. *J. Phys. Chem. Solids* 58, 133 (1997); B. Nag, K. P. Ghatak, Nonlin. *Opt. and Quan. Opt.*, 19, 1 (1998).

[17] Ghatak, K. P. and B. Nag, *Nanostruct. Mat.*, 10, 923 (1998).

[18] Bose, P. K., N. Paitya, S. Bhattacharya, D. De, S. Saha, K. M. Chatterjee, S. Pahari and K. P. Ghatak, *Quantum Matter*, 1, 89(2012); K. P. Ghatak, B. Mitra, *International journal of electronics*, 72,541(1992); M. Mondal, N. Chattopadhyay and K. P. Ghatak, *Journal of low temperature physics*, 66,131(1987); P. K. Chakraborty, G. C. Datta and K. P. Ghatak, *Physica Scripta*, 68,368(2003); K. P. Ghatak and M. Mondal, *Zeitschrift für Physik B Condensed Matter*, 69,471(1988); B. Mitra and K. P. Ghatak, *Solid-state electronics*, 32,177(1989).

[19] Adhikari, S. M., A. Sakar and K. P. Ghatak, *Quantum Matter*, 2,455(2013); K. P. Ghatak, S. Bhattacharya, S. K. Biswas, A. Dey and A. K. Dasgupta, *Physica Scripta*, 75,820(2007); M. Mondal and K. P. Ghatak, *Physics Letters A*, 131,529(1988).

[20] Chakravarti, A. N., K. P. Ghatak, K. K. Ghosh, S. Ghosh and A. Dhar, *Zeitschrift für Physik B Condensed Matter*, 47,149(1982); A. Ghoshal, B. Mitra and K. P. Ghatak, *Il Nuovo Cimento D*, 12,891(1990); A. N. Chakravarti, A. K. Chowdhury, K. P. Ghatak, S. Ghosh and A. Dhar, *Applied physics*, 25,105(1981).

[21] Mondal, M. and K. P. Ghatak, *Physica Status Solidi (b),* 135, K21(1986); M. Mondal and K. P. Ghatak, *Physica Status Solidi (b),* 123, K143(1984); K. P. Ghatak, A. Ghoshal and B. Mitra, *Il Nuovo Cimento D,* 13, 867(1991).

[22] Ghatak, K. P. and B. Mitra, *International Journal of Electronics Theoretical and Experimental,* 70, 343(1991); K. P. Ghatak, B. Mitra and A. Ghoshal, *Physica Status Solidi (b)*, 154, K121(1989); K. P.

Ghatak and M. Mondal, *Zeitschrift für Naturforschung A*, 41, 881(1986).

[23] Mondal, M. and K. P. Ghatak, *Physica Status Solidi (b),* 129, 745(1985); K. P. Ghatak, A. Ghoshal and B. Mitra, *Il Nuovo Cimento D,* 14, 903(1992); B. Mitra and K. P. Ghatak, *Solid-state electronics,* 32, 515(1989).

[24] Ghatak, K. P., N. Chatterjee and M. Mondal, *Physica Status Solidi (b),* 139, K25(1987); A. N. Chakravarti and K. P. Ghatak, K. K. Ghosh and H. M. Mukherjee, *Physica Status Solidi (b),* 116, 17(1983); A. N. Chakravarti, K. P. Ghatak, A. Dhar and K. K. Ghosh, *Czechoslovak Journal of Physics,* 33, 65(1983).

[25] Ghatak, K. P. and A. Ghosal, *Physica Status Solidi (b)*, 151, K135 (1989); S. Bhattacharya, D. De, S. M. Adhikari and K. P. Ghatak, *Superlattices and Microstructures,* 51, 203(2012); K. P. Ghatak and M. Mondal, *Physica Status Solidi (b)*, 175, 113(1993).

[26] Mondal, M. and K. P. Ghatak, *Czechoslovak Journal of Physics B*, 36, 1389(1986); M. Mondal and K. P. Ghatak, *Physica Status Solidi (b),* 126, K47(1984); D. De, S. Bhattacharya, S. M. Adhikari, A. Kumar, P. K. Bose and K. P. Ghatak, *Beilstein journal of nanotechnology,* 2, 339(2011).

[27] Mondal, M. and K. P. Ghatak, *Physica Status Solidi (b),* 139, 185(1987); S. Debbarma, A. Bhattacharjee, S. Bhattacharyya, A. Mondal and K. P. Ghatak, *Journal of Advanced Physics,* 1, 84(2012); D. Bhattacharyya and K. P. Ghatak, *Physica Status Solidi (b),* 187, 523(1995).

[28] Adhikari, S. M., D. De, J. K. Baruah, S. Chowdhury and K. P. Ghatak, *Advanced Science Focus,* 1, 57(2013); P. K. Bose, S. Bhattacharya, D. De, N. Paitya and K. P. Ghatak, *Advanced Science, Engineering and Medicine,* 5, 245(2013).

[29] Bhattacharya, S. and K. P. Ghatak, *Springer Science & Business Media,* 167(2012); K. P. Ghatak, *Acta Physica Hungarica,* 74, 257(1994); M. Mondal and K. P. Ghatak, *Acta Physica Slovaca,* 36, 325(1986); M. Mondal, K. P. Ghatak and *Acta physica Polonica.* A, 66, 47(1984).

[30] Bhattachrya, S., N. Paitya and K.P. Ghatak, *Journal of Computational and Theoretical Nanoscience*, 10, 1999(2013); N. Paitya and K. P. Ghatak, *Journal of nanoscience and nanotechnology,* 12, 8985(2012); N. Chattopadhyay, K. P. Ghatak and M. Mondal, *Acta Physica Hungarica,* 67, 163(1990); M. Mondal and K. P. Ghatak, *Acta physica Polonica.* A, 67, 983(1985).

[31] Chakraborty, P. K., G. C. Datta and K. P. Ghatak, *Physica Scripta*, 68, 368(2003); B. Nag, P. K. Chakrabarty and K. P. Ghatak, *Journal of Wave Material Interaction,* 11, 211(1996).

[32] Ghatak, K. P., D. K. Basu and B. Nag, *Journal of Wave Material Interaction*, 10, 29(1995); K. P. Ghatak and S. N Banik, *FIZIKA A,* 4, 33(1995); K. P. Ghatak and D. Bhattacharyya, *Physica Status Solidi (b)*, 179, 383(1993); K. P. Ghatak and D. Bhattacharyya, *Journal of Wave Material Interaction,* 8, 233(1993).

[33] Ghoshal, A. and K. P. Ghatak, *Acta Physica Slovaca, Czechoslovakia,* 42, 370(1992); K. P. Ghatak and S. N. Biswas, *Orlando'91, Orlando, FL,* 149(1991); K. P. Ghatak and A. Ghoshal, *Physica Status Solidi (a)*, 126, K53(1991); K. P. Ghatak, *Acta Physica Hungarica,* 70, 77(1991); M. Mondal and K. P. Ghatak, *Physica Status Solidi (b),*128, K133(1985).

[34] Hwang, J., *J. Appl. Phys.* 41, 2668 (1970); W. Sritrakool, H. R. Glyde, V. Sa Yakanit Can. *J. Phys.,* 60, 373 (1982); H. Ikoma, *J. Phys. Soc. Japan*, 27,514 (1969).

[35] Chakravarti, N., A. K. Chowdhury, K. P. Ghatak, S. Ghosh, A. Dhar, *Appl. Phys*. 25, 105 (1981); A. N. Chakravarti, K. P. Ghatak, K. K. Ghosh, S. Ghosh, A. Dhar, *Zeitschrift für Physik B Conden. Matter.* 47, 149 (1982); M. Mondal, N. Chattopadhyay, K. P. Ghatak, *Jour. of Low. Temp. Phys.* 66, 131 (1987); M. Mondal, K. P. Ghatak, *Phys. Lett. A* 131, 529 (1988); K. P. Ghatak, A. Ghoshal, B. Mitra, Il *Nuovo Cimento., D* 13, 867 (1991); K. P. Ghatak, A. Ghoshal, *Phys. Stat. Sol.* (a) 126, K53 (1991); E. A. Aruhanov, A. F. Knyazev, A. N. Nateprov, S. I. Radautsan, *Sov. Phys. Semiconduct*. 15, 828 (1981); P. N. Hai, W. M. Chen, I. A. Buyanova, H. P. Xin, C. W. Tu, *Appl. Phys. Letts*. 77, 1843 (2000); D. P. DiVincenzo, E. J. Mele, *Phys.*

Rev. B, 29, 1685 (1984); P. Perlin, E. Litwin-Staszewska, B. Suchanek, W. Knap, J. Camassel, T. Suski, R. Piotrzkowski, I. Grzegory, S. Porowski, E. Kaminska, *J. C. Chervin, Appl. Phys. Letts.* 68, 1114 (1996); F. Masia, G. Pettinari, A. Polimeni, M. Felici, A. Miriametro, M. Capizzi, A. Lindsay, S. B. Healy, E. P. O'Reilly, A. Cristofoli, G. Bais, M. Piccin, S. Rubini, F. Martelli, A. Franciosi, P. J. Klar, K. Volz, W. Stolz, *Phys. Rev. B,* 73, 073201 (2006); I. M. Tsidilkovski, *Band Structures of Semiconductors,* (Pergamon Press, London, 1982); K. P. Ghatak, M. Mondal, Z. F. *Physik B,* B69, 471 (1988); B. Mitra, A. Ghoshal, K. P. Ghatak, *Nouvo Cimento D 12D,* 891 (1990); K. P. Ghatak, S. N. Biswas, *Nonlin. Opt. Quant. Opts.*, 12, 83 (1995); K. P. Ghatak, S. N. Biswas, *Proc. SPIE*, 1484, 149 (1991); M. Mondal, K. P. Ghatak, *Graphite Intercalation Compounds: Science and Applications, MRS Proceedings*, ed. by M. Endo, M. S. Dresselhaus, G. Dresselhaus, *MRS Fall Meeting*, EA 16, 173 (1988); K. P. Ghatak, M. Mondal, Z. fur, *Nature A*, 41A, 881 (1986); C. C. Wu, C. J. Lin, *J. Low Temp. Phys.* 57, 469 (1984); M. H. Chen, C. C. Wu, C. J. Lin, *J. Low Temp. Phys.* 55, 127 (1984).

[36] Schulman, J. N., Y. C. Chang, *Phys. Rev. B,* 24, 4445 (1981).

[37] Anderson, N. G., W. D. Laidig, R. M. Kolbas, Y. C. Lo, *J. Appl. Phys.* 60, (2361) (1986); N. Paitya, K. P. Ghatak *Jour. Adv. Phys.* 1, 161(2012); S. Bhattacharya, D. De, S. M. Adhikari, K. P. Ghatak *Superlatt. Microst.* 51, 203(2012); D. De, S. Bhattacharya, S. M. Adhikari, A. Kumar, P. K. Bose, K. P. Ghatak, Beilstein *Jour. Nanotech.* 2, 339(2012); D. De, A. Kumar, S. M. Adhikari, S. Pahari, N. Islam, P. Banerjee, S. K. Biswas, S. Bhattacharya, K. P. Ghatak, *Superlatt. and Microstruct.* 47, 377(2010); S. Pahari, S. Bhattacharya, S. Roy, A. Saha, D. De, K. P. Ghatak, *Superlatt. and Microstruct.* 46, 760(2009); S. Pahari, S. Bhattacharya, K. P. Ghatak *Jour. of Comput. and Theo. Nanosci.* 6, 2088(2009); L. J. Singh, S. Choudhury, D. Baruah, S. K. Biswas, S. Pahari, K. P. Ghatak, *Phys. B: Conden. Matter*, 368, 188(2005); S. Chowdhary, L. J. Singh, K. P. Ghatak, *Phys. B: Conden. Matter,* 365, 5(2005); K. P. Ghatak, J. Mukhopadhyay J. P. Banerjee, *SPIE proceedings series*, 4746,

1292(2002); K. P. Ghatak, S. Dutta, D. K. Basu, B. Nag, Il *Nuovo Cimento. D* 20, 227(1998); K. P. Ghatak, B. De, *Mat. Resc. Soc. Proc.* 300, 513(1993); K. P. Ghatak, B. Mitra, Il *Nuovo Cimento. D* 15, 97(1993); K. P. Ghatak, *Inter. Soci. Opt. and Photon. Proc. Soc. Photo Opt. Instru. Engg.*, 1626, 115(1992); K. P. Ghatak, A Ghoshal, S. N. Biswas, M. Mondal, *Proc. Soc. Photo Opt. Instru. Engg*. 1308, 356(1990); S. N. Biswas, K. P. Ghatak, *Internat. Jour. Electronics Theo. Exp*. 70, 125(1991); B. Mitra, K. P Ghatak, *Phys. Lett. A.* 146, 357(1990); B. Mitra, K. P. Ghatak, *Phys. Lett. A.* 142, 401(1989); K. P. Ghatak, B Mitra, A Ghoshal, *Phy. Stat. Sol.* (b) 154, K121(1989); B. Mitra, K. P. Ghatak, *Phys. Stat. Sol.* (b). 149, K117(1988); S. Bhattacharyya, K. P. Ghatak, S. Biswas, OE/Fibers' 87, *Inter. Soc. Opt. Photon*. 73 (1987); K. P. Ghatak, A. N. Chakravarti, *Phys. Stat. Sol.* (b). 117, 707(1983).

[38] Capasso, F., *Semiconductors, and Semimetals* 22, 2(1985).

[39] Capasso, F., K. Mohammed, A. Y. Cho, R. Hull, A. L. Hutchinson, *Appl. Phys. Letts.,* 47, 420 (1985).

[40] Capasso, F., R. A. Kiehl, *J. Appl. Phys*. 58, 1366(1985).

[41] Ploog, K., G. H. Doheler, *Adv. Phys.* 32, 285(1983).

[42] Capasso, F., K. Mohammed, A. Y. Cho, *Appl. Phys. Lett.,* 478(1986).

[43] Grill, R., C. Metzner, G. H. Döhler, *Phys. Rev. B*, 63, 235316(2001); *Phys. Rev. B,* 61, 15614 (2000).

[44] Kost, R., M. H. Jupina, T. C. Hasenberg, E. M. Garmire, *J. Appl. Phys.*, 99, 023501(2006).

[45] Willardson, R. K., ed. by A. C. Beer, 1, *Semiconductors and Semimetals* (Academic Press, New York, 1966) p. 102.

[46] Kane, E. O., *Phys. Rev.* 131, 79 (1963); *Phys. Rev. B.*, 139, 343 (1965).

[47] Bonch Bruevich, V. L., *Sov. Phys. Sol. Stat.*, 4, 53 (1963).

[48] Logan, R. A., A. G. Chynoweth, *Phys. Rev.*, 131, 89 (1963).

[49] Hwang, J., *J. Appl. Phys.* 40, 3731 (1969).

[50] Pankove, J. I., *Phys. Rev. A*, 130, 2059 (1965).

[51] Halperin, I., M. Lax, *Phys. Rev.,* 148, 722 (1966).

[52] Abram, R. A., G. J. Rees, B. L. H. Wilson, *Adv. Phys.* 27,799 (1978).

[53] Shklovskii, B. I., A. L. Efros, *Electronic Properties of Doped Semiconductors*, Springer series in Solid State Sciences, vol.45 (Springer, Berlin, 1984).

[54] Kane, O., *Solid State Electron*, 28, 3 (1985).

[55] Chakraborty, P. K., J. C. Biswas, *J. Appl. Phys*. 82,3328 (1997).

[56] Nag, B. R., *Electron Transport in Compound Semiconductors*, Springer Series in Solid state Sciences, vol. 11, (Springer, Heidelberg, 1980).

[57] Schmid, P. E., *Phys. Rev. B*, 23, 5531 (1981).

[58] Jellison, Jr. G. E., F. A. Modine, C. W. White, R. F. Wood, R. T. Young, *Phys. Rev. Lett.*, 46, 1414 (1981).

[59] Fistul, V. I., *Heavily Doped Semiconductors* (Plenum, New York, 1969) ch 7.

[60] Hwang, C. J., *J. Appl. Phys.* 41, 2668 (1970); W. Sritrakool, H. R. Glyde, V. Sa Yakanit *Can. J. Phys.,* 60, 373 (1982); H. Ikoma, *J. Phys. Soc. Japan*, 27,514 (1969); M. H. Chen, C. C. Wu, C. J. Lin, *J. Low Temp. Phys*. 55, 127 (1984).

In: Quantum Wires: An Overview
Editor: Kamakhya Prasad Ghatak
ISBN: 978-1-53617-676-6

Chapter 3

DISPERSION RELATIONS IN QUANTUM WIRES OF NON-PARABOLIC SEMICONDUCTORS

N. Debbarma[1], M. Mitra[2], S. Das[1], S. Debbarma[1] and K. P. Ghatak[3,*]

[1]Department of Computer Science and Engineering, National Institute of Technology, Agartala, Tripura, India
[2]Department of Applied Physics, University of Calcutta, Kolkata, India
[3]Institute of Engineering and Management, Salt Lake, Kolkata, India and University of Engineering and Management, Kolkata and Jaipur, India

ABSTRACT

The concept of Dispersion Relations (DR) of the carriers in quantized structures occupies a singular position in Nano Technology and whose importance has already been established since the inception of the

* Corresponding Author's Email: kamakhyaghatakcu@gmail.com.

theory of band structure of Solid State Physics. The concept of DRs is of fundamental importance for not only the characterization of semiconductor nano-structures but also the study of the carrier transport in semiconductors and their quantized counter parts through the proper formulation of the Boltzmann Transport equation which, in turn, needs the corresponding carrier energy spectra. In this chapter we formulate the DR's of quantum wires (QWs) of n-GaP, IV-VI, Ge, Zero Gap, GaSb, $PtSb_2$, GaAs, n-InSb, $Pb_{1-x}Ga_xTe$, Te, Graphite, Sb, Bi_2Te_3, p-InSb, II-V, CuCl and Stressed p-type Semiconductors having diamond structure valance bands respectively. The influence of quantum confinement is immediately apparent from the 1D DRs since the DR depends strongly on the thickness of the quantum-confined materials in contrast with the corresponding bulk specimens. Finally, it may be noted that the basic aim of this chapter is not solely to demonstrate the influence of quantum confinement on the DR from QWs of the said materials but also to formulate the appropriate DOS in the most generalized form, since the transport and other phenomena in quantized structures having different band structures and the derivation of the expressions of many important electronic properties are based on the DR in such materials.

INTRODUCTION

The creation of Nano Electronics, the subset of the generalized set Physics, is based on following two important concepts:

The symmetry of the wave-vector space of the charge carriers in electronic materials having various band structures is being reduced from a 3D closed surface to a quantized 2D closed surface, quantized non-parabolas and fully quantized wave vector space leading to the formation of 0D systems such as ultra thin films (UFs), Doping superlattices, inversion and accumulation layers, quantum wells (QWs), quantum well super-lattices, carbon nano-tubes, nano wires (NWs), quantum wire super-lattices, magnetic quantization, magneto size quantization, quantum dots (QDs), magneto inversion and accumulation layers, magneto quantum well super-lattices, magneto NIPIs, quantum dot super-lattices and other field aided nano structures. The advent of modern experimental methods namely, Fine Line Lithography (FLL), Metallo-Organic Chemical Vapor

Deposition (MOCVD), Molecular Beam Epitaxy (MBE), etc. for fabricating the low dimensional nano-structured systems.

Quantum confined materials have gained much interest in modern physics because of their importance to unlock both new scientific revelations and multi dimensional all together unheard technological applications. In UFs, the quantization of the motion of the carriers in the direction perpendicular to the surface exhibits the two-dimensional carrier motion of the charge carriers and the third direction is being quantized. Another one dimensional structure known as NWs has been proposed to investigate the physical properties in these materials where the carrier gas is quantized in two transverse directions and they can move only in the longitudinal direction. As the concept of quantization increases from 1D to 3D, the degree of freedom of the free carriers decreases drastically and the total density-of-states (DOS) function changes from Heaviside step function to the Dirac's delta function forming QDs which, in turn, depend on the carrier energy spectra in different materials. An enormous range of important applications of such low dimensional structures for modern physics in the quantum regime together with a rapid increase in computing power, have generated considerable interest in the study of the optical properties of quantum effect devices based on various new materials of reduced dimensionality. Examples of such new applications include quantum switches, quantum registers, quantum sensors, hetero-junction field-effect, quantum logic gates, quantum well and quantum wire transistors, quantum cascade lasers, high-frequency microwave circuits, high-speed digital networks, high-resolution terahertz spectroscopy, advanced integrated circuits, super-lattice photo-oscillator, super-lattice photo-cathodes, resonant tunneling diodes and transistors, super-lattice coolers, thermoelectric devices, thin film transistors, micro-optical systems, intermediate-band solar cells, high performance infrared imaging systems, optical modulators, optical switching systems, single electron/molecule electronics, nano-tube based diodes, and other nano-electronic devices [1-14].

Although many new effects in quantized structures have already been reported, the interest for further research of other aspects of such quantum-confined materials is becoming increasingly important. One such significant concept is the Dispersion Relations (DRs) of the carriers in semiconductors and their nano-structures which occupies a singular position in the whole arena of Modern Physics and related disciplines in general and whose importance [15-36] has already been established since the inception of the theory of band structure of Solid State Physics. The concept of DRs is of fundamental importance for not only the characterization of semiconductor nano-structures but also the study of the carrier transport in semiconductors and their quantized counter parts through the proper formulation of the Boltzmann Transport equation. It is important to note that six important transport quantities namely, the effective carrier mass (ECM), density-of-states (DOS) function, the sub-band energy and the measurement of band-gap in the presence of strong light waves, intense electric field and heavy doping are in disguise in the very important concept of DR. Besides, the acoustic mobility limited momentum relaxation time is inversely proportional to the respective DOS function of a particular semiconductor and integral over the DOS function leads to carrier statistics under the condition of extreme carrier degeneracy which, in turn, is connected to the twenty five important transport topics of quantum effect devices namely the Landau Dia and Pauli's Para Magnetic Susceptibilities [37], the Einstein's Photoemission [38], the Einstein Relation [39], the Debye Screening Length [40], the Generalized Raman gain [41], the Normalized Hall coefficient [42], Dispersion Relations In Quantum Wires [43], the Gate Capacitance [44], the Thermoelectric Power [45], the Plasma Frequency [46], the Magneto-Thermal effect in Quantized Structures [47], the Activity coefficient [48], the Reflection coefficient [49], the Heat Capacity [50], the Faraday rotation [51], the Optical Effective Mass [52], the Carrier contribution to the elastic constants [53], the Diffusion coefficient of the minority carriers [54], the Nonlinear optical response [55], the Third order nonlinear optical susceptibility [56], the Righi-Leduc coefficient [57], the Electric Susceptibility [58], the Electric Susceptibility Mass [59], the Electron Diffusion Thermo-power [60] and theHydrostatic Piezo-resistance Coefficient [61] respectively. Although the DRs for various technologically important materials have been extensively studied in the literature [62-100] nevertheless there lies scopes in the investigations made while the interest for further researches of the different other materials is becoming increasingly important. Keeping this in mind, in this chapter we shall study the DRs in QWs of n-GaP, IV-VI, Ge, Zero Gap, GaSb, $PtSb_2$, GaAs, n-InSb, $Pb_{1-x}Ga_xTe$, Te, Graphite, Sb, Bi_2Te_3, p-InSb, II-V, CuCl and Stressed p-type Semiconductors having diamond structure valance bands respectively. The technological importance of the said materials is already well-known in the literature.

THEORETICAL BACKGROUND

(A) The conduction electrons of *n*-GaP obey two different dispersion laws as given literature [62, 63]. In accordance with Rees [62], the electron energy spectrum is given by

$$E = \frac{\hbar^2 k_s^2}{2m_\perp^*} + \frac{\hbar^2}{2m_{||}^*}[\wp k_s^2 + k_z^2] - [\frac{\hbar^4 k_0^2}{m_{||}^{*2}}(k_s^2 + k_z^2) + |V_G|^2]^{1/2} - |V_G| \quad (1)$$

where, $k_s^2 = k_x^2 + k_y^2$, k_0 and $|V_G|$ are constants of the energy spectrum with $m_{||}^* = 0.92m_0$, $m_\perp^* = 0.25m_0$, $k_0 = 1.7\times10^{19}m^{-1}$, $|V_G| = 0.21eV$, $g_v = 6$, $g_s = 2$ and $\wp = 1$.

The Dispersion Relation in quantum wires in this case can be written as

$$E = \frac{\hbar^2[(\frac{n_x\pi}{d_x})^2 + (\frac{n_y\pi}{d_y})^2]}{2m_\perp^*} + \frac{\hbar^2}{2m_{||}^*}[\wp((\frac{n_x\pi}{d_x})^2 + (\frac{n_y\pi}{d_y})^2) + k_z^2]$$
$$-[\frac{\hbar^4 k_0^2}{m_{||}^{*2}}(((\frac{n_x\pi}{d_x})^2 + (\frac{n_y\pi}{d_y})^2) + k_z^2) + |V_G|^2]^{1/2} - |V_G| \quad (2)$$

(i) In accordance with Ivchenko and Pikus [63], the electron dispersion law can be written as

$$E = \frac{\hbar^2 k_s^2}{2m_{||}^*} + \frac{\hbar^2 k_z^2}{2m_{\perp}^*} \mp \frac{\bar{\Delta}}{2} \pm [(\frac{\bar{\Delta}}{2})^2 + P_1 k_z^2 + D_1 k_x^2 k_y^2]^{1/2} \quad (3)$$

where, $\bar{\Delta} = 335 meV, P_1 = 2\times 10^{-10} eVm, D_1 = P_1 a_1$ *and* $a_1 = 5.4\times 10^{-10} m$

The Dispersion Relation in quantum wires in this case can be written as

$$E = \frac{\hbar^2 ((\frac{n_x \pi}{d_x})^2 + (\frac{n_y \pi}{d_y})^2)}{2m_{||}^*} + \frac{\hbar^2 k_z^2}{2m_{\perp}^*} \mp \frac{\bar{\Delta}}{2} \pm [(\frac{\bar{\Delta}}{2})^2 + P_1 k_z^2 + D_1 ((\frac{n_x \pi}{d_x})^2 (\frac{n_y \pi}{d_y})^2)]^{1/2} \quad (4)$$

(B) The dispersion relation for the conduction electrons for IV-VI compounds can also be described by the models of Dimmock [64], Bangert et al. [65] and Foley et al. [66] respectively.

(i) In accordance with Dimmock model [64], the carrier energy spectrum of IV-VI materials assumes the form

$$[\in - \frac{E_g}{2} - \frac{\hbar^2 k_s^2}{2m_t^-} - \frac{\hbar^2 k_z^2}{2m_l^-}][\in + \frac{E_g}{2} + \frac{\hbar^2 k_s^2}{2m_t^+} + \frac{\hbar^2 k_z^2}{2m_l^+}] = P_\perp^2 k_s^2 + P_{||}^2 k_z^2 \quad (5)$$

where, $\in$ is the energy as measured from the center of the band gap E_g, $m_t^{\pm}$ and $m_l^{\pm}$ represent the contribution of the transverse and longitudinal effective masses of the external L_6^+ and L_6^- bands arising from the **k.p** perturbations with the other bands taken to the second order and $g_v = 4$.

The Dispersion Relation in quantum wires in this case can be written as

$$[\in - \frac{E_g}{2} - \frac{\hbar^2 ((\frac{n_x \pi}{d_x})^2 + (\frac{n_y \pi}{d_y})^2)}{2m_t^-} - \frac{\hbar^2 k_z^2}{2m_l^-}][\in + \frac{E_g}{2} + \frac{\hbar^2 ((\frac{n_x \pi}{d_x})^2 + (\frac{n_y \pi}{d_y})^2)}{2m_t^+} + \frac{\hbar^2 k_z^2}{2m_l^+}]$$
$$= P_\perp^2 ((\frac{n_x \pi}{d_x})^2 + (\frac{n_y \pi}{d_y})^2) + P_{||}^2 k_z^2 \quad (6)$$

(ii) In accordance with Bangert et al. [65] the dispersion relation is given by

$$\Gamma(E) = F_1(E)k_s^2 + F_2(E)k_z^2 \tag{7}$$

where,

$$\Gamma(E) \equiv 2E \, , F_1(E) \equiv \frac{R_1^2}{E+E_g} + \frac{S_1^2}{E+\Delta_c'} + \frac{Q_1^2}{E+E_g} \, , F_2(E) \equiv \frac{2C_5^2}{E+E_g} + \frac{(S_1+Q_1)^2}{E+\Delta_c''} \, ,$$

$R_1^2 = 2.3\times10^{-19}(eVm)^2$, $C_5^2 = 0.83\times10^{-19}(eVm)^2$, $Q_1^2 = 1.3R_1^2$, $S_1^2 = 4.6R_1^2$, $\Delta_c' = 3.07eV$, $\Delta_c'' = 3.28eV$ and $g_v = 4$. It may be noted that under the substitution $S_1 = 0$, $Q_1 = 0$, $R_1^2 \equiv \frac{\hbar^2 E_g}{m_\perp^*}$, $C_5^2 \equiv \frac{\hbar^2 E_g}{2m_{||}^*}$, equation (7) assumes the form $E(1+\alpha E) = \frac{\hbar^2 k_s^2}{2m_\perp^*} + \frac{\hbar^2 k_z^2}{2m_{||}^*}$ which is the simplified Lax model.

The Dispersion Relation in quantum wires in this case can be written as

$$\Gamma(E) = F_1(E)((\frac{n_x\pi}{d_x})^2 + (\frac{n_y\pi}{d_y})^2) + F_2(E)k_z^2 \tag{8}$$

(iii) The carrier energy spectrum of IV-VI semiconductors in accordance with Foley et al. [66] can be written as

$$E + \frac{E_g}{2} = E_-(k) + [[E_+(k) + \frac{E_g}{2}]^2 + P_\perp^2 k_s^2 + P_{||}^2 k_z^2]^{1/2} \tag{9}$$

where, $E_+(k) = \frac{\hbar^2 k_s^2}{2m_\perp^+} + \frac{\hbar^2 k_z^2}{2m_{||}^+}, E_-(k) = \frac{\hbar^2 k_s^2}{2m_\perp^-} + \frac{\hbar^2 k_z^2}{2m_{||}^-}$ represents the contribution from the interaction of the conduction and the valance band edge states with the more distant bands and the free electron term,

$$\frac{1}{m_\perp^\pm} = \frac{1}{2}[\frac{1}{m_{tc}} \pm \frac{1}{m_{tv}}], \frac{1}{m_{||}^\pm} = \frac{1}{2}[\frac{1}{m_{1c}} \pm \frac{1}{m_{1v}}]$$

For n-PbTe,

$$P_{\perp} = 4.61 \times 10^{-10} eVm, P_{\parallel} = 1.48 \times 10^{-10} eVm, \frac{m_0}{m_{tv}} = 10.36, \frac{m_0}{m_{1v}} = 0.75, \frac{m_0}{m_{tc}} = 11.36, \frac{m_0}{m_{1c}} = 1.20 \text{ and } g_v = 4$$

The Dispersion Relation in quantum wires in this case can be written as

$$E + \frac{E_g}{2} = E_{-}(k) + [[E_{+}(k) + \frac{E_g}{2}]^2 + P_{\perp}^2((\frac{n_x \pi}{d_x})^2 + (\frac{n_y \pi}{d_y})^2) + P_{\parallel}^2 k_z^2]^{1/2} \tag{10}$$

$$\text{where, } E_{+}(k) = \frac{\hbar^2((\frac{n_x \pi}{d_x})^2 + (\frac{n_y \pi}{d_y})^2)}{2m_{\perp}^{+}} + \frac{\hbar^2 k_z^2}{2m_{\parallel}^{+}}, E_{-}(k) = \frac{\hbar^2((\frac{n_x \pi}{d_x})^2 + (\frac{n_y \pi}{d_y})^2)}{2m_{\perp}^{-}} + \frac{\hbar^2 k_z^2}{2m_{\parallel}^{-}}$$

(C) The importance of Germanium is well known since the inception of semiconductor physics. The conduction electrons of *n-Ge* obey two different dispersion laws since band non-parabolicity has been included in two different ways as given literature [67-68]. In accordance with Cardona et al. [67] and Wang et al. [68] the electron dispersion laws in Ge can respectively, be expressed as

$$E = -\frac{E_g}{2} + \frac{\hbar^2 k_z^2}{2m_{\parallel}^*} + [\frac{E_g^2}{4} + E_g k_s^2 (\frac{\hbar^2}{2m_{\perp}^*})]^{1/2} \tag{11}$$

and

$$E = a_9 k_z^2 + l_9 k_s^2 - c_9 k_s^4 - d_9 k_s^2 k_z^2 - e_9 k_z^4 \tag{12}$$

where,

$$a_9 = (\hbar^2 / 2m_{||}^*), m_{||}^* = 1.588 m_0, l_9 = (\hbar^2 / 2m_{\perp}^*), m_{\perp}^* = 0.0815 m_0, c_9 = 1.4 A_9 A_9 \equiv \frac{1}{4} (\hbar^4 / E_g m_{\perp}^{*2})[1 - \frac{m_{\perp}^*}{m_0}]^2$$

$$E_g = 2.2 eV, d_9 = 0.8 A_9, e_9 = 0.005 A_9, g_v = 4 \text{ and } g_s = 2$$

The Dispersion Relation in quantum wires in this case can be written as

$$E = -\frac{E_g}{2} + \frac{\hbar^2 k_z^2}{2m_{\parallel}^*} + [\frac{E_g^2}{4} + E_g ((\frac{n_x \pi}{d_x})^2 + (\frac{n_y \pi}{d_y})^2)(\frac{\hbar^2}{2m_{\perp}^*})]^{1/2} \tag{13}$$

and

$$E = a_9 k_z^2 + l_9 ((\frac{n_x \pi}{d_x})^2 + (\frac{n_y \pi}{d_y})^2) - c_9 ((\frac{n_x \pi}{d_x})^2 + (\frac{n_y \pi}{d_y})^2)^2 - d_9 ((\frac{n_x \pi}{d_x})^2 + (\frac{n_y \pi}{d_y})^2) k_z^2 - e_9 k_z^4 \tag{14}$$

(D) The dispersion relation of the conduction electrons of zero-gap materials (e.g., HgTe) is given by [69]

$$E = \frac{\hbar^2 k^2}{2m^*} + \frac{3e^2}{128\varepsilon_\infty} k - (\frac{2E_B}{\pi}) \ln \left| \frac{k}{k_0} \right| \tag{15}$$

where, ε_∞ is the semiconductor permittivity in the high frequency limit,

$$E_B \equiv \frac{m_0 e^2}{2\hbar^2 \varepsilon_\infty^2} \text{ and } k_0 \equiv \frac{m_0 e^2}{\hbar^2 \varepsilon_\infty}$$

The Dispersion Relation in quantum wires in this case can be written as

$$E = \frac{\hbar^2 [(\frac{n_x \pi}{d_x})^2 + (\frac{n_y \pi}{d_y})^2 + k_z^2]}{2m^*} + \frac{3e^2}{128\varepsilon_\infty} [(\frac{n_x \pi}{d_x})^2 + (\frac{n_y \pi}{d_y})^2 + k_z^2]^{1/2}$$
$$-(\frac{2E_B}{\pi}) \ln \left| \frac{[(\frac{n_x \pi}{d_x})^2 + (\frac{n_y \pi}{d_y})^2 + k_z^2]^{1/2}}{k_0} \right| \tag{16}$$

(E) The conduction electrons of n-GaSb obey the following three dispersion relations:

(i) In accordance with the model of Seiler et al. [70]

$$E = [-\frac{E_g}{2} + \frac{E_g}{2}[1 + \alpha_4 k^2]^{1/2} + \frac{\bar{\varsigma}_0 \hbar^2 k^2}{2m_o} + \frac{\bar{v}_0 f_1(k) \hbar^2}{2m_o} \pm \frac{\bar{\omega}_0 f_2(k) \hbar^2}{2m_o}] \tag{17}$$

where, $\alpha_4 \equiv 4P^2(E_g + \frac{2}{3}\Delta)[E_g^2(E_g + \Delta)]^{-1}$, P is the isotropic momentum matrix element, $f_1(k) \equiv k^{-2}[k_x^2 k_y^2 + k_y^2 k_z^2 + k_z^2 k_x^2]$ represents the warping of the Fermi surface, $f_2(k) \equiv [\{k^2(k_x^2 k_y^2 + k_y^2 k_z^2 + k_z^2 k_x^2) - 9k_x^2 k_y^2 k_z^2\}^{1/2} k^{-1}]$ represents the inversion asymmetry splitting of the conduction band and $\bar{\varsigma}_0, \bar{v}_0$ and $\bar{\omega}_0$ represent the constants of the electron spectrum in this case.

The Dispersion Relation in quantum wires in this case can be written as

$$E=[-\frac{E_g}{2}+\frac{E_g}{2}[1+\alpha_4[(\frac{n_x\pi}{d_x})^2+(\frac{n_y\pi}{d_y})^2+k_z^2]]^{1/2}+\frac{\bar{\varsigma}_0\hbar^2[(\frac{n_x\pi}{d_x})^2+(\frac{n_y\pi}{d_y})^2+k_z^2]}{2m_o}$$
$$+\frac{\bar{v}_0 f_1(k_z)\hbar^2}{2m_o}\pm\frac{\bar{\omega}_0 f_2(k_z)\hbar^2}{2m_o}] \tag{18}$$
$$where,\ f_1(k_z)\equiv[(\frac{n_x\pi}{d_x})^2(\frac{n_y\pi}{d_y})^2+(\frac{n_y\pi}{d_y})^2k_z^2+k_z^2(\frac{n_x\pi}{d_x})^2][(\frac{n_x\pi}{d_x})^2+(\frac{n_y\pi}{d_y})^2+k_z^2]^{-1},$$
$$f_2(k_z)\equiv[\{[(\frac{n_x\pi}{d_x})^2+(\frac{n_y\pi}{d_y})^2+k_z^2]((\frac{n_x\pi}{d_x})^2(\frac{n_y\pi}{d_y})^2+(\frac{n_y\pi}{d_y})^2k_z^2$$
$$+k_z^2(\frac{n_x\pi}{d_x})^2)-9(\frac{n_x\pi}{d_x})^2(\frac{n_y\pi}{d_y})^2k_z^2\}^{1/2}[(\frac{n_x\pi}{d_x})^2+(\frac{n_y\pi}{d_y})^2+k_z^2]^{-1/2}]$$

(ii) In accordance with the model of Mathur et al. [71],

$$E=\frac{\hbar^2k^2}{2m_0}+\frac{E_{g1}}{2}[(1+\frac{2\hbar^2k^2}{E_{g1}}\{\frac{1}{m^*}-\frac{1}{m_0}\})^{1/2}-1] \tag{19}$$
$$where, E_{g1}=\{E_g+[(5\times10^{-5}T^2)(2(112+T)^{-1})]\}eV$$

The Dispersion Relation in quantum wires in this case can be written as

$$E=\frac{\hbar^2[(\frac{n_x\pi}{d_x})^2+(\frac{n_y\pi}{d_y})^2+k_z^2]}{2m_0}+\frac{E_{g1}}{2}[(1+\frac{2\hbar^2[(\frac{n_x\pi}{d_x})^2+(\frac{n_y\pi}{d_y})^2+k_z^2]}{E_{g1}}\{\frac{1}{m^*}-\frac{1}{m_0}\})^{1/2}-1] \tag{20}$$

(iii) In accordance with the model of Zhang et al. [72]

$$E=[E_2^{(1)}+E_2^{(2)}K_{4,1}]k^2+[E_4^{(1)}+E_4^{(2)}K_{4,1}]k^4+k^6[E_6^{(1)}+E_6^{(2)}K_{4,1}+E_6^{(3)}K_{6,1}] \tag{21a}$$

where, $K_{4,1}\equiv\frac{5}{4}\sqrt{21}[\frac{k_x^4+k_y^4+k_z^4}{k^4}-\frac{3}{5}]$, $K_{6,1}\equiv\sqrt{\frac{639639}{32}}[\frac{k_x^2k_y^2k_z^2}{k^6}+\frac{1}{22}(\frac{k_x^4+k_y^4+k_z^4}{k^4}-\frac{3}{5})-\frac{1}{105}]$, the coefficients are in eV, the values of k are $10(\frac{a}{2\pi})$ times those of k in atomic units (a is the lattice constant), $E_2^{(1)}=1.0239620$, $E_2^{(2)}=0$, $E_4^{(1)}=-1.1320772$, $E_4^{(2)}=0.05658$, $E_6^{(1)}=1.1072073$, $E_6^{(2)}=-0.1134024$ and $E_6^{(3)}=-0.0072275$.

The Dispersion Relation in quantum wires in this case can be written as

$$E=[E_2^{(1)}+E_2^{(2)}K_{4,1}][(\frac{n_x\pi}{d_x})^2+(\frac{n_y\pi}{d_y})^2+k_z^2]+[E_4^{(1)}+E_4^{(2)}K_{4,1}][(\frac{n_x\pi}{d_x})^2+(\frac{n_y\pi}{d_y})^2+k_z^2]^2$$
$$+[(\frac{n_x\pi}{d_x})^2+(\frac{n_y\pi}{d_y})^2+k_z^2]^3[E_6^{(1)}+E_6^{(2)}K_{4,1}+E_6^{(3)}K_{6,1}] \quad (21b)$$

$$where,\ K_{4,1}\equiv\frac{5}{4}\sqrt{21}[\frac{(\frac{n_x\pi}{d_x})^4+(\frac{n_y\pi}{d_y})^4+k_z^4}{[(\frac{n_x\pi}{d_x})^2+(\frac{n_y\pi}{d_y})^2+k_z^2]^2}-\frac{3}{5}]$$

and

$$K_{6,1}\equiv\sqrt{\frac{639639}{32}}[\frac{(\frac{n_x\pi}{d_x})^2(\frac{n_y\pi}{d_y})^2k_z^2}{[(\frac{n_x\pi}{d_x})^2+(\frac{n_y\pi}{d_y})^2+k_z^2]^3}+\frac{1}{22}(\frac{(\frac{n_x\pi}{d_x})^4+(\frac{n_y\pi}{d_y})^4+k_z^4}{[(\frac{n_x\pi}{d_x})^2+(\frac{n_y\pi}{d_y})^2+k_z^2]^2}-\frac{3}{5})-\frac{1}{105}]$$

(F) The dispersion relation of the carriers in p-type Platinum antimonide ($PtSb_2$) following Emtage [73] can be written as

$$(E+\lambda_1\frac{a^2}{4}k^2-l_1k_s^2\frac{a^2}{4})(E+\delta_0'-\upsilon_1\frac{a^2}{4}k^2-n_1\frac{a^2}{4}k_s^2)=I_0\frac{a^4}{16}k^4 \quad (22)$$

where λ_1, l_1, δ_0', ν_1, n_1 and I_0 are the energy band constants and a is the lattice constant.

The Dispersion Relation in quantum wires in this case can be written as

$$(E+\lambda_1\frac{a^2}{4}[(\frac{n_x\pi}{d_x})^2+(\frac{n_y\pi}{d_y})^2+k_z^2]-l_1[(\frac{n_x\pi}{d_x})^2+(\frac{n_y\pi}{d_y})^2]\frac{a^2}{4})(E+\delta_0'-\upsilon_1\frac{a^2}{4}[(\frac{n_x\pi}{d_x})^2$$
$$+(\frac{n_y\pi}{d_y})^2+k_z^2]-n_1\frac{a^2}{4}[(\frac{n_x\pi}{d_x})^2+(\frac{n_y\pi}{d_y})^2])=I_0\frac{a^4}{16}[(\frac{n_x\pi}{d_x})^2+(\frac{n_y\pi}{d_y})^2+k_z^2]^2 \quad (23)$$

(G) In addition to the well known three band model of Kane, the conduction electrons of n-GaAs obey the following three dispersion relations:

In accordance with the model of Stillman et al. [74]

$$E = \frac{\hbar^2 k^2}{2m^*} - (1 - \frac{m^*}{m_0})^2 (\frac{\hbar^2 k^2}{2m^*})^2 [\frac{3E_g + 4\Delta + (\frac{2\Delta^2}{E_g})}{(E_g + \Delta)(2\Delta + 3E_g)}] \tag{24}$$

The Dispersion Relation in quantum wires in this case can be written as

$$E = \frac{\hbar^2[(\frac{n_x \pi}{d_x})^2 + (\frac{n_y \pi}{d_y})^2 + k_z^2]}{2m^*} - (1 - \frac{m^*}{m_0})^2 (\frac{\hbar^2}{2m^*})^2 [(\frac{n_x \pi}{d_x})^2 + (\frac{n_y \pi}{d_y})^2 + k_z^2]^2 [\frac{3E_g + 4\Delta + (\frac{2\Delta^2}{E_g})}{(E_g + \Delta)(2\Delta + 3E_g)}] \tag{25}$$

(i) In accordance with the model of Newson et al. [75]

$$E = \bar{\alpha}_0 k_z^4 + [\frac{\hbar^2}{2m^*} + (2\bar{\alpha}_0 + \bar{\beta}_0) k_s^2] k_z^2 + \frac{\hbar^2}{2m^*} k_s^2 + (2\bar{\alpha}_0 + \bar{\beta}_0) k_x^2 k_y^2 + \bar{\alpha}_0 (k_x^4 + k_y^4) \tag{26}$$

where, $\bar{\alpha}_0 (= -1.97 \times 10^{-37} eVm^4)$ is the non-parabolicity constant and $\bar{\beta}_0 A (= -2.3 \times 10^{-37} eVm^4)$ is the wrapping constant.

The Dispersion Relation in quantum wires in this case can be written as

$$E = \bar{\alpha}_0 k_z^4 + [\frac{\hbar^2}{2m^*} + (2\bar{\alpha}_0 + \bar{\beta}_0)[(\frac{n_x \pi}{d_x})^2 + (\frac{n_y \pi}{d_y})^2]] k_z^2 + \frac{\hbar^2}{2m^*}[(\frac{n_x \pi}{d_x})^2 + (\frac{n_y \pi}{d_y})^2]$$
$$+ (2\bar{\alpha}_0 + \bar{\beta}_0)(\frac{n_x \pi}{d_x})^2 (\frac{n_y \pi}{d_y})^2 + \bar{\alpha}_0 ((\frac{n_x \pi}{d_x})^4 + (\frac{n_y \pi}{d_y})^4) \tag{27}$$

(ii) In accordance with the model of Rossler [76]

$$E = \frac{\hbar^2 k^2}{2m^*} + \bar{\alpha}_{10} k^4 + \bar{\beta}_{10}[k_x^2 k_y^2 + k_y^2 k_z^2 + k_z^2 k_x^2] \pm \bar{\gamma}_{10}[k^2 (k_x^2 k_y^2 + k_y^2 k_z^2 + k_z^2 k_x^2) - 9 k_x^2 k_y^2 k_z^2]^{1/2} \tag{28}$$

where, $\bar{\alpha}_{10}=\bar{\alpha}_{11}+\bar{\alpha}_{12}k$, $\bar{\beta}_{10}=\bar{\beta}_{11}+\bar{\beta}_{12}k$ and $\bar{\gamma}_{10}=\bar{\gamma}_{11}+\bar{\gamma}_{12}k$, in which, $\bar{\alpha}_{11}=-2132\times10^{-40}eVm^4$, $\bar{\alpha}_{12}=9030\times10^{-50}eVm^5$, $\bar{\beta}_{11}=-2493\times10^{-40}eVm^4$, $\bar{\beta}_{12}=12594\times10^{-50}eVm^5$, $\bar{\gamma}_{11}=30\times10^{-30}eVm^3$ and $\bar{\gamma}_{12}=-154\times10^{-42}eVm^4$.

The Dispersion Relation in quantum wires in this case can be written as

$$E=\frac{\hbar^2[(\frac{n_x\pi}{d_x})^2+(\frac{n_y\pi}{d_y})^2+k_z^2]}{2m^*}+\bar{\alpha}_{10}[(\frac{n_x\pi}{d_x})^2+(\frac{n_y\pi}{d_y})^2+k_z^2]^2+\bar{\beta}_{10}[(\frac{n_x\pi}{d_x})^2(\frac{n_y\pi}{d_y})^2+(\frac{n_y\pi}{d_y})^2k_z^2+k_z^2(\frac{n_x\pi}{d_x})^2]$$
$$\pm\bar{\gamma}_{10}[[(\frac{n_x\pi}{d_x})^2+(\frac{n_y\pi}{d_y})^2+k_z^2]((\frac{n_x\pi}{d_x})^2(\frac{n_y\pi}{d_y})^2+(\frac{n_y\pi}{d_y})^2k_z^2+k_z^2(\frac{n_x\pi}{d_x})^2)-9(\frac{n_x\pi}{d_x})^2(\frac{n_y\pi}{d_y})^2k_z^2]^{1/2} \quad (29)$$

$$\text{where, } \bar{\alpha}_{10}=\bar{\alpha}_{11}+\bar{\alpha}_{12}[(\frac{n_x\pi}{d_x})^2+(\frac{n_y\pi}{d_y})^2+k_z^2]^{1/2}, \bar{\beta}_{10}=\bar{\beta}_{11}+\bar{\beta}_{12}[(\frac{n_x\pi}{d_x})^2+(\frac{n_y\pi}{d_y})^2+k_z^2]^{1/2}$$
$$\text{and } \bar{\gamma}_{10}=\bar{\gamma}_{11}+\bar{\gamma}_{12}[(\frac{n_x\pi}{d_x})^2+(\frac{n_y\pi}{d_y})^2+k_z^2]^{1/2}$$

(H) In addition to the well known three band model of Kane, the conduction electrons of n-InSb obey the following three dispersion relations:

(i) To the fourth order effective mass theory, and taking into account the interactions of the conduction, the heavy hole, the light hole and the split-off bands, the electron energy spectrum in n-InSb is given by [77]

$$E=\frac{\hbar^2k^2}{2m^*}+\bar{b}_1k^4 \quad (30)$$

$$\text{where, } \bar{b}_1\equiv\frac{K_2\hbar^4}{4E_g(m^*)^2}, K_2\equiv-[\frac{(1+\frac{1}{2}x_1^2)}{1+\frac{1}{2}x_1}](1-y_1)^2, x_1\equiv[1+(\frac{\Delta}{E_g})]^{-1} \text{ and } y_1\equiv\frac{m^*}{m_0}$$

The Dispersion Relation in quantum wires in this case can be written as

$$E=\frac{\hbar^2[(\frac{n_x\pi}{d_x})^2+(\frac{n_y\pi}{d_y})^2+k_z^2]}{2m^*}+\bar{b}_1[(\frac{n_x\pi}{d_x})^2+(\frac{n_y\pi}{d_y})^2+k_z^2]^2 \quad (31)$$

(ii) In accordance with Johnson and Dickey [78], the electron energy spectrum assumes the form

$$E = -\frac{E_g}{2} + \frac{\hbar^2 k^2}{2}[\frac{1}{m_0} + \frac{1}{m_{\gamma b}}] + \frac{E_g}{2}[1 + 4\frac{\hbar^2 k^2}{2m_c'}\frac{\bar{f}_1(E)}{E_g}]^{1/2} \tag{32}$$

$$where,\ \frac{m_0}{m_c'} \equiv P^2[\frac{(E_g + \frac{2\Delta}{3})}{E_g(E_g + \Delta)}], \bar{f}_1(E) \equiv \frac{(E_g + \Delta)(E + E_g + \frac{2\Delta}{3})}{(E_g + \frac{2\Delta}{3})(E + E_g + \Delta)}, m_c' = 0.139 m_0$$

$$and\ m_{\gamma b} = [\frac{1}{m_c'} - \frac{2}{m_0}]^{-1}$$

The Dispersion Relation in quantum wires in this case can be written as

$$E = -\frac{E_g}{2} + \frac{\hbar^2[(\frac{n_x\pi}{d_x})^2 + (\frac{n_y\pi}{d_y})^2 + k_z^2]}{2}[\frac{1}{m_0} + \frac{1}{m_{\gamma b}}] + \frac{E_g}{2}[1 + 4\frac{\hbar^2[(\frac{n_x\pi}{d_x})^2 + (\frac{n_y\pi}{d_y})^2 + k_z^2]}{2m_c'}\frac{\bar{f}_1(E)}{E_g}]^{1/2} \tag{33}$$

(iii) In accordance with Agafonov et al. [79], the electron energy spectrum can be written as

$$E = \frac{\bar{\eta} - E_g}{2}[1 - \frac{\hbar^2 k^2}{2\bar{\eta} m^*}\{\frac{D\sqrt{3} - 3\bar{B}}{2(\frac{\hbar^2}{m^*})}\}[\frac{k_x^4 + k_y^4 + k_z^4}{k^4}]] \tag{34}$$

$$where,\ \bar{\eta} = (E_g^2 + \frac{8}{3}P^2 k^2)^{1/2}, B \equiv -21\frac{\hbar^2}{2m_0}\ and\ D \equiv -40(\frac{\hbar^2}{2m_0})$$

The Dispersion Relation in quantum wires in this case can be written as

$$E = \frac{\bar{\eta}_1 - E_g}{2}[1 - \frac{\hbar^2[(\frac{n_x\pi}{d_x})^2 + (\frac{n_y\pi}{d_y})^2 + k_z^2]}{2\bar{\eta}_1 m^*}\{\frac{D\sqrt{3} - 3\bar{B}}{2(\frac{\hbar^2}{m^*})}\}\frac{[(\frac{n_x\pi}{d_x})^4 + (\frac{n_y\pi}{d_y})^4 + k_z^4]}{[(\frac{n_x\pi}{d_x})^2 + (\frac{n_y\pi}{d_y})^2 + k_z^2]^2}] \tag{35}$$

$$where,\ \bar{\eta}_1 = (E_g^2 + \frac{8}{3}P^2[(\frac{n_x\pi}{d_x})^2 + (\frac{n_y\pi}{d_y})^2 + k_z^2])^{1/2},$$

The dispersion relation of the carriers in n-type $Pb_{1-x}Ga_xTe$ with x=0.01 following Vassilev [80] can be written as

$$[E-0.606k_s^2-0.0722k_z^2][E+\bar{E}_g+0.411k_s^2+0.0377k_z^2]=0.23k_s^2+0.02k_z^2$$
$$\pm[0.06\bar{E}_g+0.061k_s^2+0.0066k_z^2]k_s \tag{36}$$

where, $\bar{E}_g(=0.21eV)$ is the energy gap for the transition point, the zero of the energy E is at the edge of the conduction band of the Γ point of the Brillouin zone and is measured positively upwards, k_x, k_y and k_z are in the units of $10^9 m^{-1}$.

The Dispersion Relation in quantum wires in this case can be written as

$$[E-0.606[(\frac{n_x\pi}{d_x})^2+(\frac{n_y\pi}{d_y})^2]-0.0722k_z^2][E+\bar{E}_g+0.411k_s^2+0.0377k_z^2]=0.23[(\frac{n_x\pi}{d_x})^2+(\frac{n_y\pi}{d_y})^2]+0.02k_z^2$$
$$\pm[0.06\bar{E}_g+0.061[(\frac{n_x\pi}{d_x})^2+(\frac{n_y\pi}{d_y})^2]+0.0066k_z^2][(\frac{n_x\pi}{d_x})^2+(\frac{n_y\pi}{d_y})^2]^{1/2} \tag{37}$$

(J) The charge carriers of Tellurium obey two different dispersion laws as given literature [80, 81].

(i) The dispersion relation of the conduction electrons in Tellurium, following Bouat, [81] can be written as

$$E=A_6k_z^2+B_6k_s^2\pm[\bar{\vartheta}k_z^2+\bar{\ell}k_s^2]^{1/2} \tag{38a}$$

where, $A_6(=6.7\times10^{-16}meVm^2), B_6(=4.2\times10^{-16}meVm^2), \bar{\vartheta}(=(6\times10^{-8}meVm)^2)$ *and* $\bar{\ell}(=(3.8\times10^{-8}meVm)^2)$ are the band constants.

The Dispersion Relation in quantum wires in this case can be written as

$$E=A_6k_z^2+B_6[(\frac{n_x\pi}{d_x})^2+(\frac{n_y\pi}{d_y})^2]\pm[\bar{\vartheta}k_z^2+\bar{\ell}[(\frac{n_x\pi}{d_x})^2+(\frac{n_y\pi}{d_y})^2]]^{1/2} \tag{38b}$$

(ii) The energy spectrum of the carriers in the two higher valance bands and the single lower valance band of Te can respectively be expressed as [82]

$$\bar{E} = A_{10}k_z^2 + B_{10}k_s^2 \pm [\Delta_{10}^2 + (\beta_{10}k_z)^2]^{1/2} \tag{39}$$

and

$$\bar{E} = \Delta_{||} + A_{10}k_z^2 + B_{10}k_s^2 \pm \beta_{10}k_z \tag{40}$$

where, $\bar{E}$ is the energy measured within the valance bands, $A_{10} = 3.77\times10^{-19} eVm^2$, $B_{10} = 3.57\times10^{-19} eVm^2$, $\Delta_{10} = 0.628eV$, $(\beta_{10})^2 = 6\times10^{-20}(eVm)^2$ and $\Delta_{||} = 1004\times10^{-5}eV$ are the spectrum constants.

The Dispersion Relation in quantum wires in this case can be written as

$$\bar{E} = A_{10}k_z^2 + B_{10}[(\frac{n_x\pi}{d_x})^2 + (\frac{n_y\pi}{d_y})^2] \pm [\Delta_{10}^2 + (\beta_{10}k_z)^2]^{1/2} \tag{41}$$

and

$$\bar{E} = \Delta_{||} + A_{10}k_z^2 + B_{10}[(\frac{n_x\pi}{d_x})^2 + (\frac{n_y\pi}{d_y})^2] \pm \beta_{10}k_z \tag{42}$$

(K) The dispersion relation for the electrons in graphite can be written following Brandt [83] as

$$E = \frac{1}{2}[E_2 + E_3] \pm [\frac{1}{4}(E_2 - E_3)^2 + \eta_2^2 k^2]^{1/2} \tag{43}$$

where, $E_2 \equiv \bar{\Delta} - 2\bar{\gamma}_1\cos\phi_0 + 2\bar{\gamma}_5\cos^2\phi_0, \phi_0 \equiv \frac{c_6 k_z}{2}, E_3 \equiv 2\bar{\gamma}_2\cos^2\phi_0, \eta_2 \equiv (\frac{\sqrt{3}}{2})a_6(\bar{\gamma}_0 + 2\bar{\gamma}_4\cos\phi_0)$

in which the band constants are

$$\bar{\Delta} = -0.0002eV, \bar{\gamma}_0 = 3eV, \bar{\gamma}_1 = 0.392eV, \bar{\gamma}_2 = -0.019eV,$$

$$\bar{\gamma}_4 = 0.193eV, \bar{\gamma}_5 = 0.194eV, a_6 = 2.46A^\circ \text{ and } c_6 = 6.74A^\circ$$

The Dispersion Relation in quantum wires in this case can be written as

$$E = \frac{1}{2}[E_2 + E_3] \pm [\frac{1}{4}(E_2 - E_3)^2 + \eta_2^2[(\frac{n_x\pi}{d_x})^2 + (\frac{n_y\pi}{d_y})^2 + k_z^2]]^{1/2} \quad (44)$$

(L) The dispersion relation of the conduction electrons in Antimony (Sb) in accordance with Ketterson [84] can be written as

$$2m_0E = \alpha_{11}p_x^2 + \alpha_{22}p_y^2 + \alpha_{33}p_z^2 + 2\alpha_{23}p_yp_z \quad (45)$$

and

$$2m_0E = a_1p_x^2 + a_2p_y^2 + a_3p_z^2 + a_4p_yp_z \pm a_5p_xp_z \pm a_6p_xp_y \quad (46a)$$

where, $a_1 = \frac{1}{4}(\alpha_{11} + 3\alpha_{22}), a_2 = \frac{1}{4}(\alpha_{22} + 3\alpha_{11}), a_3 = \alpha_{33}, a_4 = \alpha_{33}, a_5 = \sqrt{3}$ *and* $a_6 = \sqrt{3}(\alpha_{22} - \alpha_{11})$

in which $\alpha_{11} = 16.7, \alpha_{22} = 5.98, \alpha_{33} = 11.61$ *and* $\alpha_{23} = 7.54$ are the system constants.

The Dispersion Relation in quantum wires in this case can be written as

$$2m_0E = \hbar^2[\alpha_{11}(\frac{n_x\pi}{d_x})^2 + \alpha_{22}(\frac{n_y\pi}{d_y})^2 + \alpha_{33}k_z^2 + 2\alpha_{23}(\frac{n_y\pi}{d_y})k_z] \quad (46b)$$

$$2m_0E = \hbar^2[a_1(\frac{n_x\pi}{d_x})^2 + a_2(\frac{n_y\pi}{d_y})^2 + a_3k_z^2 + a_4(\frac{n_y\pi}{d_y})k_z \pm a_5(\frac{n_x\pi}{d_x})k_z \pm a_6(\frac{n_x\pi}{d_x})(\frac{n_y\pi}{d_y})] \quad (46c)$$

(M) The dispersion relation of the holes in p- Bi_2Te_3 can be written [85] as

$$E(1 + \frac{E}{E_g}) = \frac{\hbar^2}{2m_0}[\alpha_{11}k_x^2 + \alpha_{22}k_y^2 + \alpha_{33}k_z^2 + 2\alpha_{23}k_yk_z] \quad (47)$$

where, x, y and z are parallel to binary, bisectrix and trigonal axes respectively, $E_g = 0.145eV$, $\alpha_{11} = 32.5$, $\alpha_{22} = 4.81$, $\alpha_{33} = 9.02$, $\alpha_{23} = 4.15$, $g_s = 2$ and $g_v = 6$.

The Dispersion Relation in quantum wires in this case can be written as

$$E(1+\frac{E}{E_g})=\frac{\hbar^2}{2m_0}[\alpha_{11}(\frac{n_x\pi}{d_x})^2+\alpha_{22}(\frac{n_y\pi}{d_y})^2+\alpha_{33}k_z^2+2\alpha_{23}(\frac{n_y\pi}{d_y})k_z] \quad (48)$$

(N) The dispersion relation of the holes in p- InSb in accordance with Cunningham [86] can be written as

$$\bar{E}=c_4\left(1+\gamma_4 f_4\right)k^2 \pm \frac{1}{3}[2\sqrt{2}\sqrt{c_4}\sqrt{16+5\gamma_4}\sqrt{E_4}g_4 k] \quad (49)$$

where, $\bar{E}$ is the energy of the hole as measured from the top of the valance and within it, $c_4 \equiv \frac{\hbar^2}{2m_0}+\theta_4$, $\theta_4 \equiv 4.7\frac{\hbar^2}{2m_0}$, $\gamma_4 \equiv \frac{b_4}{c_4}$, $b_4 \equiv \frac{3}{2}b_5+2\theta_4$, $b_5 \equiv 2.4\frac{\hbar^2}{2m_0}$, $f_4 \equiv \frac{1}{4}[\sin^2 2\theta+\sin^4\theta\sin^2 2\phi]$, θ is measured from the positive z- axis, ϕ is measured from positive x-axis, $g_4 \equiv \sin\theta[\cos^2\theta+\frac{1}{4}\sin^4\theta\sin^2 2\phi]$ and $E_4=5\times10^{-4}eV$.

The Dispersion Relation in quantum wires in this case can be written as

$$\bar{E}=c_4(1+\gamma_4 f_4)[(\frac{n_x\pi}{d_x})^2+(\frac{n_y\pi}{d_y})^2+k_z^2]\pm\frac{1}{3}[2\sqrt{2}\sqrt{c_4}\sqrt{16+5\gamma_4}\sqrt{E_4}g_4[(\frac{n_x\pi}{d_x})^2+(\frac{n_y\pi}{d_y})^2+k_z^2]^{1/2}] \quad (50)$$

(O) The dispersion relation of the valance bands of II-V compounds in accordance with Yamada [87] can be written as

$$E=\frac{1}{2}(\bar{t}_1+\bar{t}_2)k_x^2+\frac{1}{2}(\bar{t}_3+\bar{t}_4)k_y^2+\frac{1}{2}(\bar{t}_5+\bar{t}_6)k_z^2+\frac{1}{2}(\bar{t}_7+\bar{t}_8)k_x$$
$$\pm[\{\frac{1}{2}(\bar{t}_1-\bar{t}_2)k_x^2+\frac{1}{2}(\bar{t}_3-\bar{t}_4)k_y^2+\frac{1}{2}(\bar{t}_5-\bar{t}_6)k_z^2+\frac{1}{2}(\bar{t}_7-\bar{t}_8)k_x\}^2+t_9^2k_y^2+t_{10}^2]^{1/2} \quad (51)$$

where, $\bar{t}_i (i=1 \text{ to } 8), t_9$ *and* t_{10} are the constants of the energy spectra.

For p-CdSb, $\bar{t}_1=-32.3\times10^{-20}eVm^2$, $\bar{t}_2=-60.7\times10^{-20}eVm^2$, $\bar{t}_3=-1.63\times10^{-19}eVm^2$, $\bar{t}_4=-2.44\times10^{-19}eVm^2$, $\bar{t}_5=-9.19\times10^{-19}eVm^2$,

$\bar{t}_6 = -10.5\times10^{-19} eVm^2$, $\bar{t}_7 = 2.97\times10^{-10} eVm$, $\bar{t}_8 = -3.47\times10^{-10} eVm$, $t_9 = 1.3\times10^{-10} eVm$ and $t_{10} = 0.070 eV$.

The Dispersion Relation in quantum wires in this case can be written as

$$E = \frac{1}{2}(\bar{t}_1+\bar{t}_2)(\frac{n_x\pi}{d_x})^2 + \frac{1}{2}(\bar{t}_3+\bar{t}_4)(\frac{n_y\pi}{d_y})^2 + \frac{1}{2}(\bar{t}_5+\bar{t}_6)k_z^2 + \frac{1}{2}(\bar{t}_7+\bar{t}_8)(\frac{n_x\pi}{d_x})$$
$$\pm[\{\frac{1}{2}(\bar{t}_1-\bar{t}_2)(\frac{n_x\pi}{d_x})^2 + \frac{1}{2}(\bar{t}_3-\bar{t}_4)(\frac{n_y\pi}{d_y})^2 + \frac{1}{2}(\bar{t}_5-\bar{t}_6)k_z^2 + \frac{1}{2}(\bar{t}_7-\bar{t}_8)(\frac{n_x\pi}{d_x})\}^2 + t_9^2(\frac{n_y\pi}{d_y})^2 + t_{10}^2]^{1/2} \quad (52)$$

(P) The energy spectrum of the valance bands of CuCl in accordance with Yekimov et al. [88] can be written as

$$E_h = (\gamma_6 - 2\gamma_7)\frac{\hbar^2k^2}{2m_0} \quad (53)$$

and

$$E_{l,s} = (\gamma_6+\gamma_7)\frac{\hbar^2k^2}{2m_0} - \frac{\Delta_1}{2} \pm [\frac{\Delta_1^2}{4} + \gamma_7\Delta_1\frac{\hbar^2k^2}{2m_0} + 9(\frac{\gamma_7\hbar^2k^2}{2m_0})^2]^{1/2} \quad (54)$$

where, $\gamma_6 = 0.53$, $\gamma_7 = 0.07$, $\Delta_1 = 70meV$

The Dispersion Relation in quantum wires in this case can be written as

$$E_h = (\gamma_6 - 2\gamma_7)\frac{\hbar^2[(\frac{n_x\pi}{d_x})^2 + (\frac{n_y\pi}{d_y})^2 + k_z^2]}{2m_0} \quad (55)$$

and

$$E_{l,s} = (\gamma_6+\gamma_7)\frac{\hbar^2[(\frac{n_x\pi}{d_x})^2 + (\frac{n_y\pi}{d_y})^2 + k_z^2]}{2m_0} - \frac{\Delta_1}{2}$$
$$\pm[\frac{\Delta_1^2}{4} + \gamma_7\Delta_1\frac{\hbar^2[(\frac{n_x\pi}{d_x})^2 + (\frac{n_y\pi}{d_y})^2 + k_z^2]}{2m_0} + 9(\frac{\gamma_7\hbar^2[(\frac{n_x\pi}{d_x})^2 + (\frac{n_y\pi}{d_y})^2 + k_z^2]}{2m_0})^2]^{1/2} \quad (56)$$

where, $\gamma_6 = 0.53$, $\gamma_7 = 0.07$, $\Delta_1 = 70meV$

(Q) In the presence of stress, χ_6 along <001> and <111> directions, the energy spectra of the holes in semiconductors having diamond structure valance bands can be respectively expressed following Roman [89] et al. as

$$E = A_6 k^2 \pm [\bar{B}_7^2 k^4 + \delta_6^2 + B_7 \delta_6 (2k_z^2 - k_s^2)]^{1/2} \tag{57}$$

and

$$E = A_6 k^2 \pm [\bar{B}_7^2 k^4 + \delta_7^2 + \frac{D_6}{\sqrt{3}} \delta_7 (2k_z^2 - k_s^2)]^{1/2} \tag{58}$$

where, A_6, B_7, D_6 and C_6 are inverse mass band parameters in which $\delta_6 \equiv l_7(\bar{S}_{11} - \bar{S}_{12})\chi_6$, $\bar{S}_{ij}$ are the usual elastic compliance constants, $\bar{B}_7^2 \equiv (B_7^2 + \frac{c_6^2}{5})$ and $\delta_7 \equiv (\frac{d_8 S_{44}}{2\sqrt{3}})\chi_6$. For gray tin, $d_8 = -4.1eV$, $l_7 = -2.3eV$, $A_6 = 19.2\frac{\hbar^2}{2m_0}$, $B_7 = 26.3\frac{\hbar^2}{2m_0}$, $D_6 = 31\frac{\hbar^2}{2m_0}$ and $c_6^2 = -1112\frac{\hbar^2}{2m_0}$. The Dispersion Relation in quantum wires in this case can be written as

$$E = A_6[(\frac{n_x\pi}{d_x})^2 + (\frac{n_y\pi}{d_y})^2 + k_z^2] \pm [\bar{B}_7^2[(\frac{n_x\pi}{d_x})^2 + (\frac{n_y\pi}{d_y})^2 + k_z^2]^2 + \delta_6^2 + B_7\delta_6(2k_z^2 - [(\frac{n_x\pi}{d_x})^2 + (\frac{n_y\pi}{d_y})^2])]^{1/2} \tag{59}$$

and

$$E = A_6[(\frac{n_x\pi}{d_x})^2 + (\frac{n_y\pi}{d_y})^2 + k_z^2] \pm [\bar{B}_7^2[(\frac{n_x\pi}{d_x})^2 + (\frac{n_y\pi}{d_y})^2 + k_z^2]^2 + \delta_7^2 + \frac{D_6}{\sqrt{3}}\delta_7(2k_z^2 - [(\frac{n_x\pi}{d_x})^2 + (\frac{n_y\pi}{d_y})^2])]^{1/2} \tag{60}$$

CONCULSION

The influence of quantum confinement is immediately apparent from the 1D DRs since the DR depends strongly on the thickness of the quantum-confined materials in contrast with the corresponding bulk specimens. Finally, it may be noted that the basic aim of this chapter is not solely to demonstrate the influence of quantum confinement on the DR from QWs of n-GaP, IV-VI, Ge, Zero Gap, GaSb, $PtSb_2$, GaAs, n-InSb, $Pb_{1-x}Ga_xTe$, Te, Graphite, Sb, Bi_2Te_3, p-InSb, II-V, CuCl and Stressed p-type semiconductors having diamond structure valance bands respectively, but also to formulate the appropriate DOS in the most generalized form, since the transport and other phenomena in quantized structures having different band structures and the derivation of the expressions of many important electronic properties are based on the DR in such materials.

REFERENCES

[1] Handbook of computational and theoretical nanoscience, edited by M. Rieth and W. Schommers, American Scientific Publishers, Los Angeles (2006), Vols. 1-10; M. Rieth and W. Schommers, *Jour. Nanosci. and Nanotechnol.* 2, 679 (2002).

[2] *Encyclopedia of Nanoscience and Nanotechnology*, edited by H. S. Nalwa, American Scientific Publishers, Los Angeles (2004/2011), Vols. 1-25; *Soft Nanomaterials*, edited by H. S. Nalwa, American Scientific Publishers, Los Angeles (2009), Vols. 1-2; *Magnetic Nanostructures*, edited by H. S. Nalwa, American Scientific Publishers, Los Angeles (2002).

[3] *Nanomaterials for Energy Storage Applications*, edited by H. S. Nalwa, American Scientific Publishers, Los Angeles (2009); *Handbook of Nanostructured Biomaterials and Their Applications in Nanobiotechnology*, edited by H. S. Nalwa, American Scientific Publishers, Los Angeles (2005), Vols. 2; *Quantum Dots and Nanowires*, edited by S. Bandyopadhyay and H. S. Nalwa, American Scientific Publishers, Los Angeles (2003), Vol. 5.

[4] *Handbook of Nanoceramics and Their Based Nanodevices*, edited by T. Y. Tseng and H. S. Nalwa, American Scientific Publishers, Los Angeles (2009), Vols. 1-5; *Bottom-up Nanofabrication* edited by K. Ariga and H. S. Nalwa, American Scientific Publishers, Los Angeles (2009), Vols. 1-6; K. P. Ghatak and S. N. Biswas, *Acta physica Slovaca* 43, 425 (1993).

[5] *Handbook of Electrochemical Nanotechnology*, edited by Y. Lin and H. S. Nalwa, American Scientific Publishers, Los Angeles (2009),

Vol. 1-2; *Packaging Nanotechnology,* edited by A. K. Mohanty, M. Misra and H. S. Nalwa, American Scientific Publishers, Los Angeles (2009).

[6] Yan S., Y. Shi, Z. Xiao, M. Zhou, W. Yan, H. Shen and D. Hu, *J. Nanosci. Nanotechnol.* 12, 6873 (2012); J. R. C. Smirnov, E. Jover, R. Amade, G. Gabriel, R. Villa and E. Bertran, *J. Nanosci. Nanotechnol.* 12, 6941 (2012).

[7] Mashapa M. G., N. Chetty and S. S. Ray, *J. Nanosci. Nanotechnol.* 12, 7030 (2012); D. H. Kim, J. H. You, J. H. Kim, K. H. Yoo and T. W. Kim, *J. Nanosci. Nanotechnol.* 12, 5687 (2012); A. Ooba, Y. Fujimura, K. Takahashi, T. Komine and R. Sugita, *J. Nanosci. Nanotechnol.* 12, 7411 (2012); C. W. Tan and B. K. Tay, *J. Nanosci. Nanotechnol.* 12, 707 (2012).

[8] Noh S. J., Y. Miyamoto, M. Okuda, N. Hayashi and Y. K. Kim, *J. Nanosci. Nanotechnol.* 12, 428 (2012); S. Padalkar, K. Schroeder, Y. H. Won, H. S. Jang and L. Stanciu, *J. Nanosci. Nanotechnol.* 12, 227 (2012); K. Y. Lee, B. Kumar, H. K. Park, W. M. Choi, J. Y. Choi and S. W. Kim, *J. Nanosci. Nanotechnol.* 12, 1551 (2012).

[9] Wu J. M., *J. Nanosci. Nanotechnol.* 12, 1434 (2012); Y. Zhang, L. Su, K. Lubonja, C. Yu, Y. Liu, D. Huo, C. Hou and Y. Lei, *Sci. Adv. Mater.* 4, 825 (2012).

[10] Tamang R., B. Varghese, E. S. Tok, S. Mhaisalkar and C. H. Sow, *Nanosci. Nanotechnol. Lett.* 4, 716 (2012); Y. Yue, M. Chen and Y. J, *Nanosci. Nanotechnol. Lett.* 4, 414 (2012).

[11] Adhikari S. M., A. Sakar and K. P. Ghatak, *Quantum Matter* 2, 455 (2013); K. P. Ghatak, P. K. Bose, S. Bhattacharya, A. Bhattacharjee, D. De and S. Ghosh, *Quantum Matter* 2, 83 (2013); K. P. Ghatak, S. Bhattacharya, A. Mondal, S. Debbarma, P. Ghorai and A. Bhattacharjee, *Quantum Matter* 2, 25 (2013);

[12] Paitya N., S. Bhattacharya, D. De and K. P. Ghatak, *Quantum Matter* 1, 63 (2012); S. Bhattacharya, N. Paitya and K. P. Ghatak, *Jour. Comp. and Theo. Nanosci.*10, 1999 (2013).

[13] Choudhury S., D. De, S. Mukherjee, A. Neogi, A. Sinha, M. Pal, S. K. Biswas, S. Pahari, S. Bhattacharya and K. P. Ghatak, *Jour. Comp.*

and Theo. Nanosci. 5, 375 (2008); K. P. Ghatak, S. Bhattacharya, D. De, R. Sarkar, S. Pahari, A. Dey, A. K. Dasgupta and S. N. Biswas, *Jour. Comp. and Theo. Nanosci.* 5, 1345 (2008); S. Bhattacharya, R. Sarkar, D. De, S. Mukherjee, S. Pahari, A. Saha, S. Roy, N. C. Paul, S. Ghosh and K. P. Ghatak, *Jour. of Comput. and Theo. Nanosci.* 6, 112 (2009).

[14] Bhattacharya S., S. Chowdhury, S. Ghoshal, S. K. Biswas, D. De and K. P. Ghatak, *Jour. Comp. and Theo. Nanosci.* 3, 423 (2006); S. Bhattacharya, D. De, S. Chowdhury, S. Karmakar, D. K. Basu, S. Pahari and K. P. Ghatak, *Jour. Comp. and Theo. Nanosci.* 3, 280(2006); K. P. Ghatak, S. Karmakar, D. De, S. Pahari, S. K. Charaborty, S. K. Biswas and S. Chowdhury, *Jour. Comp. and Theo. Nanosci.* 3, 153 (2006); K. P. Ghatak, P. K. Bose, A. R. Ghatak, D. De, S. Pahari, S. K. Chakraborty and S. K. Biswas, *Jour. Comp. and Theo. Nanosci.* 2, 423 (2005).

[15] Paitya N., K. P. Ghatak, *J. Adv. Phys.* 1, 161 (2012); S. Debbarma, A. Bhattacharjee, S. Bhattacharyya, A. Mondal and K. P. Ghatak, *J. Adv. Phys.* 1, 84 (2012); S. Bhattacharya, D. De, S. M. Adhikari and K. P. Ghatak, *Superlattices and Microstructures* 51, 203 (2012); D. De, S. Bhattacharya, S. M. Adhikari, A. Kumar, P. K. Bose and K. P. Ghatak, Beilstein *Jour. of Nanotechnol.* 2, 339 (2011).

[16] Bhattacharya S., S. Choudhury and K. P. Ghatak, *Superlattices and Microstructures* 48, 257 (2010); S. Pahari, S. Bhattacharya, S. Roy, A. Saha, D. De and K. P. Ghatak, *Superlattices and Microstructures* 46, 760 (2009); D. De and K. P. Ghatak, *Journal of Materials Science: Materials in Electronics* 20, 185 (2009).

[17] Adhikari S. M., D. De, J. K. Baruah, S. Chowdhury, K. P. Ghatak, *Adv. Science Focus,* 1, 54 (2013); S. Bhattacharya, S. Pahari, R. Sarkar, S. Ghosh, K. P. Ghatak, *Physica B: Condensed Matter,* 403, 3635 (2008).

[18] Singh L. J., S. Choudhury, D. Baruah, S. K. Biswas, S. Pahari, K. P. Ghatak, *Physica B: Condensed Matter* 368, 188 (2005); P. K. Chakraborty, S. Bhattacharya, K. P. Ghatak, *Jour. Appl. Phys.* 98, 053517 (2005).

[19] Chakraborty P. K., S. Ghoshal and K. P. Ghatak, *Physica B: Condensed Matter,* 382, 26 (2006); S. Chowdhary, L. J. Singh, K. P. Ghatak, *Physica B: Condensed Matter,* 365, 5 (2005); K. P. Ghatak, D. K. Basu, S. Bhattacharya, S. Ghosal, S. K. Biswas, A. Mallik and G. Majumdar, *Nonlinear Optics Quantum Optics,* 33, 221 (2005); L. J. Singh, S. Choudhury, S. S. Roy and K. P. Ghatak, *Electrical Engineering,* 87, 1(2005).

[20] Ghatak K. P., S. K. Biswas, D. De, S. Ghosal and S. Chatterjee, *Physica B: Condensed Matter,* 353, 127 (2004); P. K. Chakraborty, S. K. Biswas and K. P. Ghatak, *Physica B: Condensed Matter,* 352, 111 (2004); P. K. Chakraborty, L. J. Singh and K. P. Ghatak, *Jour. Appl. Phys.* 95, 5311 (2004); P. K Chakraborty, G. C. Datta and K. P. Ghatak, *Physica B: Condensed Matter,* 339, 198 (2003).

[21] Chakraborty P. K., K. P. Ghatak, *J. Phys. Chem. Solids,* 62, 1061 (2001); K. P. Ghatak, J. Y. Siddiqui, B. Nag, *Phys. Lett.* A, 282, 428 (2001); P. K. Chakraborty, K. P. Ghatak, *Jour. Appl. Phys.* 89, 1075 (2001).

[22] Chakraborty P. K., K. P. Ghatak, *Jour. of Phys. D Appl. Phys.* 32, 2438 (1999); B. Nag, K. P. Ghatak, *J. Phys. Chem. Solids* 59, 713 (1998); K. P. Ghatak, B. Nag, *J. Phys. Chem. Solids* 59, 411 (1998), K. P. Ghatak, B. Nag, *Phys. Stat. Sol. (b)* 205, 519 (1998).

[23] Ghatak K. P., J. P. Banerjee, B. Nag, *Jour. Appl. Phys.* 83, 1420 (1998); B. Nag, K. P. Ghatak, *Nonlinear Optics* 19, 1 (1998); J. P. Banerjee, S. Ghosh, B. Nag, K. P. Ghatak, *Internat. Jour. Elect.* 82, 335 (1997).

[24] Ghatak K. P., D. K Basu and B. Nag, *J. Phys. Chem. Solids* 58, 133 (1997); B. Mitra, D. K. Basu, B. Nag and K. P. Ghatak, *Nonlinear Optics* 17, 171 (1997); K. P. Ghatak, B. Mitra and B. Nag, *Phys. Stat. Sol. (b)* 199, 95 (1997); B. Nag, K. P. Ghatak, *Phys. Scr.* 54, 657 (1996).

[25] Ghatak K. P., D. K. Basu, D. Basu and B. Nag, *Il Nuovo Cimento D* 18, 947 (1996); K. P. Ghatak, S. Beta, A. Ali and B. Nag, *Fizika* A-5, 111 (1996); K. P. Ghatak, B. De, B. Nag and P. K. Chakraborty,

Nonlinear Optics 16, 221 (1996); K. P. Ghatak, J. P Banerjee, B. Goswami, B. Nag, *Nonlinear Optics* 16, 241 (1996).

[26] Ghatak K. P., J. P. Banerjee, M. Mitra and B. Nag, *Nonlinear optics* 16, 193 (1996); K. P. Ghatak, M. Mitra, B. Goswami and B. Nag, *Nonlinear optics* 16, 167 (1996); K. P. Ghatak and B. Nag, *Nanostructured materials* 5, 769 (1995); K. P. Ghatak, B. Nag and D. Bhattacharyya, *Jour. Low Temp. Phys.* 101, 983 (1995).

[27] Ghatak K. P., D. Bhattacharya, D. Basu and B. Nag, *Phys. Stat. Sol. (b)* 191, 141 (1995); K. P. Ghatak and D. Bhattacharyya, *Phys. Scr.* 52, 343 (1995); D. Bhattacharyya and K. P. Ghatak, *Phys. Stat. Sol. (b)* 187, 523 (1995); K. P. Ghatak and S. N. Banik, *Fizika* A 4, 33 (1995).

[28] Ghatak K. P., and M. Mondal, *Phys. Stat Sol. (b)* 185, K5 (1994); K. P. Ghatak and S. N. Biswas, *Solid-state electronics* 37, 1437 (1994); K. P. Ghatak, *Acta Physica Hungarica* 74, 167 (1994); K. P. Ghatak and D. Bhattacharyya, *Phys. Lett.* A 184, 366 (1994). S. Banik, K. P. Ghatak and S. Biswas, *Fizika* A 3, 77 (1994); K. P. Ghatak and S. N. Banik, *Fizika* A 3, 127 (1994); K. P. Ghatak and S. N. Banik, *Fizika* A 3, 155 (1994); K. P. Ghatak and D. Bhattacharyya, *Phys. Stat Sol.(b)* 181, 127 (1994); K. P. Ghatak and S. Biswas, *Fizika* A 3, 7 (1994); K. P. Ghatak and D. Bhattacharya, *Phys. Stat. Sol.* (b) 179, 383 (1993); P. K. Chakraborty and K. P. Ghatak, *Jour. Appl. Phys.* 74, 3246 (1993).

[29] Ghatak K. P., and S. N. Biswas, *Nanostructured materials* 2, 91 (1993); K. P. Ghatak, *Fizika* A. 2, 133 (1993); K. P. Ghatak and S. N. Biswas, *Il Nuovo Cimento* D 15, 87 (1993).

[30] Ghatak K. P., A. Ghoshal and S. N. Biswas, *Il Nuovo Cimento* D 15, 39 (1993); K. P. Ghatak, *Fizika* A 2, 59 (1993); K. P. Ghatak and M. Mondal, *Phys. Stat Sol.* (b) 175, 113 (1993); K. P. Ghatak and B. Mitra, *Il Nuovo Cimento* D 15, 97 (1993).

[31] Ghatak K. P., *Acta Physica Hungarica* 72, 147 (1992); K. P. Ghatak, *Il Nuovo Cimento* D 14, 1187 (1992); P. K. Chakraborty, K. P. Ghatak, *Jour. Appl. Phys.* 72, 2086 (1992); K. P. Ghatak, A. Ghoshal, B. Mitra, *Il Nuovo Cimento* D 14, 903 (1992).

[32] Biswas S. N., and K. P. Ghatak, *Internat. Jour. Electron.* 73, 287 (1992); K. P. Ghatak, B. Mitra, *Phys. Scr.* 46, 182 (1993); K. P. Ghatak, B. Mitra, *Internat. Jour. Elect.* 72, 541 (1992); K. P. Ghatak, M. Mondal, *Phys. Stat. Sol. (b)* 170, 57 (1992).

[33] Ghatak K. P., and A. Ghoshal, *Phys. Stat. Sol. (b)* 170, K27 (1992); K. P. Ghatak and M. Mondal, *Jour. Appl. Phys.* 71, 1277 (1992); K. P. Ghatak and M. Mondal, *Fizika* A 1, 113 (1992); K. P. Ghatak, *Fizika* A 1, 197 (1992).

[34] Ghatak K. P., *Fizika* A 1, 207 (1992); M. Mondal, A. Ghoshal and K. P. Ghatak, *Il Nuovo Cimento* D 14, 63 (1992); K. P. Ghatak and S. N. Biswas, *Jour. Appl. Phys.* 70, 4309 (1991); K. P. Ghatak, *Il Nuovo Cimento D* 13, 1321 (1991).

[35] Mohammad S. N., *J. Appl. Phys.*95, 4856 (2004); V. K. Arora, *Appl. Phys. Lett.* 80, 3763 (2002); S. N. Mohammad, *J. Appl. Phys.* 95, 7940 (2004); S. N. Mohammad, *Philos. Mag.* 84, 2559 (2004).

[36] Mondal M., K. P. Ghatak, *Phys. Stat. Sol.* (b) 126, K41 (1984); D. R. Choudhury, A. K. Chowdhury, K. P. Ghatak, A. N. Chakravarti, *Phys. Stat. Sol.* (b) 98, K141 (1980); K. P. Ghatak, M. Mondal, *Phys. Stat. Sol.* (b) 135, K119 (1986); K. P. Ghatak, S. Biswas, Society of Photo-Optical Instrumentation Engineers (SPIE) Conference Series, 818, 115 (1987), E. V. Rozenfeld, Yu. P. Irkhin and P. G. Guletskii, *Sov. Phys. Solid State* 29, 1134 (1987).

[37] Ghatak K. P., D. De, S. Bhattacharya, *Photoemission from optoelectronic materials and their nanostructures*, Springer Series in Nanostructure Science and Technology, Springer (2009); R. K. Pathria, Statistical Mechanics, 2nd ed. Butterworth-Heinemann, Oxford (1996); L. Torres, L. Lopez-Diaz, O. Alejos, J. Iniguez, *J. Appl. Phys.* 85, 6208 (1999); S. Tiwari, S. Tiwari, *Cryst. Res. Technol.* 41, 78 (2006); R. Houdré, C. Hermann, G. Lampel, P. M. Frijlink, A. C. Gossard, *Phys. Rev. Lett.* 55, 734 (1985); K. P. Ghatak, S. Bhattacharya, K. M. Singh, S. Choudhury, S. Pahari, *Physica B* 403, 2116 (2008); H. Shimoda, B. Gao, X. P. Tang, A. Kleinhammes, L. Fleming, Y. Wu, O. Zhou, *Phys. Rev. Lett.* 88,

015502 (2002); M. Maillard, P. Monchicourt, M. P. Pileni, *Chem. Phys. Lett.* 380, 704 (2003).

[38] Chakravarti A. N., B. R. *Nag, Phys. Stat. Sol.* (a) 14, K55 (1972); *Int. J. Elect.* 37, (281) (1974); *Phys. Stat. Sol.* (a) 14, K23 (1972); A. N. Chakravarti, D. P. Parui, *Canad. J. Phys.* 51, (451) (1973). D. Mukherjee, A. N. Chakravarti, B. R. Nag, *Phys. Stat. Sol* (a) 26, K27 (1974).

[39] Ghatak K. P., S.Bhattacharya, *J. Appl. Phys.* 102, 073704 (2007); M. Mondal, K. P. Ghatak, *Phys. Lett.* 102A, 54 (1984); P. K. Chakraborty, G. C. Datta, K. P. Ghatak, *Phys. Scr.* 68, 368 (2003). E. O. Kane, *Solid-State Electron.* 28, 3 (1985); W. Zawadzki, in *Handbook on Semiconductors*, edited by W. Paul, North Holland, New York, (1982), 1, p. 715.

[40] Wherreff B. S., W. Wolland, C. R. Pidgeon, R. B. Dennis, S. D. Smith, in *Proceedings of the 12th International Conference of the Physics of the Semiconductors*, ed. by M. H. Pilkahn, R. G. Tenbner (Staffgard,1978), p.793.

[41] Blatt F. J., *J. Phys. Chem. Solids* 17, 177 (1961).

[42] Fowler R. H., L. Nordheim, *Proc. of Royal Society of London*, Series-A, 119,173 (1928) A. Van Der Ziel, *Solid State Physical Electronics*, Prentice-Hall, Inc., Englewood Cliffs U.S.A., p.176, (1957); B. Mitra, K. P. Ghatak, *Phys. Lett.* A, 357, 146 (1990); K. P. Ghatak, M. Mondal, *Jour. Mag. Mag. Mat.*, 74, 203 (1988); K. P. Ghatak, B. Mitra, *Phys. Lett.*, 156 A,. 233, (1991); K. P. Ghatak, A. Ghosal, S. N. Biswas, M. Mondal, *Proc. of SPIE*, USA, 1308, 356 (1990).

[43] Ghatak K. P., S. Biswas, *J. Vac. Sci. Technol.* B7, 104 (1989); K. P. Ghatak, M. Mondal, *Z. Phys. B-Condensed Matter* 64, 223 (1986); D. R. Choudhury, A. K. Chowdhury, A. N. Chakravarti, *Phys. Scr.* 22, 656 (1981); D. R. Choudhury, A. K. Chowdhury, A. N. Chakravarti, *Czech. J. of Phys. B,* 30 (1980); D. R. Choudhury, A. K. Chowdhury, A. N. Chakravarti, B. R. Nag, *Phys. Stst. Sol. (a)* 58, K51 (1980); B. Mitra, A. Ghoshal, K. P. Ghatak, *Phys. Stat. Sol. (b)* 153, K209 (1989); M. Mondal, K. P. Ghatak, *Phys. Stat. Sol. (a)* 93, 377 (1986);

D. R. Choudhury, A. K. Chowdhury, A. N. Chakravarti, *Appl. Phys.* 22, 145 (1980).

[44] Klitzing K. v., G. Dorda, M. Pepper, *Phys. Rev. Lett.* 45, 494 (1980); K. v. Klitzing, *Rev. Mod. Phys.* 58, 519 (1986); W. Zawadzki, In: *Two-Dimensional Systems, Heterostructures and Superlattices*, Springer Series in Solid State Sciences, Vol. 53, edited by G. Bauer, F. Kuchar, H. Heinrich (Springer, Berlin, Heidelberg, 1984).

[45] Peschke C. H., *Phys. Stat. Sol.* (b) 191, 325 (1995).

[46] Isihara A., Y. Shiwa, *J. Phys. C*, 18, 4703 (1985); *Semiconducting Lead Chalocogenides,* V. G. I. Ravish, B. A. Efimova and I. A. Smirnov, Plenum Press, New York, p163, 1970).

[47] Hwang C. J., and J. R. Brews, *J. Phys. Chem. Solids,* 32, 837 (1971); P. T. Landsberg and H. C. Cheng, *Phys. Rev.* 32B, 8021 (1985); H. Reiss, C. S. Fuller, and F. J. Morin, *Bell Syst. Tech.*J., 35, 535 (1956); P. T. Landsberg and A. G. Guy, *Phys. Rev.* B, 28, 1187 (1983).

[48] Roy D. K., *Quantum Mechanical Tunneling and its applications*, p. 309 1986, World Scientific Publication Co.Ptc.Ltd.

[49] Zawadzki W., *Physica* 127, 388 (1980).

[50] Piller H., *Semiconductors and Semimetals*, Ed. by R. K. Willardson and A. C. Beer, 124, 8 (1972).

[51] Madelung O., *Physics of III-V compounds.* John Wiley and Sons, Inc. New York, (1966).

[52] Sreedhar A. K., and S. C. Gupta, *Phys. Rev.* B 5, 3160 (1972); R. W. Keyes, IBM. *J. Res. Devlopm.* 5, 266 (1961); *Solid State Phys.* 20, 37, (1967); B. Nag, and K. P. Ghatak, *Nonlin. Opt.*, 19, 1 (1998); K. P. Ghatak and B. Nag, *Nanostruc. Material.*, 10, 923 (1998); B. Nag and K. P. Ghatak, *J. Phys. Chem. Sol.,* 58, 427, (1997).

[53] Ghatak K. P., S. Bhattacharya, D. De, Einstein Relation in Compound Semiconductors and Their Nanostructures, *Springer Series in Materials Science*, Vol. 116 (Springer-Verlag, Germany, 2008).

[54] Filipchenko A. S., I. G. Lang, D. N. Nasledov, S. T. Pavlov, L. N. Radaikine, *Phys. Stat. Sol.* (b) 66, 417 (1974).

[55] Wegener M., *Extreme Nonlinear Optics* (Springer-Verlag, Germany, 2005).

[56] Askerov B. M., N. F. Gushimzade and M. M. Panakhov, *Sov. Phys. Sol. Stat.* 29, 465 (1987).

[57] Mitra B., A. Ghoshal and K. P. Ghatak, *Phys. Stat Sol.* (b) 155, K23 (1989).

[58] Biswas S. N., N. Chattopadhyay and K. P. Ghatak, *Phys. Stat Sol.* (b) 141, K47 (1987).

[59] Woollam J. A., *Phys. Rev.* 3, 1148 (1971).

[60] Kaminoskii V. V., N. N. Stepanav, l. M. Smirnov, *Sov. Phys. Sol. Stat.* 27, 1295 (1985).

[61] Rees G. *J., Phys. of Compounds, Proc. of the 13th Inter. Nat. Conf.* Ed. F. G. Fumi, 1166, North Holland Company (1976).

[62] Ivchenko E. L., and G. E. Pikus, *Sov. Phys. Semicond.* 13, 579 (1979).

[63] Dimmock J. O., In *the Physics of Semimetals and Narrow Gap Compounds* by D. L. Carter and R. T. Bates, Pergamon Press, Oxford pp. 319 (1971).

[64] Bangert E., and P. Kastner, *Phys. Stat. Sol.* (b) 61, 503 (1974).

[65] Foley G. M. T., and P. N. Langenberg, *Phys. Rev. B,* 15B, 4850 (1977).

[66] Cardona M., W. Paul and H. Brooks Helv, *Acta Physica*, 33, 329 (1960) and A. F. Gibson in *"Proceeding of International school of physics "ENRICO FERMI"* course XIII, Ed. R. A Smith, Academic Press, pp.171 (1963).

[67] Wang C. C., and N. W. Ressler, *Phys. Rev.* 2, 1827 (1970).

[68] Ivanov-Omskii V. I., A. Sh. Mekhtisev, S. A. Rustambekova and E. N. Ukraintsev, *Phys. Stat. Sol. (b)* 119, 159 (1983).

[69] Seiler D. G., W. M. Beeker and L. M. Roth, *Phys. Rev.* 1, 764 (1970).

[70] Mathur P. C., and S. Jain, *Phys. Rev.* 19, 3159 (1979).

[71] Zhang H. I., *Phys. Rev.* 1B, 3450 (1970).

[72] Emtage P. R., *Phys. Rev. A* 246, 138 (1965).

[73] Stillman G. E., C. M. Wolfe and J. C. Dinmockin, in *Semiconductors and Semimetals*, Ed. R. K. Willardson and A. C. Beer, 12, 169 (1977).

[74] Newson D. J., and A. Kurobe, *Semicond. Sci. Tech.* 3, 786 (1988).

[75] Rossler U., *Solid State Communication,* 49, 943 (1984).

[76] Palik E. D., G. S. Picus, S. Teither and R. F. Wallis, *Phys. Rev.* 122, 475 (1961).

[77] Johnson E. J., and D. H. Dicley, *Phys. Rev.* 1, 2676 (1970).

[78] Agafonov V. G., P. M. Valov, B. S. Ryvkin and I. D. Yaroshetskii, *Sov. Phys. Semicond.* 12, 1182 (1978).

[79] Vassilev L. A., *Phys. Stat. Sol.* (b) 121, 203 (1984).

[80] Bouat J., and J. C. Thuillier, *Surface Science* 73, 528 (1978).

[81] Averkiev N. S., V. M. Asnin, A. A. Bakun, A. M. Danishevskii, E. L. Ivchenko, G. E. Pikus and A. A. Rogachev, *Sov. Phys. Semicond.* 18, 379 and 402 (1984).

[82] Brandt N. B., V. N. Davydov, V. A. Kulbachinskii and O. M. Nikitina, *Sov. Phys. Sol. Stat.* 29, 1014, (1987).

[83] Ketterson J. B., *Phys. Rev.* 129, 18 (1963).

[84] Stordeur M., and W. Kuhnberger, *Phys. Stat. Sol.* (b) 69, 377 (1975); H. Kohler, *Phys. Stat. Sol.* (b) 73, 95 (1976).

[85] Cunningham R. W., *Phys. Rev.* 167, 761 (1968).

[86] Yamada Y., *J. Phys. Soc. Japan,* 37, 606 (1974).

[87] Yekimov A. I., A. A. Onushchenko, A. G. Plyukhin and Al. L. Efros, *J. Expt. Theor. Phys.,* 88, 1490 (1985).

[88] Roman B. J., A. W. Ewald, *Phys Rev,* B5, 3914 (1972).

[89] Ghatak K. P., S Bhattacharya, *Journal of Applied Physics,* 102, 73704, (2007); K. P. Ghatak, S. Bhattacharya, S. K. Biswas, A. Dey, A. K. Dasgupta, *Physica Scripta,* 75, 820, (2007); S. Mukherjee, S. N. Mitra, P. K. Bose, A. R. Ghatak, A. Neogi, J. P. Banerjee, A. Sinha, M. Pal, S. Bhattacharya, K. P. Ghatak, *Journal of Computational and Theoretical Nanoscience,* 4, 550, (2007).

[90] Mukherjee S., D. De, D. J. Mukherjee, S. Bhattacharya, A. Sinha, K. P. Ghatak, *Physica B: Condensed Matter,* 393, 347,(2007); A. Sinha, A. K. Sharma, R. Barui, A. R. Ghatak, S. Bhattacharya, K. P. Ghatak,

Physica B: Condensed Matter, 391, 141, (2007); P. K. Chakraborty, A. Sinha, S Bhattacharya, K. P. Ghatak, *Physica B: Condensed Matter,* 390, 325-339, (2007).

[91] Biswas S. K., A. R. Ghatak, A. Neogi, A. Sharma, S. Bhattacharya, K. P. Ghatak, *Physica E: Low-Dimensional Systems and Nanostructures,* 36, 163, (2007); P. K. Chakraborty, S. Choudhury, K. P. Ghatak, *Physica B: Condensed Matter,* 387, 333, (2007); D Baruah, S. Choudhury, K. M. Singh, K. P. Ghatak, *Journal of Physics: Conference Series,* 61, 80, (2007).

[92] Ghatak K. P., S. Bhattacharya, H. Saikia, D. Baruah, A. Saikia, K. M. Singh, A. Ali, S. N. Mitra, P. K. Bose, A. Sinha, *Jour. of Compu. and Theo. Nanosci.,* 3, 727, (2006); P. K. Chakraborty, S. Ghoshal, K. P. Ghatak, *Physica B: Condensed Matter,* 382, 26, (2006); S. Bhattacharya, S. Chowdhury, S. Ghoshal, S. K. Biswas, D. De, K. P. Ghatak, *Journal of Computational and Theoretical Nanoscience*, 3, 423, (2006).

[93] Bhattacharya S., D. De, S. Chowdhury, S. Karmakar, D. K. Basu, S. Pahari, K. P. Ghatak, *Jour. of Compu. and Theo. Nanosci.,* 3, 280, (2006); K. P. Ghatak, S. Karmakar, D. De, S. Pahari, S. K. Charaborty, S. K. Biswas, S. Chowdhury, *Jour. of Compu. and Theo. Nanosci.,* 3, 153, (2006).

[94] Singh L. J., S. Choudhury, D. Baruah, S. K. Biswas, S. Pahari, K. P. Ghatak, *Physica B: Condensed Matter,* 368, 188, (2005); K. P. Ghatak, P. K. Bose, A. R. Ghatak, D. De, S. Pahari, S. K. Chakraborty, S. K. Biswas, *Jour. of Compu. and Theo. Nanosci.,* 2, 423, (2005); P. K. Chakraborty, S. Bhattacharya, K. P. Ghatak, *Jour. of appl. Phys.,* 98, 53517, (2005).

[95] Chowdhary S., L. J. Singh, K. P. Ghatak, *Physica B: Condensed Matter,* 365, 5, (2005); L. J. Singh, S. Choudhary, A. Mallik, K. P. Ghatak, *Jour. of Compu. and Theo. Nanosci.,* 2, 287, (2005); K. P. Ghatak, D. K. Basu, S. Bhattacharya, S. Ghosal, S. K. Biswas, A. Mallik, G. Majumdar, *Nonlinear Optics Quantum Optics*, 33, 221, (2005); L. J. Singh, S. Choudhury, S. S. Roy, K. P. Ghatak, *Electrical Engineering,* 87, 1, (2005).

[96] Ghatak K. P., S. K. Biswas, D. De, S. Ghosal, S. Chatterjee, *Physica B: Condensed Matter,* 353, 127, (2004); P. K. Chakraborty, S. K. Biswas, K. P. Ghatak, *Physica B: Condensed Matter,* 352, 111, (2004); P. K. Chakraborty, L. J. Singh, K. P. Ghatak, *Jour. of appl. Phys.*, 95, 5311, (2004).

[97] Chakraborty P. K., G. C. Datta, K. P. Ghatak, *Physica B: Condensed Matter,* 339, 198, (2003); P. K. Chakraborty, B. Nag, K. P. Ghatak, *Jour. of Phys. and Chem. of Solids,* 64,2191, (2003); S Choudhury, L. J. Singh, K. P. Ghatak, *Nanotechnology,* 15, 180, (2003).

[98] Sourangshu Mukhopadhyay, Nirmalya Pahari, Archan Kumar Das, Partha Pratime Das, Abhijit Sinha, Harihar Bhowmik, Tushar Kanti Dalal, Ajitesh Biswas, Sugata Ghosh, Piyasi Mukherjee, Pradip Dhak, Subhomoy Singha Ray, Tushar Kanti Tapadar, S. Karmakar, J. P. Banerjee, Kamakhya Prasad Ghatak, Tridib Tripathy, B. R. De, Manabesh Bhattacharya, W. G. Devi, S. Mondal, M. Maiti, Amiya K. Shyamal, Madhumangal Pal, Susanta K. Khan, B. Mondal, S. Kar, A. K. Bhunia, Kalishankar Tiwari, S. C. Debnath, Ram Kumar Bhakat, *Journal of Physical Science,* Vol. 8, (2003); P. K. Chakraborty, G. C. Datta, K. P. Ghatak, *Physica Scripta,* 68, 368, (2003); P. K. Chakraborty, G. C. Datta, K. P. Ghatak, *Physica Scripta,* 68, 368,(2003); P. K. Chakraborty, K. P. Ghatak, *Physics Letters A*, 288, 335, (2001).

[99] Chakraborty P. K., K. P. Ghatak, *Jour. of Physics and Chem. of Solids,* 62,1061, (2001); K. P. Ghatak, J. Y. Siddiqui, B. Nag, *Physics Letters A,* 282, 428, (2001); P. K. Chakraborty, K. P. Ghatak, *Journal of Applied Physics,* 89, 1075-1081, (2001).

In: Quantum Wires: An Overview
Editor: Kamakhya Prasad Ghatak ISBN: 978-1-53617-676-6

Chapter 4

THE DIFFUSIVITY MOBILITY RATIO IN QUANTUM WIRE SUPERLATTICES

S. K. Biswas[1,*], M. Mitra[2], P. K. Das[3], S. Chakrabarti[4] and N. Debbarma[5]

[1]Department of Physics, University of Calcutta, Kolkata, India
[2]Department of Applied Physics, University of Calcutta, Kolkata, India
[3]Department of Basic Science Institute of Engineering & Management, Salt Lake, Kolkata, West Bengal, India
[4]Department of Computer Science and Engineering, University of Engineering and Management and Institute of Engineering and Management, Kolkata and Jaipur, Salt Lake Electronics Complex, Kolkata, and Gurukul, Jaipur, Rajasthan, India
[5]Department of Computer Science and Engineering, National Institute of Technology, Agartala, Tripura, India

[*] Corresponding Author's Email: skbiswas23@gmail.com.

ABSTRACT

The importance of superlattices of non-parabolic semiconductors has been well known in the field of Nano technology since its inception. We study the diffusivity mobility ratio (DMR) in quantum Wire $GaAs / Ga_{1-x}Al_xAs, CdS / CdTe, PbTe / PbSnTe\ and\ HgTe / CdTe$ super-lattices with graded interfaces by formulating the appropriate electron statistics per unit length in each case. We also investigate the DMR in Quantum Wire Effective Mass Super-lattices of the said materials by deriving appropriate 1D Dispersion Relation. It appears that the DMR increases for both types of SLs as considered in this chapter with decreasing film thickness and increasing electron concentration per unit length respectively, although the numerical values are totally band structure dependent.

INTRODUCTION

It is well known that Keldysh [1] first suggested the fundamental concept of a superlattice (SL), although it was successfully experimental realized by Esaki and Tsu [2]. The importance of SLs in the field of nano-electronics have already been described in [3-5]. The most extensively studied III-V SL is the one consisting of alternate layers of GaAs and $Ga_{1-x}Al_xAs$ owing to the relative ease of fabrication. The GaAs layers forms quantum wells and $Ga_{1-x}Al_xAs$ form potential barriers. The III-V SL's are attractive for the realization of high speed electronic and optoelectronic devices [6]. In addition to SLs with usual structure, SLs with more complex structures such as II-VI [7], IV-VI [8] and HgTe/CdTe [9] SL's have also been proposed. The IV-VI SLs exhibit quite different properties as compared to the III-V SL due to the peculiar band structure of the constituent materials [10]. The epitaxial growth of II-VI SL is a relatively recent development and the primary motivation for studying the mentioned SLs made of materials with the large band gap is in their potential for optoelectronic operation in the blue [10]. HgTe/CdTe SL's have raised a great deal of attention since 1979 as promising new materials for long wavelength infrared detectors and other electro-optical applications [11]. Interest in Hg-based SL's has been further increased as new properties with potential device applications were revealed [11, 12]. These features arise from the unique zero band gap material HgTe [13] and the direct band gap semiconductor CdTe which can be described by the three band mode of Kane [14]. The combination of the aforementioned materials with specified dispersion relation makes HgTe/CdTe SL very attractive, especially because of the possibility to tailor the material properties for various applications by varying the energy band constants of the SLs. In addition to it, for effective mass SLs, the electronic subbands appear continually in real space [15].

We note that all the aforementioned SLs have been proposed with the assumption that the interfaces between the layers are sharply defined, of zero thickness, i.e., devoid of any interface effects. The SL potential distribution may be then considered as a one dimensional array of rectangular potential wells. The aforementioned advanced experimental techniques may produce SLs with physical interfaces between the two materials crystallo graphically abrupt; adjoining their interface will change at least on an atomic scale. As the potential form changes from a well (barrier) to a barrier (well), an intermediate potential region exists for the electrons. The influence of finite thickness of the interfaces on the electron dispersion law is very important, since, the electron energy spectrum governs the electron transport in SLs.

It is well known that the Einstein relation for the diffusivity-mobility ratio of the carriers in semiconductors (DMR) occupies a central position in the whole field of semiconductor science and technology [16] since the diffusion constant (a quantity very useful for device analysis where exact experimental determination is rather difficult) can be obtained from this ratio by knowing the experimental values of the mobility. The classical value of the DMR is equal to $(k_B T/|e|)$,(k_B, T, and $|e|$ are Boltzmann's constant, temperature and the magnitude of the carrier charge respectively).

This relation in this form was first introduced to study the diffusion of gas particles and is known as the Einstein relation [17, 18]. Therefore, it appears that the DMR increases linearly with increasing T and is independent of electron concentration. This relation holds for both types of charge carrier only under non-degenerate carrier concentration although its validity has been suggested erroneously for degenerate materials [19]. Landsberg first pointed out that the DMR for semiconductors having degenerate electron concentration are essentially determined by their energy band structures [20, 21]. This relation is useful for semiconductor homostructures [22, 23], semiconductor-semiconductor heterostructures [24, 25], metals-semiconductor heterostructures [26-34] and insulator-semiconductor heterostructures [35-38]. The nature of the variations of the DMR under different physical conditions has been studied in the literature [16-18, 20, 21, 39-65] and some of the significant features, which have emerged from these studies, are:

(a) the ratio increases monotonically with increasing electron concentration in bulk materials and the nature of these variations are significantly influenced by the energy band structures of different materials;
(b) the ratio increases with the increasing quantizing electric field as in inversion layers;
(c) the ratio oscillates with the inverse quantizing magnetic field under magnetic quantization due to the Shubnikov de Hass effect;

In this chapter, we shall study the Diffusivity Mobility Ratio (DMR) in quantum wire superlattices of III-V, II-VI, IV-VI and HgTe/CdTe SLs with graded interfaces and in the same chapter we shall also investigate the DMR in III-V, II-VI, IV-VI and HgTe/CdTe effective mass SLs respectively.

THEORETICAL BACKGROUND

The Diffusivity Mobility Ratio in Quantum Wire III-V Superlattices with Graded Interfaces

The energy spectrum of the conduction electrons in bulk specimens of the constituent materials of III-V SLs whose energy band structures are defined by three band model of Kane can be written as

$$(\hbar^2 k^2)/(2m_i^*) = EG(E, E_{gi}, \Delta_i) \tag{1}$$

$$i = 1,2,..., G(E, E_{gi}, \Delta_i) \equiv \frac{(E_{gi} + \frac{2}{3}\Delta_i)(E + E_{gi} + \Delta_i)(E + E_{gi})}{[E_{gi}(E_{gi} + \Delta_i)(E + E_{gi} + \frac{2}{3}\Delta_i)]}$$

where, $\hbar$ is planck constant, m_i^* is the effective electron mass at the edge of the conduction band, E is the total electron energy, E_{gi} is the band gap Δ_i is te spin orbit splitting constant of the valance band.

Therefore, the dispersion law of the electrons of III-V SLs with graded interfaces can be expressed, following Jiang and Lin [16], as

$$\cos(L_o k) = \frac{1}{2}\Phi(E, k_s) \tag{2}$$

where $L_0(\equiv a_0 + b_0)$ is the period length, a_0 and b_0 are the widths of the barrier and the well respectively,

$$\begin{aligned}
\Phi(E,k_s) \equiv\ & [2\cosh\{\beta(E,k_s)\}\cos\{\gamma(E,k_s)\} + \varepsilon(E,k_s)\sinh\{\beta(E,k_s)\}\sin\{\gamma(E,k_s)\} \\
& + \Delta_0[(\frac{K_1^2(E,k_s)}{K_2(E,k_s)} - 3K_2(E,k_s))\cosh\{\beta(E,k_s)\}\sin\{\gamma(E,k_s)\} + (3K_1(E,k_s) \\
& - \frac{\{K_2(E,k_s)\}^2}{K_1(E,k_s)})\sinh\{\beta(E,k_s)\}\cos\{\gamma(E,k_s)\}] \\
& + \Delta_0[2(\{K_1(E,k_s)\}^2 - \{K_2(E,k_s)\}^2)\cosh\{\beta(E,k_s)\}\cos\{\gamma(E,k_s)\} \\
& + \frac{1}{12}[\frac{5\{K_2(E,k_s)\}^3}{K_1(E,k_s)} + \frac{5\{K_1(E,k_s)\}^3}{K_2(E,k_s)} - 34K_2(E,k_s)K_1(E,k_s)]\sinh\{\beta(E,k_s)\}\sin\{\gamma(E,k_s)\}]]
\end{aligned}$$

$$\varepsilon(E,k_s) \equiv [\frac{K_1(E,k_s)}{K_2(E,k_s)} - \frac{K_2(E,k_s)}{K_1(E,k_s)}],$$

$\beta(E,k_s) \equiv K_1(E,k_s)[a_0 - \Delta_0]$, Δ_o is the interface width, $\gamma(E,k_s) = K_2(E,k_s)[b_0 - \Delta_0]$, $K_1(E,k_s) \equiv [\frac{2m_2^* E'}{\hbar^2} G(E - V_o, \alpha_2, \Delta_2) + k_s^2]^{1/2}$, $E' \equiv V_0 - E$, V_0 is the potential barrier encountered by the electron $(V_0 \equiv |E_{g_2} - E_{g_1}|)$, $\alpha_i \equiv 1/E_{gi}$, $K_2(E,k_s) \equiv [\frac{2m_1^* E}{\hbar^2} G(E, \alpha_1, \Delta_1) - k_s^2]^{1/2}$ and $k_s^2 = k_x^2 + k_y^2$.

The electron dispersion law in III-V quantum wire superlattices (QWSLs), can be written following equation (2) as

$$k_z^2 = [\frac{1}{L_0^2}\{\rho_8(n_x,n_y,E)\} - \phi(n_x,n_y)] \tag{3}$$

where, $\rho_8(n_x,n_y,E) = [\cos^{-1}\{\frac{1}{2}\psi_8(n_x,n_y,E)\}]^2, \phi(n_x,n_y) = (n_x\pi/d_x)^2 + (n_y\pi/d_y)^2$,

$n_x(=1,2,3,...)$ and $n_y(=1,2,3,...)$ are the size quantum numbers along the x- and y- direction respectively and d_x and d_y are the nano-thickness along the respective directions.

$$\psi_8(n_x,n_y,E) = [2\cosh\{\beta_8(n_x,n_y,E)\}\cos\{\gamma_8(n_x,n_y,E)\} + \varepsilon_8(n_x,n_y,E)$$
$$\sinh\{\beta_8(n_x,n_y,E)\}$$
$$\sin\{\gamma_8(n_x,n_y,E)\} + \Delta_0[(\frac{\{K_9(n_x,n_y,E)\}^2}{K_{10}(n_x,n_y,E)}$$
$$-3K_{10}(n_x,n_y,E))\cosh\{\beta_8(n_x,n_y,E)\}\sin\{\gamma_8(n_x,n_y,E)\} + (3K_9(n_x,n_y,E)$$
$$-\frac{\{K_{10}(n_x,n_y,E)\}^2}{K_9(n_x,n_y,E)})\sinh\{\beta_8(n_x,n_y,E)\}\cos\{\gamma_8(n_x,n_y,E)\}$$
$$+\Delta_0[2(K_9(n_x,n_y,E) - \{K_{10}(n_x,n_y,E)\}^2)\cosh\{\beta_8(n_x,n_y,E)\}\cos\{\gamma_8(n_x,n_y,E)\}$$

$$+\frac{1}{12}(\frac{5\{K_9(n_x,n_y,E)\}^3}{K_{10}(n_x,n_y,E)}+\frac{5\{K_{10}(n_x,n_y,E)\}^3}{K_9(n_x,n_y,E)}$$
$$-\{34K_{10}(n_x,n_y,E)K_9(n_x,n_y,E)\})\sinh\{\beta_8(n_x,n_y,E)\}$$
$$\sin\{\gamma_8(n_x,n_y,E)\}]]$$
$$\varepsilon_8(n_x,n_y,E)\equiv[\frac{K_9(n_x,n_y,E)}{K_{10}(n_x,n_y,E)}-\frac{K_{10}(n_x,n_y,E)}{K_9(n_x,n_y,E)}],$$
$$\beta_8(n_x,n_y,E)\equiv K_9(n_x,n_y,E)[a_0 - \Delta_0]$$
$$K_9(n_x,n_y,E)\equiv[\frac{2m_2^*E'}{\hbar^2}G(E-V_0,\alpha_2,\Delta_2)+\phi(n_x,n_y)]^{1/2},$$
$$\gamma_8(n_x,n_y,E)=K_{10}(n_x,n_y,E)[b_0-\Delta_0]$$

and

$$K_{10}(n_x,n_y,E)\equiv[\frac{2m_1^*E}{\hbar^2}G(E,\alpha_1,\Delta_1)-\phi(n_x,n_y)]^{1/2}$$

The effective momentum mass (EMM) in this case assumes the form

$$m^*(n_x,n_y,E_{FQWSL})=\frac{\hbar^2}{2L_0^2}\{\rho_8(n_x,n_y,E_{FQWSL})\}' \tag{4}$$

where, E_{FQWSL} is the Fermi energy in this case. Thus, the EMM in III-V QWSLs is a function of both the Fermi energy and the size quantum numbers.

The subband energies $(E_{n_{SL9}})$ can be expressed as

$$\rho_8(n_x,n_y,E_{n_{SL9}})=[\phi(n_x,n_y)] \tag{5}$$

Considering only the lowest mini band, since in an actual SL only the lowest mini band is significantly populated at low temperatures, where the quantum effects become prominent, the relation between the 1D electron concentration (n_{1D}) and the Fermi energy in the present case can be written as

$$n_{1D} = (\frac{2g_v}{\pi})\sum_{n_x=1}^{n_{x\max}}\sum_{n_y=1}^{n_{y\max}} [T_{917}(n_x, n_y, E_{FQWSL}) + T_{918}(n_x, n_y, E_{FQWSL})] \tag{6a}$$

where, $T_{917}(n_x, n_y, E_{FQWSL}) \equiv [\rho_8(n_x, n_y, E_{FQWSL}) - \phi(n_x, n_y)]^{1/2}$,

$T_{918}(n_x, n_y, E_{FQWSL}) \equiv \sum_{r=1}^{s} L(r)[T_{917}(n_x, n_y, E_{FQWSL})]$,

$L(r) = 2(1-2^{1-2r})\xi(2r)(k_B T)^{2r}$,

k_B Boltzmann constant and T is temperature.

The DMR can in general be expressed as

$$\frac{D}{\mu} = \frac{n_{1D}}{e}(\frac{dn_{1D}}{dE_{FQWSL}})^{-1} \tag{6b}$$

where, E_{FQWSL} is the Fermi energy in this case.

The use of equations (6a) and (6b) leads to the expression of the DMR in this case as

$$\frac{D}{\mu} = \frac{1}{|e|}\frac{\sum_{n_x=1}^{n_{x\max}}\sum_{n_y=1}^{n_{y\max}} [T_{917}(n_x, n_y, E_{FQWSL}) + T_{918}(n_x, n_y, E_{FQWSL})]}{\sum_{n_x=1}^{n_{x\max}}\sum_{n_y=1}^{n_{y\max}} [\{T_{917}(n_x, n_y, E_{FQWSL})\}' + \{T_{918}(n_x, n_y, E_{FQWSL})\}']} \tag{7}$$

The Diffusivity Mobility Ratio in Quantum Wire III-V Superlattices with Graded Interfaces

The energy spectrum of the conduction electrons of the constituent materials of II-VI SLs are given by [17].

$$E = \frac{\hbar^2 k_s^2}{2m_{\perp,1}^*} + \frac{\hbar^2 k_z^2}{2m_{\parallel,1}^*} + \bar{\lambda}_0 k_s \tag{8}$$

and

$$\frac{\hbar^2 k^2}{2m_2^*} = EG(E, E_{g2}, \Delta_2) \tag{9}$$

where $m_{\perp,1}^*$ and $m_{\parallel,1}^*$ are the transverse and longitudinal effective electron masses respectively at the edge of the conduction band for the first material and $\bar{\lambda}_0$ is the splitting of the two spin-states. By the spin orbit coupling and the crystalline field. The electron dispersion law in II-VI SLs with graded interfaces can be expressed as

$$\cos(L_o k) = \frac{1}{2}\Phi_1(E, k_s) \tag{10}$$

where,

$$\Phi_1(E,k_s) \equiv \left[\frac{2\cosh\{\beta_1(E,k_s)\}\cos\{\gamma_1(E,k_s)\} + \varepsilon_1(E,k_s)}{\sinh\{\beta_1(E,k_s)\}\sin\{\gamma_1(E,k_s)\}}\right.$$
$$+\Delta_0[(\frac{\{K_3(E,k_s)\}^2}{K_4(E,k_s)} - 3K_4(E,k_s))\cosh\{\beta_1(E,k_s)\}\sin\{\gamma_1(E,k_s)\}$$
$$+(3K_3(E,k_s) - \frac{\{K_4(E,k_s)\}^2}{K_3(E,k_s)})\sinh\{\beta_1(E,k_s)\}\cos\{\gamma_1(E,k_s)\}]$$
$$+\Delta_0[2(\{K_3(E,k_s)\}^2 - \{K_4(E,k_s)\}^2)\cosh\{\beta_1(E,k_s)\}\cos\{\gamma_1(E,k_s)\}$$
$$+\frac{1}{12}[\frac{5\{K_3(E,k_s)\}^3}{K_4(E,k_s)} + \frac{5\{K_4(E,k_s)\}^3}{K_3(E,k_s)} - 34K_4(E,k_s)K_3(E,k_s)]$$
$$\sinh\{\beta_1(E,k_s)\}\sin\{\gamma_1(E,k_s)\}]]$$
$$\varepsilon_1(E,k_s) \equiv [\frac{K_3(E,k_s)}{K_4(E,k_s)} - \frac{K_4(E,k_s)}{K_3(E,k_s)}], \beta_1(E,k_s) \equiv K_3(E,k_s)$$
$$[a_0 - \Delta_0], \gamma_1(E,k_s) = K_4(E,k_s)[b_0 - \Delta_0]$$
$$K_3(E,k_s) \equiv [\frac{2m_2^* E'}{\hbar^2} G(E - V_o, \alpha_2, \Delta_2) + k_s^2]^{1/2} \; and \; K_4(E,k_s)$$
$$\equiv [\frac{2m_{\parallel,1}^*}{\hbar^2}[E - \frac{\hbar^2 k_s^2}{2m_{\perp,1}^*} \mp \bar{\lambda}_0 k_s]]^{1/2}$$

The electron dispersion law in II-VI QWSLs, can be written following equation (10) as

$$k_z^2 = [\frac{1}{L_0^2}\{\rho_9(n_x, n_y, E)\} - \phi(n_x, n_y)] \quad (11)$$

where,

$$\rho_9(n_x, n_y, E) = [\cos^{-1}\{\frac{1}{2}\psi_9(n_x, n_y, E)\}]^2,$$

$$\psi_9(n_x, n_y, E) = [2\cosh\{\beta_9(n_x, n_y, E)\}\cos\{\gamma_9(n_x, n_y, E)\}$$
$$+\varepsilon_9(n_x, n_y, E)\sinh\{\beta_9(n_x, n_y, E)\}$$
$$\sin\{\gamma_9(n_x, n_y, E)\} + \Delta_0[(\frac{\{K_{11}(n_x, n_y, E)\}^2}{K_{12}(n_x, n_y, E)} - 3K_{12}(n_x, n_y, E))\cosh\{\beta_9(n_x, n_y, E)\}$$
$$\sin\{\gamma_9(n_x, n_y, E)\} + (3K_{11}(n_x, n_y, E) - \frac{\{K_{12}(n_x, n_y, E)\}^2}{K_{11}(n_x, n_y, E)})$$
$$\sinh\{\beta_9(n_x, n_y, E)\}\cos\{\gamma_9(n_x, n_y, E)\}$$
$$+\Delta_0[2(K_{11}(n_x, n_y, E) - \{K_{12}(n_x, n_y, E)\}^2)\cosh\{\beta_9(n_x, n_y, E)\}\cos\{\gamma_9(n_x, n_y, E)\}$$
$$+\frac{1}{12}(\frac{5\{K_{11}(n_x, n_y, E)\}^3}{K_{12}(n_x, n_y, E)} + \frac{5\{K_{12}(n_x, n_y, E)\}^3}{K_{11}(n_x, n_y, E)}$$
$$-\{34K_{12}(n_x, n_y, E)K_{11}(n_x, n_y, E)\})\sinh\{\beta_9(n_x, n_y, E)\}\sin\{\gamma_9(n_x, n_y, E)\}]],$$

$$\varepsilon_9(n_x, n_y, E) \equiv [\frac{K_{11}(n_x, n_y, E)}{K_{12}(n_x, n_y, E)} - \frac{K_{12}(n_x, n_y, E)}{K_{11}(n_x, n_y, E)}],$$

$$\beta_9(n_x, n_y, E) \equiv K_{11}(n_x, n_y, E)[a_0 - \Delta_0]$$

$$\gamma_9(n_x, n_y, E) = K_{12}(n_x, n_y, E)\left[b_0 - \Delta_0\right], K_{11}(n_x, n_y, E) \equiv$$
$$[\frac{2m_2^*}{\hbar^2}E'G(E - V_0, \alpha_2, \Delta_2) + \phi(n_x, n_y)]^{1/2}$$

and

$$K_{12}(n_x,n_y,E)\equiv[\frac{2m^*_{\parallel,1}}{\hbar^2}[E-\frac{\hbar^2}{2m^*_{\perp,1}}\phi(n_x,n_y)\mp\bar{\lambda}_0\{\phi(n_x,n_y)\}^{1/2}]]^{1/2}$$

The EMM in this case assumes the form

$$m^*(n_x,n_y,E_{FQWSL})=\frac{\hbar^2}{2L_0^2}\{\rho_9(n_x,n_y,E_{FQWSL})\}' \tag{12}$$

The EMM in II-VI QWSLs depends on both the Fermi energy and size quantum numbers, which is the intrinsic property of such QWSLs.

The subband energies $(E_{n_{SL10}})$ can be written as

$$\rho_9(n_x,n_y,E_{n_{SL10}})=[\phi(n_x,n_y)] \tag{13}$$

The electron concentration in this case can be expressed as

$$n_{1D}=(\frac{2g_v}{\pi})\sum_{n_x=1}^{n_{x_{mas}}}\sum_{n_y=1}^{n_{y_{max}}}[T_{919}(n_x,n_y,E_{FQWSL})+T_{920}(n_x,n_y,E_{FQWSL})] \tag{14}$$

where, $T_{919}(n_x,n_y,E_{FQWSL})\equiv[\rho_9(n_x,n_y,E_{FQWSL})-\phi(n_x,n_y)]^{1/2}$

and $T_{920}(n_x,n_y,E_{FQWSL})\equiv\sum_{r=1}^{s}L(r)[T_{919}(n_x,n_y,E_{FQWSL})]$

The use of equations (6b) and (14) leads to the expression of the DMR as

$$\frac{D}{\mu}=\frac{1}{|e|}\frac{\sum_{n_x=1}^{n_{x_{mas}}}\sum_{n_y=1}^{n_{y_{max}}}[T_{919}(n_x,n_y,E_{FQWSL})+T_{920}(n_x,n_y,E_{FQWSL})]}{\sum_{n_x=1}^{n_{x_{mas}}}\sum_{n_y=1}^{n_{y_{max}}}[\{T_{919}(n_x,n_y,E_{FQWSL})\}'+\{T_{920}(n_x,n_y,E_{FQWSL})\}']} \tag{15}$$

The Diffusivity Mobility Ratio in Quantum Wire IV-VI Superlattices with Graded Interfaces

The **E-k** dispersion relation of the conduction electrons of the constituent materials of the IV-VI SLs can be expressed [18] as

$$E = a_i k_s^2 + b_i k_z^2 + [[c_i k_s^2 + d_i k_z^2] + (e_i k_s^2 + f_i k_y^2 + \frac{E_{g_i}}{2})^2]^{1/2} - \frac{E_{g_i}}{2} \tag{16}$$

where, $a_i \equiv [\frac{\hbar^2}{2m_{\perp,i}^-}], b_i \equiv (\frac{\hbar^2}{2m_{\|,i}^-})^2, c_i \equiv P_{\perp,i}^2, d_i \equiv P_{\perp,i}^2, e_i \equiv [\frac{\hbar^2}{2m_{\perp,i}^+}]$ *and* $f_i \equiv (\frac{\hbar^2}{2m_{\|,i}^+})^2$

The electron dispersion law in IV-VI SLs with graded interfaces can be expressed as

$$\cos(L_o k) = \frac{1}{2}\Phi_2(E, k_s) \tag{17}$$

where,

$$\Phi_2(E,k_s) \equiv \begin{bmatrix} 2\cosh\{\beta_2(E,k_s)\}\cos\{\gamma_2(E,k_s)\} + \varepsilon_2(E,k_s) \\ \sinh\{\beta_2(E,k_s)\}\sin\{\gamma_2(E,k_s)\} \end{bmatrix}$$
$$+\Delta_0[(\frac{\{K_5(E,k_s)\}^2}{K_6(E,k_s)} - 3K_6(E,k_s))\cosh\{\beta_2(E,k_s)\}\sin\{\gamma_2(E,k_s)\}$$
$$+(3K_5(E,k_s) - \frac{\{K_6(E,k_s)\}^2}{K_5(E,k_s)})\sinh\{\beta_2(E,k_s)\}\cos\{\gamma_2(E,k_s)\}]$$
$$+\Delta_0[2(\{K_5(E,k_s)\}^2 - \{K_6(E,k_s)\}^2)\cosh\{\beta_2(E,k_s)\}\cos\{\gamma_2(E,k_s)\}$$
$$+\frac{1}{12}[\frac{5\{K_5(E,k_s)\}^3}{K_6(E,k_s)} + \frac{5\{K_6(E,k_s)\}^3}{K_5(E,k_s)} - 34K_6(E,k_s)$$
$$K_5(E,k_s)]\sinh\{\beta_2(E,k_s)\}\sin\{\gamma_2(E,k_s)\}]]$$
$$\varepsilon_2(E,k_s) \equiv [\frac{K_5(E,k_s)}{K_6(E,k_s)} - \frac{K_6(E,k_s)}{K_5(E,k_s)}], \beta_2(E,k_s) \equiv$$
$$K_5(E,k_s)[a_0 - \Delta_0], \gamma_2(E,k_s) = K_6(E,k_s)[b_0 - \Delta_0]$$
$$K_6(E,k_x,k_y) \equiv [[EH_{11} + H_{21}(k_x,k_y)] - [H_{31}E^2 + EH_{41}(k_x,k_y)$$
$$+ H_{51}(k_x,k_y)^{1/2}]^{1/2}, H_{1i} = (b_i^2 - f_i^2)^{-1}$$

and

$$H_{2i}(k_x,k_y)=[2H_{1i}]^{-1}[E_{g}b_i+d_i+f_iE_{g_i}+2(e_if_i-a_ib_i)(k_x^2+k_y^2)],\ H_{3i}=\frac{f_i^2}{(b_i^2-f_i^2)^2},$$
$$H_{4i}(k_x,k_y)=[4H_{1i}^2]^{-1}[4b_id_i+4b_if_iE_{g_i}+4f_i^2E_{g_i}+8(k_x^2+k_y^2)[e_if_ib_i-a_if_i^2]]$$
$$H_{5i}(k_x,k_y)\equiv[4H_{1i}^2][(k_x^2+k_y^2)^2[-8a_ib_ie_if_i+4b_i^2e_i^2+4f_i^2a_i^2]+(k_x^2+k_y^2)[4e_if_iE_{g}b_i$$
$$-4e_if_id_i+4e_if_i^2E_{g_i}-4a_ib_i^2E_{g}-4a_ib_id_i-4a_ib_if_iE_{g_i}+4b_i^2e_iE_{g}+4b_i^2c_i+4b_i^2E_{g_i}a_i-4f_i^2e_iE_{g_i}-4f_i^2c_i-4f_i^2E_{g}a_i]$$
$$+[E_{g_i}^2b_i^2+d_i^2+f_i^2g_i^2+2E_{g_i}b_id_i+2E_{g_i}^2b_if_i+2d_if_iE_{g}]]$$

and

$$K_5(E,k_x,k_y)\equiv[[(E-V_0)^2H_{32}+(E-V_0)H_{42}(k_x,k_y)+H_{52}(k_x,k_y)]^{1/2}-[(E-V_0)H_{12}+H_{22}(k_x,k_y)]]^{1/2}$$

The electron dispersion law in IV-VI QWSLs can be written from equation (17a) as

$$k_z^2=[\frac{1}{L_0^2}\{\rho_{10}(n_x,n_y,E)\}-\phi(n_x,n_y)] \tag{18}$$

where, $\rho_{10}(n_x,n_y,E)=[\cos^{-1}\{\frac{1}{2}\psi_{10}(n_x,n_y,E)\}]^2,$

$$\psi_{10}(n_x,n_y,E)=[2\cosh\{\beta_{10}(n_x,n_y,E)\}\cos\{\gamma_{10}(n_x,n_y,E)\}+\varepsilon_{10}(n_x,n_y,E)\sinh\{\beta_{10}(n_x,n_y,E)\}$$
$$\sin\{\gamma_{10}(n_x,n_y,E)\}+\Delta_0[(\frac{\{K_{11}(n_x,n_y,E)\}^2}{K_{12}(n_x,n_y,E)}-3K_{12}(n_x,n_y,E))\cosh\{\beta_{10}(n_x,n_y,E)\}$$
$$\sin\{\gamma_{10}(n_x,n_y,E)\}+(3K_{11}(n_x,n_y,E)-\frac{\{K_{12}(n_x,n_y,E)\}^2}{K_{11}(n_x,n_y,E)})\sinh\{\beta_{10}(n_x,n_y,E)\}\cos\{\gamma_{10}(n_x,n_y,E)\}$$
$$+\Delta_0[2(K_{11}(n_x,n_y,E)-\{K_{12}(n_x,n_y,E)\}^2)\cosh\{\beta_{10}(n_x,n_y,E)\}\cos\{\gamma_{10}(n_x,n_y,E)\}$$
$$+\frac{1}{12}(\frac{5\{K_{11}(n_x,n_y,E)\}^3}{K_{12}(n_x,n_y,E)}+\frac{5\{K_{12}(n_x,n_y,E)\}^3}{K_{11}(n_x,n_y,E)}-\{34K_{12}(n_x,n_y,E)K_{11}(n_x,n_y,E)\})\sinh\{\beta_{10}(n_x,n_y,E)\}\sin\{\gamma_{10}(n_x,n_y,E)\}]]$$

$$\varepsilon_{10}(n_x,n_y,E)\equiv[\frac{K_{11}(n_x,n_y,E)}{K_{12}(n_x,n_y,E)}-\frac{K_{12}(n_x,n_y,E)}{K_{11}(n_x,n_y,E)}],\beta_{10}(n_x,n_y,E)\equiv K_{11}(n_x,n_y,E)[a_0-\Delta_0]$$
$$K_{12}(n_x,n_y,E)\equiv[[EH_{11}+H_{21}(n_x,n_y)]-[E^2H_{31}+EH_{41}(n_x,n_y)+H_{51}(n_x,n_y)]^{1/2}]^{1/2}$$

and

$$H_{2i}(n_x,n_y)=[2H_{1i}]^{-1}[E_{g_i}b_i+d_i+f_iE_{g_i}+2(e_if_i-a_ib_i)\phi(n_x,n_y)],$$
$$H_{4i}(n_x,n_y)=[4H_{1i}^2]^{-1}[4b_id_i+4b_if_iE_{g_i}+4f_i^2E_{g_i}+8\phi(n_x,n_y)[e_if_ib_i-a_if_i^2]],$$
$$H_{5i}(n_x,n_y)\equiv[4H_{1i}^2]^{-1}[\{\phi(n_x,n_y)\}^2[-8a_ib_ie_if_i+4b_i^2e_i^2+4f_i^2a_i^2]+\{\phi(n_x,n_y)\}[4e_if_iE_{g_i}b_i$$
$$-4e_if_id_i+4e_if_i^2E_{g_i}-4a_ib_i^2E_{g_i}-4a_ib_id_i-4a_ib_if_iE_{g_i}+4b_i^2e_iE_{g_i}+4b_i^2c_i+4b_i^2E_{g_i}a_i-4f_i^2e_iE_{g_i}-4f_i^2c_i-4f_i^2E_{g_i}a_i]$$
$$+[E_{g_i}^2b_i^2+d_i^2+f_i^2g_i^2+2E_{g_i}b_id_i+2E_{g_i}^2b_if_i+2d_if_iE_{g_i}]]$$
$$K_{11}(n_x,n_y,E)\equiv[[(E-V_0)^2H_{32}+(E-V_0)H_{42}(n_x,n_y)+H_{52}(n_x,n_y)]^{1/2}-[(E-V_0)H_{12}+H_{22}(n_x,n_y)]]^{1/2}$$

The EMM in this case assumes the form

$$m^*(n_x,n_y,E_{FQWSL})=\frac{\hbar^2}{2L_0^2}\{\rho_{10}(n_x,n_y,E_{FQWSL})\}' \quad (19)$$

The EMM in IV-VI QWSLs under magnetic quantization depends on both the Fermi energy and magnetic quantum number, which is the intrinsic property of such SLs.

The Landau Subband energies $(E_{n_{SL11}})$ can be written as

$$\rho_{10}(n_x,n_y,E_{n_{SL11}})=\{\phi(n_x,n_y,E_{n_{SL11}})\}' \quad (20)$$

The electron concentration in this case can be expressed as

$$n_{1D}=(\frac{2g_v}{\pi})\sum_{n_x=1}^{n_{x_{\max}}}\sum_{n_y=1}^{n_{y_{\max}}}[T_{921}(n_x,n_y,E_{FQWSL})+T_{922}(n_x,n_y,E_{FQWSL})] \quad (21)$$

$$\textit{where, } T_{921}(n_x,n_y,E_{FQWSL})\equiv[\rho_{10}(n_x,n_y,E_{FQWSL})-\{\phi(n_x,n_y)\}]^{1/2}$$

$$\textit{and } T_{922}(n_x,n_y,E_{FQWSL})\equiv\sum_{r=1}^{s}L(r)[T_{921}(n_x,n_y,E_{FQWSL})]$$

The use of equations (6b) and (21) leads to the expression of the DMR as

$$\frac{D}{\mu}=\frac{1}{|e|}\frac{\sum_{n_x=1}^{n_{x_{\max}}}\sum_{n_y=1}^{n_{y_{\max}}}[T_{921}(n_x,n_y,E_{FQWSL})+T_{922}(n_x,n_y,E_{FQWSL})]}{\sum_{n_x=1}^{n_{x_{\max}}}\sum_{n_y=1}^{n_{y_{\max}}}[\{T_{921}(n_x,n_y,E_{FQWSL})\}'+\{T_{922}(n_x,n_y,E_{FQWSL})\}']} \quad (22)$$

The Diffusivity Mobility Ratio in Quantum Wire HgTe/CdTe Superlattices with Graded Interfaces

The dispersion relation of the conduction electrons of the constituent materials of HgTe/CdTe SLs can be expressed [13] as

$$E = \frac{\hbar^2 k^2}{2m_1^*} + \frac{3|e|^2 k}{128\varepsilon_{sc}} \quad (23)$$

$$\frac{\hbar^2 k^2}{2m_2^*} = EG(E_1 E_{g2}, \Delta_2) \quad (24)$$

where, ε_{sc} is the semiconductor permittivity.

The electron energy dispersion law in HgTe/CdTe SL is given by

$$\cos(L_o k) = \frac{1}{2}\Phi_3(E, k_s) \quad (25)$$

where, $\Phi_3(E,k_s) \equiv [2\cosh\{\beta_3(E,k_s)\}\cos\{\gamma_3(E,k_s)\} + \varepsilon_3(E,k_s)$

$$\sinh\{\beta_3(E,k_s)\}\sin\{\gamma_3(E,k_s)\}$$
$$+\Delta_0[(\frac{\{K_7(E,k_s)\}^7}{K_8(E,k_s)} - 3K_8(E,k_s))\cosh\{\beta_3(E,k_s)\}\sin\{\gamma_3(E,k_s)\}$$
$$+(3K_7(E,k_s) - \frac{\{K_8(E,k_s)\}^2}{K_7(E,k_s)})$$
$$\sinh\{\beta_3(E,k_s)\}\cos\{\gamma_3(E,k_s)\}]$$
$$+\Delta_0[2(\{K_7(E,k_s)\}^2 - \{K_8(E,k_s)\}^2)\cosh\{\beta_3(E,k_s)\}\cos\{\gamma_3(E,k_s)\}$$
$$+\frac{1}{12}[\frac{5\{K_8(E,k_s)\}^3}{K_7(E,k_s)} + \frac{5\{K_7(E,k_s)\}^3}{K_8(E,k_s)} - 34K_7(E,k_s)K_8(E,k_s)]$$
$$\sinh\{\beta_3(E,k_s)\}\sin\{\gamma_3(E,k_s)\}]],$$

$$\varepsilon_3(E,k_s) \equiv [\frac{K_7(E,k_s)}{K_8(E,k_s)} - \frac{K_8(E,k_s)}{K_7(E,k_s)}], \beta_3(E,k_s) \equiv K_7(E,k_s)$$
$$[a_0 - \Delta_0], \gamma_3(E,k_s) = K_8(E,k_s)[b_0 - \Delta_0],$$
$$K_8(E,k_x,k_y) \equiv [\frac{B_0^2 + 2AE - B_0\sqrt{B_0^2 + 4AE}}{2A^2} - k_s^2]^{1/2},$$
$$B_0 = \frac{3|e|^2}{128\varepsilon_{sc}}, A = \frac{\hbar^2}{2m_1^*}$$
$$and\ K_7(E,k_s) \equiv [\frac{2m_2^* E'}{\hbar^2} G(E - V_0, E_{g_2}, \Delta_2) + k_s^2]^{1/2}$$

The electron dispersion law in HgTe/CdTe QWSLs can be written from equation (24) as

$$k_z = [\frac{1}{L_0^2}\{\rho_{11}(n_x,n_y,E)\} - \{\phi(n_x,n_y,E)\}]^{1/2} \tag{26}$$

$$where,\ \rho_{11}(n_x,n_y,E) = [\cos^{-1}\{\frac{1}{2}\psi_{11}(n_x,n_y,E)\}]^2,$$
$$\psi_{11}(n_x,n_y,E) = [2\cosh\{\beta_{11}(n_x,n_y,E)\}\cos\{\gamma_{11}(n_x,n_y,E)\} + \varepsilon_{11}(n_x,n_y,E)\sinh\{\beta_{11}(n_x,n_y,E)\},$$
$$\sin\{\gamma_{11}(n_x,n_y,E)\} + \Delta_0[(\frac{\{K_{13}(n_x,n_y,E)\}^2}{K_{14}(n_x,n_y,E)} - 3K_{14}(n_x,n_y,E))\cosh\{\beta_{11}(n_x,n_y,E)\}$$

$$\sin\{\gamma_{11}(n_x,n_y,E)\} + (3K_{13}(n_x,n_y,E) - \frac{\{K_{14}(n_x,n_y,E)\}^2}{K_{13}(n_x,n_y,E)})$$
$$\sinh\{\beta_{11}(n_x,n_y,E)\}\cos\{\gamma_{11}(n_x,n_y,E)\}$$
$$+\Delta_0\left[2(\{K_{13}(n_x,n_y,E)\}^2 - \{K_{14}(n_x,n_y,E)\}^2)\right.$$
$$\cosh\{\beta_{11}(n_x,n_y,E)\}\cos\{\gamma_{11}(n_x,n_y,E)\}$$
$$+\frac{1}{12}(\frac{5\{K_{13}(n_x,n_y,E)\}^3}{K_{14}(n_x,n_y,E)} + \frac{5\{K_{14}(n_x,n_y,E)\}^3}{K_{13}(n_x,n_y,E)} -$$
$$-\{34K_{14}(n_x,n_y,E)\{K_{13}(n_x,n_y,E)\})$$

$\sinh\{\beta_{11}(n_x,n_y,E)\}\sin\{\gamma_{11}(n_x,n_y,E)\}]]$

$$\varepsilon_{11}(n_x,n_y,E) \equiv [\frac{K_{13}(n_x,n_y,E)}{K_{14}(n_x,n_y,E)} - \frac{K_{14}(n_x,n_y,E)}{K_{13}(n_x,n_y,E)}],$$

$$\beta_{11}(n_x,n_y,E) \equiv K_{13}(n_x,n_y,E)[a_0 - \Delta_0],$$

$$\gamma_{11}(n_x,n_y,E) = K_{14}(n_x,n_y,E)[b_0 - \Delta_0]$$

$$K_{14}(n_x,n_y,E) \equiv [\frac{B_0^2 + 2AE - B_0\sqrt{B_0^2 + 4AE}}{2A^2} - \{\phi(n_x,n_y)\}]^{1/2}$$

$$and\ K_{13}(n_x,n_y,E) \equiv [(\frac{2m_2^* E'}{\hbar^2})G(E - V_0,\alpha_2,\Delta_2) + \{\phi(n_x,n_y)\}]^{1/2}$$

The EMM in this case assumes the form

$$m^*(n_x,n_y,E_{FQWSL}) = \frac{\hbar^2}{2L_0^2}\{\rho_{11}(n_x,n_y,E_{FQWSL})\}' \qquad (27)$$

Thus the EMM in this case depends on both the Fermi energy and size quantum numbers.

The subband energies $(E_{n_{SL12}})$ can be written as

$$\rho_{11}(n_x,n_y,E_{n_{SL12}}) = [\phi(n_x,n_y,E_{n_{SL12}})] \qquad (28)$$

The electron concentration in this case can be expressed as

$$n_{1D} = (\frac{2g_v}{\pi})\sum_{n_x=1}^{n_{\max}}\sum_{n_y=1}^{n_{y\max}} [T_{923}(n_x,n_y,E_{FQWSL}) + T_{924}(n_x,n_y,E_{FQWSL})] \qquad (29)$$

$$where,\ T_{923}(n_x,n_y,E_{FQWSL}) \equiv [\rho_{11}(n_x,n_y,E_{FQWSL}) - \{\phi(n_x,n_y)\}]^{1/2}$$

$$and\ T_{924}(n_x,n_y,E_{FQWSL}) \equiv \sum_{r=1}^{s} L(r)[T_{923}(n_x,n_y,E_{FQWSL})]$$

The use of equations (6b) and (29) leads to the expression of the DMR as

$$\frac{D}{\mu}=\frac{1}{|e|}\frac{\sum_{n_x=1}^{n_{\max}}\sum_{n_y=1}^{n_{y\max}}[T_{923}(n_x,n_y,E_{FQWSL})+T_{924}(n_x,n_y,E_{FQWSL})]}{\sum_{n_x=1}^{n_{\max}}\sum_{n_y=1}^{n_{y\max}}[\{T_{923}(n_x,n_y,E_{FQWSL})\}'+\{T_{924}(n_x,n_y,E_{FQWSL})\}']} \quad (30)$$

The Diffusivity Mobility Ratio in Quantum Wire III-V Effective Mass Superlattices

Following Sasaki [15], the electron dispersion law in III-V effective mass superlattices (EMSLs) can be written as

$$k_x^2=[\frac{1}{L_0^2}\{\cos^{-1}(f(E,k_y,k_z))\}^2-k_\perp^2] \quad (31)$$

in which, $f(E,k_y,k_z)=a_1\cos[a_0C_1(E,k_\perp)+b_0D_1(E,k_\perp)]$
$-a_2\cos[a_0C_1(E,k_\perp)-b_0D_1(E,k_\perp)]$,

$k_\perp^2=k_y^2+k_z^2, a_1=[\sqrt{\frac{m_2^*}{m_1^*}}+1]^2[4(\frac{m_2^*}{m_1^*})^{1/2}]^{-1}$,

$a_2=[-1+\sqrt{\frac{m_2^*}{m_1^*}}]^2[4(\frac{m_2^*}{m_1^*})^{1/2}]^{-1}, C_1(E,k_\perp)\equiv[(\frac{2m_1^*E}{\hbar^2})G(E,E_{g_1},\Delta_1)-k_\perp^2]^{1/2}$

and $D_1(E,k_\perp)\equiv[(\frac{2m_2^*E}{\hbar^2})G(E,E_{g_2},\Delta_2)-k_\perp^2]^{1/2}$

The electron dispersion law in III-V, effective mass quantum wire superlattices (EMQWSLs) can be written following equation (31) as

$$k_x^2=[\rho_{12}(n_y,n_z,E)] \quad (32)$$

in which, $\rho_{12}(n_y,n_z,E)=\frac{1}{L_0^2}[\cos^{-1}(f_{12}(n_y,n_z,E))]^2$

$-\{\phi(n_y,n_z)\},\phi(n_y,n_z)=\{(\frac{n_y\pi}{d_y})^2+(\frac{n_z\pi}{d_z})^2\},$

$f_{12}(n_y,n_z,E)=a_7\cos[a_0C_5(n_y,n_z,E)+b_0D_5(n_y,n_z,E)]$

$-a_8\cos[a_0C_5(n_y,n_z,E)-b_0D_5(n_y,n_z,E)],$

$C_5(n_y,n_z,E)\equiv[(\frac{2m_1^*E}{\hbar^2})G(E,E_{g_1},\Delta_1)-\{\phi(n_y,n_z)\}]^{1/2}$

and

$$D_5(n_y,n_z,E)\equiv[(\frac{2m_2^*E}{\hbar^2})G(E,E_{g_1},\Delta_1)-\{\phi(n_y,n_z)\}]^{1/2}$$

The EMM in this case assumes the form

$$m^*(n_y,n_z,E_{FQWSL})=\frac{\hbar^2}{2}\{\rho_{12}(n_y,n_z,E_{FQWSL})\}' \tag{33}$$

The EMM in III-V EMQWSLs depends on both the Fermi energy and size quantum numbers, which is the intrinsic property of such SLs.

The subband energies $(E_{n_{SL13}})$ can be written as

$$\rho_{12}(n_y,n_z,E_{n_{SL13}})=0 \tag{34}$$

The electron concentration in this case can be expressed as

$$n_{1D}=(\frac{2g_v}{\pi})\sum_{n_y=1}^{n_{y\max}}\sum_{n_z=1}^{n_{z\max}}[T_{925}(n_y,n_z,E_{FQWSL})+T_{926}(n_y,n_z,E_{FQWSL})] \tag{35}$$

where, $T_{925}(n_y,n_z,E_{FQWSL})\equiv[\rho_{12}(n_y,n_z,E_{FQWSL})]^{1/2}$

and $T_{926}(n_y,n_z,E_{FQWSL})\equiv\sum_{r=1}^{s}L(r)[T_{925}(n_y,n_z,E_{FQWSL})]$

The use of equations (6b) and (35) leads to the expression of the DMR as

$$\frac{D}{\mu}=\frac{1}{|e|}\frac{\sum_{n_y=1}^{n_{y_{\max}}}\sum_{n_z=1}^{n_{z_{\max}}}[T_{925}(n_y,n_z,E_{FQWSL})+T_{926}(n_y,n_z,E_{FQWSL})]}{\sum_{n_y=1}^{n_{y_{\max}}}\sum_{n_z=1}^{n_{z_{\max}}}[\{T_{925}(n_y,n_z,E_{FQWSL})\}'+\{T_{926}(n_y,n_z,E_{FQWSL})\}']}$$

(36)

The Diffusivity Mobility Ratio in Quantum Wire II-VI Effective Mass Superlattices

Following Sasaki [15], the electron dispersion law in II-VI EMSLs can be written as

$$k_z^2=[\frac{1}{L_0^2}\{\cos^{-1}(f_1(E,k_x,k_y))\}^2-k_s^2] \tag{37}$$

in which,, $f_1(E,k_x,k_y)=a_3\cos[a_0C_2(E,k_s)+b_0D_2(E,k_s)]-a_4\cos[a_0C_2(E,k_s)-b_0D_2(E,k_s)]$, $k_s^2=k_x^2+k_y^2$

$$a_3=[\sqrt{\frac{m_2^*}{m_{\|,1}^*}}+1]^2[4(\frac{m_2^*}{m_{\|,1}^*})^{1/2}]^{-1}, a_4=[-1+\sqrt{\frac{m_2^*}{m_{\|,1}^*}}]^2[4(\frac{m_2^*}{m_{\|,1}^*})^{1/2}]^{-1}$$

and $D_2(E,k_s)\equiv[(\frac{2m_2^*}{\hbar^2})EG(E,E_{g_2},\Delta_2)-k_s^2]^{1/2}$

The dispersion law in II-VI, EMQWSLs can be written following equation (37) as

$$k_z^2=[\rho_{13}(n_x,n_y,E)] \tag{38}$$

in which, $\rho_{13}(n_x,n_y,E)=\frac{1}{L_0^2}\left[\cos^{-1}(f_{13}(n_x,n_y,E))\right]^2-\{\phi(n_x,n_y)\}$,

$f_{13}(n_x,n_y,E)=a_9\cos[a_0C_6(n_x,n_y,E)+b_0D_6(n_x,n_y,E)]$
$-a_{10}\cos[a_0C_6(n_x,n_y,E)-b_0D_6(n_x,n_y,E)]$,

$$C_6(n_x,n_y,E)\equiv(\frac{2m^*_{\parallel,1}}{\hbar^2})^{1/2}[E-\{\frac{\hbar^2}{2m^*_{\perp,1}}\phi(n_x,n_y)\}\mp\bar{\lambda}_0\{\phi(n_x,n_y)\}^{1/2}]^{1/2}$$

and

$$D_6(n_x,n_y,E)\equiv[(\frac{2m^*_2}{\hbar^2})EG(E,E_{g_2},\Delta_2)-\phi(n_x,n_y)]^{1/2}$$

The EMM in this case assumes the form

$$m^*(n_x,n_y,E)=\frac{\hbar^2}{2}\{\rho_{13}(n_x,n_y,E)\}' \quad (39)$$

Thus, we observe that the EMM in this case depends on both the Fermi energy and size quantum numbers, which is the intrinsic property of such SLs.

The subband energies $(E_{n_{SL14}})$ can be written as

$$\rho_{13}(n_x,n_y,E)=0 \quad (40)$$

The electron concentration in this case can be expressed as

$$n_{1D}=(\frac{2g_v}{\pi})\sum_{n_x=1}^{n_{x_{\max}}}\sum_{n_y=1}^{n_{y_{\max}}}[T_{927}(n_x,n_y,E_{FQWSL})+T_{928}(n_x,n_y,E_{FQWSL})] \quad (41)$$

where, $T_{927}(n_x,n_y,E_{FQWSL})\equiv[\rho_{13}(n_x,n_y,E_{FQWSL})]^{1/2}$

and $T_{928}(n_x,n_y,E_{FQWSL})\equiv\sum_{r=1}^{s}L(r)[T_{927}(n_x,n_y,E_{FQWSL})]$

Thus, using equation (6b) and (41) leads to the expression of the DMR as

$$\frac{D}{\mu}=\frac{1}{|e|}\frac{\sum_{n_x=1}^{n_{x\max}}\sum_{n_y=1}^{n_{y\max}}[T_{927}(n_x,n_y,E_{FQWSL})+T_{928}(n_x,n_y,E_{FQWSL})]}{\sum_{n_x=1}^{n_{x\max}}\sum_{n_y=1}^{n_{y\max}}[\{T_{927}(n_x,n_y,E_{FQWSL})\}'+\{T_{928}(n_x,n_y,E_{FQWSL})\}']} \tag{42}$$

The Diffusivity Mobility Ratio in Quantum Wire IV-VI Effective Mass Superlattices

Following Sasaki [15], the electron dispersion law in IV-VI, EMSLs can be written as

$$k_x^2=[\frac{1}{L_0^2}\{\cos^{-1}(f_2(E,k_y,k_z))\}^2-k_\perp^2] \tag{43}$$

in which,

$f_2(E,k_y,k_z)=a_5\cos[a_0C_3(E,k_y,k_z)+b_0D_3(E,k_y,k_z)]-a_6\cos[a_0C_3(E,k_y,k_z)$
$-b_0D_3(E,k_y,k_z)]$,

$a_5=[\sqrt{\frac{m_2^*}{m_1^*}}+1]^2[4(\frac{m_2^*}{m_1^*})^{1/2}]^{-1}$,

$m_i^*=[\frac{2\hbar^2}{\{a_i^2-b_i^2\}}][a_i+[a_iC_i+a_ie_iE_{g_i}-e_i^2E_{g_i}][E_{g_i}^2a_i^2+C_i^2+e_i^2E_{g_i}^2+2C_ie_iE_{g_i}-2E_{g_i}a_iC_i$
$-2e_ia_iE_{g_i}^2]^{1/2}]$

$C_3(E,k_y,k_z)\equiv[[EH_{11}+H_{21}(k_y,k_z)]-[E^2H_{31}+EH_{41}(k_y,k_z)+H_{51}(k_y,k_z)]^{1/2}]^{1/2}$,

$D_3(E,k_y,k_z)\equiv[[EH_{12}+H_{22}(k_y,k_z)]-[E^2H_{32}+EH_{42}(k_y,k_z)+H_{52}(k_y,k_z)]^{1/2}]^{1/2}$

and

$$a_6=[-1+\sqrt{\frac{m_2^*}{m_1^*}}]^2[4(\frac{m_2^*}{m_1^*})^{1/2}]^{-1}$$

The dispersion law in IV-VI, EMQWSLs can be written following equation (43) as

$$k_x^2 = [\rho_{14}(n_y, n_z, E)] \tag{44}$$

in which, $\rho_{14}(n_y, n_z, E) = \frac{1}{L_0^2}[\cos^{-1}(f_{14}(n_y, n_z, E))]^2 - \{\phi(n_y, n_z)\}$,

$f_{14}(n_y, n_z, E) = a_9 \cos[a_0 C_7(n_y, n_z, E) + b_0 D_7(n_y, n_z, E)] - a_{10} \cos[a_0 C_7(n_y, n_z, E) - b_0 D_7(n_y, n_z, E)]$,

$C_7(n_y, n_z, E) \equiv [[EH_{11} + H_{21}(n_y, n_z)] - [E^2 H_{31} + EH_{41}(n_y, n_z) + H_{51}(n_y, n_z)]^{1/2}]^{1/2}$

$D_7(n_y, n_z, E) \equiv [[EH_{12} + H_{22}(n_y, n_z)] - [E^2 H_{32} + EH_{42}(n_y, n_z) + H_{52}(n_y, n_z)]^{1/2}]^{1/2}$

The EMM in this case assumes the form

$$m^*(n_y, n_z, E_{FQWSL}) = \frac{\hbar^2}{2}\{\rho_{14}(n_y, n_z, E_{FQWSL})\}' \tag{45}$$

Thus, we observe that the EMM in this case depends on both the Fermi energy and size quantum numbers, which is the intrinsic property of such SLs.

The subband energies $(E_{n_{SL15}})$ can be written as

$$\rho_{14}(E_{n_{SL15}}, n_y, n_z) = 0 \tag{46}$$

The electron concentration in this case can be expressed as

$$n_{1D} = \left(\frac{2g_v}{\pi}\right)\sum_{n_y=1}^{n_{y_{\max}}}\sum_{n_z=1}^{n_{z_{\max}}} [T_{929}(n_y, n_z, E_{FQWSL}) + T_{930}(n_y, n_z, E_{FQWSL})] \tag{47}$$

where, $T_{929}(n_y, n_z, E_{FQWSL}) \equiv [\rho_{14}(n_y, n_z, E_{FQWSL})]^{1/2}$

and $T_{930}(n_y, n_z, E_{FQWSL}) \equiv \sum_{r=1}^{s} L(r)[T_{929}(n_y, n_z, E_{FQWSL})]$

Thus, using equation (6b) and (47) leads to the expression of the DMR as

$$\frac{D}{\mu}=\frac{1}{|e|}\frac{\sum_{n_y=1}^{n_{y_{\max}}}\sum_{n_z=1}^{n_{z_{\max}}}[T_{929}(n_y,n_z,E_{FQWSL})+T_{930}(n_y,n_z,E_{FQWSL})]}{\sum_{n_y=1}^{n_{y_{\max}}}\sum_{n_z=1}^{n_{z_{\max}}}\{T_{929}(n_y,n_z,E_{FQWSL})\}'+\{T_{930}(n_y,n_z,E_{FQWSL})\}'} \quad (48)$$

The Diffusivity Mobility Ratio in Quantum Wire HgTe/CdTe Effective Mass Superlattices

Following Sasaki [15], the electron dispersion law in HgTe/CdTe EMSLs can be written as

$$k_x^2=[\frac{1}{L_0^2}\{\cos^{-1}(f_3(E,k_y,k_z))\}^2-k_\perp^2] \quad (49)$$

in which, $f_3(E,k_\perp)=a_7\cos[a_0C_4(E,k_\perp)+b_0D_4(E,k_\perp)]$
$-a_8\cos[a_0C_4(E,k_\perp)-b_0D_4(E,k_\perp)],$

$$a_1=[\sqrt{\frac{m_2^*}{m_1^*}}+1]^2[4(\frac{m_2^*}{m_1^*})^{1/2}]^{-1}, a_2=[-1+\sqrt{\frac{m_2^*}{m_1^*}}]^2[4(\frac{m_2^*}{m_1^*})^{1/2}]^{-1}, C_4(E,k_\perp)$$
$$\equiv[\frac{B_0^2+2AE-B_0\sqrt{B_0^2+4AE}}{2A^2}-k_\perp^2]^{1/2}$$

and $D_4(E,k_\perp)\equiv[(\frac{2m_2^*E}{\hbar^2})G(E,E_{g_2},\Delta_2)-k_\perp^2]^{1/2}$

The dispersion law in HgTe/CdTe, EMQWSLs can be written following equation (49) as

$$k_x^2 = [\rho_{15}(n_y, n_z, E)] \tag{50}$$

in which, $\rho_{15}(n_y, n_z, E) = \frac{1}{L_0^2}[\cos^{-1}(f_{15}(n_y, n_z, E))]^2 - \{\phi(n_y, n_z)\}$,

$f_{15}(n_y, n_z, E) = a_{11}\cos[a_0 C_8(n_y, n_z, E) + b_0 D_8(n_y, n_z, E)] - a_{12}\cos[a_0 C_8(n_y, n_z, E) - b_0 D_8(n_y, n_z, E)]$,

$C_8(n_y, n_z, E) \equiv [\frac{B_0^2 + 2AE - B_0\sqrt{B_0^2 + 4AE}}{2A^2} - \{\phi(n_y, n_z)\}]^{1/2}$

and $D_8(n_y, n_z, E) \equiv [(\frac{2m_2^* E}{\hbar^2})G(E, E_{g_2}, \Delta_2) - \{\phi(n_y, n_z)\}]^{1/2}$

The EMM in this case assumes the form

$$m^*(n_y, n_z, E_{FQWSL}) = \frac{\hbar^2}{2}\{\rho_{15}(n_y, n_z, E_{FQWSL})\}' \tag{51}$$

The EMM in HgTe/CdTe QWEMSLs depends on both the Fermi energy and size quantum numbers.

The subband energies $(E_{n_{SL16}})$ can be written as

$$\rho_{15}(n_y, n_z, E_{n_{SL16}}) = 0 \tag{52}$$

The electron concentration in this case can be expressed as

$$n_{1D} = (\frac{2g_v}{\pi})\sum_{n_y=1}^{n_{y\max}}\sum_{n_z=1}^{n_{z\max}}[T_{931}(n_y, n_z, E_{FQWSL}) + T_{932}(n_y, n_z, E_{FQWSL})] \tag{53}$$

where, $T_{931}(n_y, n_z, E_{FQWSL}) \equiv [\rho_{15}(n_y, n_z, E_{FQWSL})]^{1/2}$

$$T_{932}(n_y, n_z, E_{FQWSL}) \equiv \sum_{r=1}^{s} L(r)[T_{931}(n_y, n_z, E_{FQWSL})]$$

The use of equations (6b) and (53) leads to the expression of the DMR as

$$\frac{D}{\mu}=\frac{1}{|e|}\frac{\sum_{n_y=1}^{n_{y_{\max}}}\sum_{n_z=1}^{n_{z_{\max}}}[T_{931}(n_y,n_z,E_{FQWSL})+T_{932}(n_y,n_z,E_{FQWSL})]}{\sum_{n_y=1}^{n_{y_{\max}}}\sum_{n_z=1}^{n_{z_{\max}}}[\{T_{931}(n_y,n_z,E_{FQWSL})\}'+\{T_{932}(n_y,n_z,E_{FQWSL})\}']} \tag{54}$$

RESULT AND DISCUSSION

Using the appropriate equations we have plotted the normalized 1D DMR as function of film thickness for GaAs/$Ga_{1-x}Al_xAs$, CdS/CdTe, PbTe/PbSnTe and HgTe/CdTe quantum wire superlattices with graded interfaces as shown by curves (a), (b), (c) and (d) of Figure 1. In Figure 2, we have plotted all cases of Figure 1 as function of electron concentration per unit length. It appears from both the Figures 1 and 2 that the DMR increases with decreasing film thickness and increasing electron concentration per unit length respectively, although the numerical values are totally band structure dependent. Using the appropriate equations we have plotted the normalized 1D DMR in quantum wire effective mass superlattices of the aforementioned materials as functions of film thickness and electron concentration per unit length as shown by curves (a), (b), (c) and (d) in Figures 3 and 4 respectively. It appears that the DMR in this case decreases with film thickness and increases with electron concentration for all the cases.

It appears from the said figures that the 1D DMR decreases with the increasing film thickness in a step like manner as considered here although the numerical values vary widely and determined by the constants of the energy spectra. The oscillatory dependence is due to the crossing over of the Fermi level by the quantized levels. For each coincidence of a quantized level with the Fermi level, there would be a discontinuity in the density-of-states function resulting in a peak of oscillations. With large values of film thickness, the height of the steps decreases and the DMR decreases with increasing film thickness in non-oscillatory manner and exhibit monotonic decreasing dependence. The height of step size and the rate of decrement are totally dependent on the band structure.

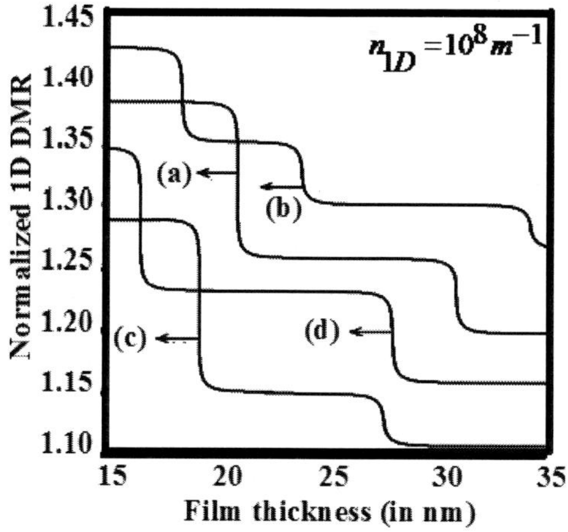

Figure 1. The plot of the 1D DMR as a function of film thickness for (a) GaAs/ $Ga_{1-x}Al_xAs$ (b) CdS/CdTe (c) PbTe/PbSnTe and (d) HgTe/CdTe superlattices with graded interfaces.

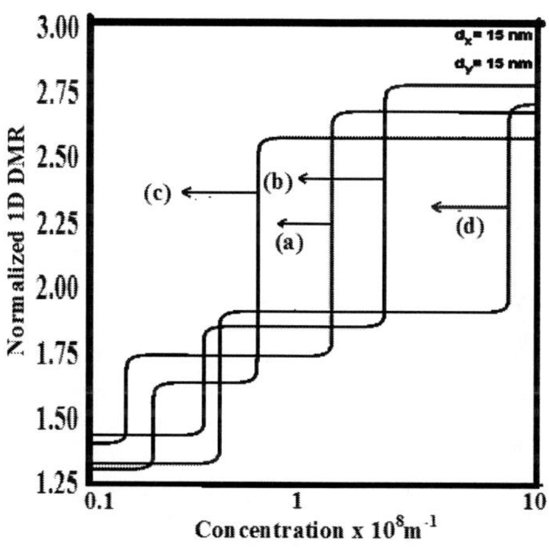

Figure 2. The plot of the 1D DMR as a function of surface electron concentration for (a) $GaAs/Ga_{1-x}Al_xAs$ (b) CdS/CdTe (c) PbTe/PbSnTe and (d) HgTe/CdTe superlattices with graded interfaces.

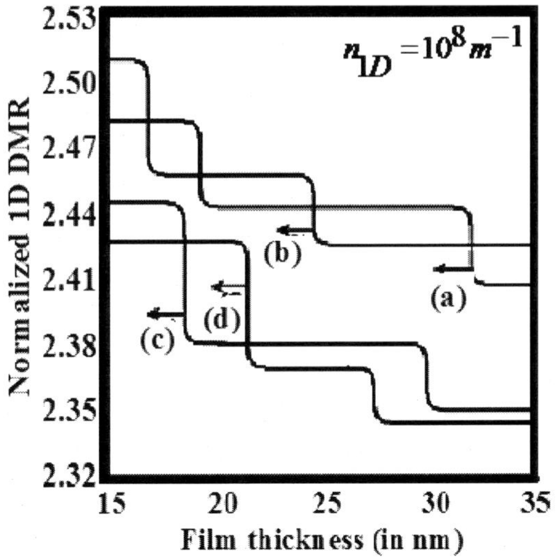

Figure 3. The plot of the 1D DMR as a function of film thickness for (a) $GaAs/Ga_{1-x}Al_xAs$ (b) CdS/CdTe (c) PbTe/PbSnTe and (d) HgTe/CdTe effective mass superlattices.

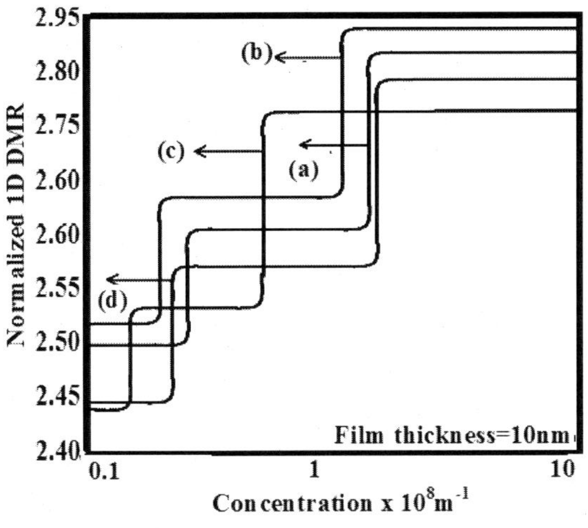

Figure 4. The plot of the 1D DMR as a function of surface electron concentration for (a) $GaAs/Ga_{1-x}Al_xAs$ (b) CdS/CdTe (c) PbTe/PbSnTe and (d) HgTe/CdTe effective mass superlattices.

REFERENCES

[1] Keldysh, LV. *Soviet Phys. Solid State*, 4, 1658, (1962).

[2] Esaki, L; Tsu, R. *IBM J. Research and Develop.*, 14, 61, (1970).

[3] Tsu, R. *Superlattices to nanoelectronics*, Elsevier, (2005). Ivchenko, EL; Pikus, G. *Superlattices and other heterostructures*, Springer-Berlin, (1995). Bastard, G. *Wave mechanics applied to heterostructures*, Editions de Physique, Les Ulis, France, (1990).

[4] Bose, PK. *Theory of optical process in semiconductors, bulk and microstructures*, Oxford University Press, (1997).

[5] Ghatak, KP; Mitra, B. *Int. J. Electronics*, 72, 541, (1992). Biswas, SN; Ghatak, KP. *Int. J. Electronics*, 70, 125, (1991). Mitra, B; Ghatak, KP. *Phys. Stat. Sol.* (b), 149, K117, (1988). Ghatak, KP; Mitra, B; Ghoshal, A. *Phys. Stat. Sol.* (b), 154, K121, (1989).

[6] Vaidyanathan, KV; Jullens, RA; Anderson, CL; Dunlap, HL. *Solid State Electron.*, 26, 717, (1983).

[7] Wilson, BA. *IEEE, J. Quantum Electron.*, 24, 1763, (1988).

[8] Krichbaum, M; Kocevar, P; Pascher, H; Bauer, G. *IEEE, J. Quantum Electron.*, 24, 717, (1988).

[9] Schulman, JN; McGill, TC. *Appl. Phys. Letts.*, 34, 663, (1979).

[10] Kinoshita, H; Sakashita, T; Fajiyasu, H. *J. Appl. Phys.*, 52, 2869, (1981).

[11] Ghenin, L; Mani, RG; Anderson, JR; Cheung, JT. *Phys. Rev. B*, 39, 1419, (1989).

[12] Hoffman, CA; Mayer, JR; Bartoli, FJ; Han, JW; Cook, JW; Schetzina, JF; Schubman, JM. *Phys. Rev. B.*, 39, 5208, (1989).

[13] Yakovlev, VA. *Sov. Phys. Semicon.*, 13, 692, (1979).

[14] Kane, EO. *J. Phys. Chem. Solids*, 1, 249, (1957).

[15] Sasaki, H. *Phys. Rev. B*, 30, 7016, (1984).

[16] Kroemer, H. *IEEE Trans, Electron Devices*, 25, 850, (1978).

[17] Einstein, A. *Ann. der Physik*, 17, 549, (1905). Nernst, WZ. *Phys. Chem. Leipzig*, 2, 613, (1888). Townsend, JS. *Trans. Roy. Soc.*, 193A, 129, (1900).

[18] Wagner, C; Physik, Z. *Chem. B*, 21, 24, (1933). Herring, C; Nichols, MH. *Rev. Mod. Phys.*, 21, 185, (1949). Landsberg, PT. *Thermodynamics and Statistical Mechanics*, Oxford University Press, Oxford (1978); *In Recombination in Semiconductors*, Cambridge University Press, U.K. (1991).

[19] Lade, RW. *Proc. IEEE*, 52, 743, (1965).

[20] Landsberg, PT. *Proc R. Soc. A*, 213, 226, (1952). *Proc. of Phys. Soc. A*, 62, 806, (1949).

[21] Landsberg, PT. *Eur. J. Phys*, 2, 213, (1981).

[22] Wang, CH; Neugroschel, A. *IEEE Electron. Dev. Lett.*, ED-11, 576, (1990).

[23] Leu, IY; Neugroschel, A. *IEEE Trans. Electron. Dev.*, ED-40, 1872, (1993).

[24] Stengel, F; Mohammad, SN; Morkoç, H. *J. Appl. Phys.*, 80, 3031, (1996).

[25] Pan, HJ; Wang, WC; Thai, KB; Cheng, CC; Yu, KH; Lin, KW; Wu, CZ; Liu, WC. *Semicond. Sci. Technol.*, 15, 1101, (2000).

[26] Mohammad, SN. *J. Appl. Phys.*, 95, 4856, (2004).

[27] Arora, VK. *Appl. Phys., Lett.*, 80, 3763, (2002).

[28] Mohammad, SN. *J. Appl. Phys.*, 95, 7940, (2004).

[29] Mohammad, SN. *Philos. Mag.*, 84, 2559, (2004).

[30] Mohammad, SN. *J. Appl. Phys.*, 97, 063703, (2005).

[31] Suzue, K; Mohammad, SN; Fan, ZF; Kim, W; Aktas, O; Botchkarev, AE; Morkoç, H. *J. Appl. Phys.*, 80, 4467, (1996).

[32] Mohammad, SN; Fan, ZF; Kim, W; Aktas, O; Botchkarev, AE; Salvador, A; Morkoç, H. *Electron. Lett.*, 32, 598, (1996).

[33] Fan, Z; Mohammad, SN; Kim, W; Aktas, O; Botchkarev, AE; Suzue, K; Morkoç, H. *J. Electron. Mater.*, 25, 1703, (1996).

[34] Lu, C; Chen, H; Lv, X; Xia, X; Mohammad, SN. *J. Appl. Phys.*, 91, 9216, (2002).

[35] Dmitriev, SG; Yu, V. Markin. *Semiconductors*, 34, 931. (2000).

[36] Tao, M; Park, D; Mohammad, SN; Li, D; Botchkerav, AE; Morkoç, H. *Philos. Mag. B*, 73, 723, (1996).

[37] Park, DG; Tao, M; Li, D; Botchkarev, AE; Fan, Z; Mohammad, SN; Morkoç, H. *J. Vac. Sci. Technol. B*, 14, 2674, (1996).
[38] Chen, Z; Park, DG; Mohammad, SN; Morkoç, H. *Appl. Phys. Lett.*, 69, 230, (1996).
[39] Landsberg, PT. *J. of Appl. Phys.*, 56, 1696, (1984). Landsberg, PT; Guy, AG. *Phys. Rev. B*, 28, 1187, (1983). Landsberg, PT. *Phys. Rev. B*, 33, 8321, (1986). Hope, SA; Feat, G; Landsberg, PT. *J. Appl. Phys.*, 61, 4909, (1987). Roichman, Y; Tessler, N. *Appl. Phys. Lett.*, 80, 1948, (2002). Peters, JMH. *Eur. J. Phys.*, 3, 19, (1982). Van Cong, H; Brunet, S; Charar, S. *Phys. Stat. Solidi B*, 109, K1, (1982). Van Cong, H. *Phys. Stat. Solidi A*, 56, 395, (1979). Van Cong, H. *Solid State Electron.*, 24, 495, (1981).
[40] Mohammad, SN; Abidi, STH. *J. Appl. Phys.*, 61, 4909, (1987). *Solid State Electron.*, 27, 1153, (1985). *J. Appl. Phys.*, 56, 3341, (1984). Sobhan, MA; Mohammad, SN. *J. Appl. Phys.*, 58, 2634, (1985). Mohammad, SN; Bemis, AV. *IEEE Trans. Electron. Dev.*, ED-39, 282, (1992). Mohammad, SN; Carter, RL. *Philos. Mag. B*, 72, 13, (1995). Mohammad, SN. *Solid State Electron.*, 46, 203, (2002). Mohammad, SN; Chen, J; Chyi, JI; Morkoç, H. *Appl. Phys. Lett.*, 56, 937, (1990).
[41] Landsberg, PT; Hope, SA. *Solid State Electronics*, 20, 421, (1977). Hope, SA; Feat, G; Landsberg, PT. *J. Phys. A. math. Gen.*, 14, 2377, (1981).
[42] Elsäber, W; Göbel, EO. *Electron. Lett.*, 19, 335, (1983). Hilfer, R; Blumen, A. *Phys. Rev. A*, 37, 578, (1988). Castner, TG. *Phys. Rev. B*, 55, 4003, (1997). Barkai, E; Fleurov, VN. *Phys. Rev. E.*, 58, 1296, (1998). Roichman, YY; Tessler, N. *Appl. Phys. Lett.*, 80, 1948, (2002). Nguyen, TH; O'Leary, SK. *Appl. Phys. Lett.*, 83, 1998, (2003). Nguyen, TH; O'Leary, SK. *J. Appl. Phys.*, 98, 076102, (2005). Rodrigues, CG; Vasconcellos, ÁR; Luzzi, R. *J. Appl. Phys.*, 99, 073701, (2006).
[43] Jain, RK. *Phys. Stat. Sol.*, (a) 42, K221, (1977). Aronzon, BA; Meilikhov, EZ. *Phys. Stat. Sol.*, (a) 19, 313, (1973).

[44] Chakravarti, AN; Nag, BR. *Int. J. Elect.*, 37, 281 (1974). Butcher, PN; Chakravarti, AN; Swaminathan, B. *Phys. Stat. Sol.*, (a) 2219, K153, (1974).

[45] Nag, BR; Chakravarti, AN. *Solid State Electron.*, 18, 109 (1975), *Phys. Stal Sol.*, (a) 22, K153, (1974).

[46] Nag, BR; Chakravarti, AN; Basu, PK. *Phys. Stat. Sol.*, (a) 68, K75, (1981).

[47] Nag, BR; Chakravarti, AN. *Phys. Stal Sol.* (a) 67, K113 (1981).

[48] Chakravarti, AN; Parui, DP. *Phys. Letts*, 40A, 113, (1972), 43A, 60 (1973). *Phys. Stat. Sol.* (a) 14, K23 (1972). *Phys. Stat. Sol.* (a) 14, K55 (1972). Chakravarti, AN; Parui, DP. *Canad. J. Phys.*, 51, 451, (1973). Mukherjee, D; Chakravarti, AN; Nag, BR. *Phys. Stat. Sol* (a) 26, K27 (1974). Ghosh, S; Chakravarti, AN. *Phys. Stat. Sol.* (b), 147, 355, (1988).

[49] Choudhury, S; De, D; Mukherjee, S; Neogi, A; Sinha, A; Pal, M; Biswas, SK; Pahari, S; Bhattacharya, S; Ghatak, KP. *J. Comp. Theo. Nanosc.*, 5, 375, (2008). Mukherjee, S; Mitra, SN; Bose, PK; Ghatak, AR; Neogi, A; Banerjee, JP; Sinha, A; Pal, M; Bhattacharya, S; Ghatak, KP. *J. Comp. Theo. Nanosc.*, 4, 550, (2007). Khan, A; Das, A. *Appl. Phys. A*, 89¸ 695, (2007).

[50] Bouchaud, JP; Georges, A. *Phys. Rep.*, 195, 127 (1996). Blickle, V; Speck, T; Lutz, C; Seifert, U; Bechinger, C. *Phys. Rev. Lett.*, 98, 210601, (2007). Kang, Y; Jean, E; Fortmonn, CM. *Appl. Phys. Lett.*, 88, 112110, (2006). Neumann, F; Genenko, YA; Seggern, HV. *J. Appl. Phys.*, 99, 013704, (2006). van de Lagemaat, J. *Phys. Rev. B.*, 73, 235319, (2005). Gu, Q; Schiff, EA; Grneber, S; Wang, F; Schwarz, R. *Phys. Rev. Lett.*, 76, 3196, (1996). Azbel, MY. *Phys. Rev. B.*, 46, 15004, (1992).

[51] Marshak, AH. *Solid State Electron.*, 30, 1089, (1987). Marshak, AH; Vliet, CMV. *Proc. IEEE*, 72, 148, (1984). Vliet, CMV; van der Zeil, A. *Solid State Electron.*, 20, 931, (1977).

[52] Chakravarti, AN; Ghatak, KP; Dhar, A; Ghosh, KK; Ghosh, S. *Appl. Phys. A*, 26, 169, (1981).

[53] Nag, BR. *Electron Transport in Compound Semiconductor*, Springer-Verlag, Germany, (1980). Madelung, O. *Semiconductors: Data handbook*, 3rd Ed. Springer, (2004). Krieehbaum, M; Kocevar, P; Pascher, H; Bauer, G. *IEEE QE*, 24, 1727, (1988).

[54] Ghatak, KP; Ghoshal, A; Biswas, SN. *Nouvo Cimento*, 15D, 39, (1993). Ghatak, KP; Biswas, SN. *Nanostructured Materials*, 2, 91, (1993). Ghatak, KP; Biswas, SN. *J. of Appl. Phys.*, 70, 4309, (1991). Ghatak, KP; Mitra, B; Mondal, M. *Ann. der Physik*, 48, 283, (1991). Mitra, B; Ghatak, KP. *Physica Scripta*, 42, 103, (1990). Mitra, B; Ghatak, KP. *in Phys. Letts.*, 135A, 397, (1989). Mondal, M; Ghatak, KP. *Ann. der Physik*, 46, 502, (1989). Ghatak, KP; Chattopadhyay, N; Mondal, M. *J. Appl. Phys.*, 63, 4536, (1988). *J. Low Temp. Phys.*, 73, 321, (1988).

[55] Ghatak, KP; Ghoshal, A; Biswas, SN. *Nouvo Cimento*, 15D, 39, (1993).

[56] Ghatak, KP; Bhattacharyya, D. *Phys. Letts. A*, 184, 366, (1994).

[57] Ghatak, KP; Bhattacharyya, D. *Physica Scripta*, 52, 343, (1995). Mondal, M; Ghatak, KP. *J. Mag. Mag. Materials*, 62, 115, (1986).

[58] Ghatak, KP; Nag, B; Bhattacharyya, D. *J. of Low Temp. Phys.*, 14, 1, (1995).

[59] Ghatak, KP; Mondal, M. *Thin Solid Films*, 148, 219, (1987).

[60] Ghatak, KP; Choudhury, AK; Ghosh, S; Chakravarti, AN. *Appl. Phys.*, 23, 241, (1980).

[61] Ghatak, KP. *Influence of Band Structure on Some Quantum Processes in Tetragonal Semiconductors, D. Eng. Thesis*, (1991), Jadavpur University, Kolkata, India.

[62] Ghatak, KP; Chattropadhyay, N; Mondal, M. *Appl. Phys. A*, 44, 305, (1987).

[63] Biswas, SN; Ghatak, KP. Proceedings of the Society of Photo-optical and Instrumentation Engineers (SPIE), *Quantum Well and Superlattice Physics*, USA, (1987), Vol. 792, p. 239. Ghatak, KP; Mondal, M; Bhattacharyya, S. *SPIE*, USA, (1990), Vol. 1284, p. 113. Ghatak, KP; Bhattacharyya, S; Mondal, M. *SPIE*, USA, (1990), Vol. 1307, p. 588. Ghatak, KP; De, B. Defect Engineering in

Semiconductor Growth, *Processing and Device technology Materials Research Society Proceedings*, (MRS) Spring meeting, (1992), Vol. 262, p. 911. Ghatak, KP; De, B. The Proc. of Wide Band Gap Semiconductors Symposium, *MRS*, (1992) p. 377. Bhattacharya, S; Ghatak, KP; Biswas, SN. *SPIE*, 836, Optoelectronic materials, Devices, Packaging Interconnects, pp. 72-87, (1988).

[64] Mondal, M; Ghatak, KP. *J. Phys. C, (Sol. State.)*, 20, 1671, (1987). Mondal, M; Banik, SN; Ghatak, KP. *Canad. J. Phys.*, 67, 72, (1989).

[65] Ghatak, KP; Mondal, M. *J. Appl. Phys.*, 70, 1277, (1992).

In: Quantum Wires: An Overview
Editor: Kamakhya Prasad Ghatak
ISBN: 978-1-53617-676-6

Chapter 5

THE MAGNETO THERMOELECTRIC POWER IN QUANTUM WIRES IN THE PRESENCE OF MICROWAVE RADIATION

M. Mitra[1,], R. Paul[2] and S. Das[3]*
[1]Department of Electronics and Telecommunication Engineering, IIEST, Shibpur, Botanic Garden, Howrah, India
[2]Department of Computer Science and Engineering, University of Engineering and Management, University Area, Kolkata, India
[3]Department of Applied Physics, University of Calcutta, Kolkata, India

ABSTRACT

With the advent of Quantum Hall Effect there has been considerable interest in studying the Magneto Thermoelectric power (MTP) in low dimensional materials whose wave vector space are being quantized forming 1D, 2D and 0D systems. In this chapter an attempt is made to

---

* Corresponding Author's Email: monojit_m1@yahoo.co.in.

study the MTP in quantum wires of optoelectronic materials in the presence of microwave radiation by formulating the new dispersion relations for such technologically important compounds. We have investigated MTP taking quantum wires of $InSb, GaAs, Hg_{1-x}Cd_xTe$ and $In_{1-x}Ga_xAs_yP_{1-x}$ as functions of film thickness, electron concentration, light intensity and wavelength respectively. It appears that MTP decreases with increasing electron concentration in oscillatory manners. For large values of electron concentration per unit length, the MTP exhibits very quick oscillatory decrement whereas for small values of the carrier degeneracy, the MTP shows the converging tendencies. It may be noted that the MTP increases with increasing light intensity and due to both quantum limits, the variation is again non-oscillatory. Because of the quantum limits, the MTP changes slowly with wavelength in a non-oscillatory manner although the wide difference in the MTP versus wave length plot for the perturbed two and three band models of Kane together with parabolic energy bands exhibit the influence of energy band models.

1. INTRODUCTION

It is well known that the Magneto Thermoelectric Power (MTP) is a very important quantity [1], since the change in entropy (a vital concept in thermodynamics) can be known from this relation by determining the experimental values of the change of electron concentration. The analysis of MTP generates information regarding the effective mass of the carriers in materials which occupies a central position in the whole field of materials science in general [2-4]. The classical MTP (G_0) equation is valid only under the non-degenerate carrier concentration and the magnitude of the MTP is given by $G_0 = (\pi^2 k_B / 3e)$ (k_B and e are Boltzmann's constant and the magnitude of the carrier charge, respectively; [2-4]). From this equation, it is readily inferred that this conventional form is a function of three fundamental constants only, being independent of the signature of the charge carriers in materials.

In recent years, with the advent of Quantum Hall Effect (QHE) [5], there has been considerable interest in studying the thermoelectric power under strong magnetic field in various types of nano-structured materials having quantum confinement of their charge carriers in one, two and three dimensions of the respective wave vector space leading to different carrier energy spectra [6-31]. The MTP can also be connected to the Debye screening length and the carrier contribution to the elastic constants, the two important physical quantities often used in the analysis of the carrier transport and the mechanical properties of technologically important optoelectronic nano structured devices. The classical MTP equation as mentioned earlier is valid only under the condition of carrier non-degeneracy, being independent of carrier concentration and reflects the fact that the signature of the band structure of any material is totally absent in the same.

The significant work of Zawadzki [4] reflects the fact that the MTP for materials having degenerate electron concentration is independent of scattering mechanisms and is exclusively determined by the dispersion laws of the respective carriers. It will, therefore, assume different values for different systems and varies with the doping, the magnitude of the reciprocal quantizing magnetic field under magnetic quantization, the nano thickness in ultrathin films, quantum wires and dots, the quantizing electric field as in inversion layers, the carrier statistics in various types of quantum confined superlattices having different carrier energy spectra and other types of low-dimensional field assisted systems.

With the advent of semiconductor optoelectronics, there has been a considerable interest in studying the optical processes in semiconductors and their nanostructures [1]. It appears from the literature that the investigations have been carried out on the assumption that the carrier energy spectra are invariant quantities in the presence of intense light waves, which is not fundamentally true. The physical properties of semiconductors in the presence of light waves which change the basic dispersion relation have been relatively less investigated in the literature [2-4]. In this chapter, we shall study the magneto thermoelectric power under micro wave radiation in in NWs of III-V, ternary and quaternary materials on the basis of electron dispersion laws.

2. THEORETICAL BACKGROUND

2.1. The MTP in III-V Quantum Wire HD Superlattices with Graded Interfaces in the Presence of Microwave Radiation

The Hamiltonian $(\hat{H})$ of an electron in the presence of light wave characterized by the vector potential $\vec{A}$ can be written as

$$\hat{H} = [\left|(\hat{p} + |e|\vec{A})\right|^2 / 2m] + V(\vec{r}) \tag{1}$$

in which, $\hat{p}$ is the momentum operator, $V(\vec{r})$ is the crystal potential and m is the free electron mass.

Equation (1) can be expressed as

$$\hat{H} = \hat{H}_0 + \hat{H}' \tag{2}$$

where, $\hat{H}_0 = \dfrac{\hat{p}^2}{2m} + V(\vec{r})$

and

$$\hat{H}' = \frac{|e|}{2m}\vec{A} \cdot \hat{p} \tag{3}$$

The perturbed Hamiltonian $\hat{H}'$ can be written as

$$\hat{H}' = \left(\frac{-i\hbar|e|}{2m}\right)(\vec{A} \cdot \nabla) \tag{4}$$

where, $i = \sqrt{-1}$ *and* $\hat{p} = -i\hbar\nabla$

The vector potential ($\vec{A}$) of the monochromatic light of plane wave can be expressed as

$$\vec{A} = A_0 \vec{\varepsilon}_s \cos(\vec{s}_0 \cdot \vec{r} - \omega t) \tag{5}$$

where A_0 is the amplitude of the light wave, $\vec{\varepsilon}_s$ is the polarization vector, $\vec{s}_0$ is the momentum vector of the incident photon, $\vec{r}$ is the position vector, ω is the angular frequency of light wave and t is the time scale.

The matrix element of $\hat{H}'_{nl}$ between initial state, $\psi_1(\vec{q},\vec{r})$ and final state $\psi_n(\vec{k},\vec{r})$ in different bands can be written as

$$\hat{H}'_{nl} = \frac{|e|}{2m} \left\langle n\vec{k} \left| \vec{A} \bullet \hat{p} \right| l\vec{q} \right\rangle \tag{6}$$

Using equations (4) and (5), we can re-write equation (6) as

$$\hat{H}'_{nl} = \left(\frac{-i\hbar |e| A_0}{4m}\right) \vec{\varepsilon}_s [\{ \left\langle n\vec{k} \left| e^{(i\vec{s}_0 \cdot \vec{r})} \nabla \right| l\vec{q} \right\rangle e^{-i\omega t} \} + \{ \left\langle n\vec{k} \left| e^{(-i\vec{s}_0 \cdot \vec{r})} \nabla \right| l\vec{q} \right\rangle e^{i\omega t} \}] \tag{7}$$

The first matrix element of equation (7) can be written as

$$\left\langle n\vec{k} \left| e^{(i\vec{s}_0 \cdot \vec{r})} \nabla \right| l\vec{q} \right\rangle = \int e^{\left(i\left[\vec{q} + \vec{s}_0 - \vec{k}\right] \cdot \vec{r}\right)} i\vec{q} u_n^* \left(\vec{k},\vec{r}\right) u_l \left(\vec{q},\vec{r}\right) d^3 r + \int e^{\left(i\left[\vec{q} + \vec{s}_0 - \vec{k}\right] \cdot \vec{r}\right)} u_n^* \left(\vec{k},\vec{r}\right) \nabla u_l \left(\vec{q},\vec{r}\right) d^3 r \tag{8}$$

The functions $u_n^* u_l$ and $u_n^* \nabla u_l$ are periodic. The integral over all space can be separated into a sum over unit cells times an integral over a single unit cell. It is assumed that the wave length of the electromagnetic wave is sufficiently large so that if $\vec{k}$ and $\vec{q}$ are within the Brillouin zone, $(\vec{q}+\vec{s}_0-\vec{k})$ is not a reciprocal lattice vector.

Therefore, we can write equation (8) as

$$< n\vec{k}\left|e^{(i\vec{s}_0\cdot\vec{r})}\nabla\right|l\vec{q} > = [\frac{(2\pi)^3}{\Omega}]\{i\vec{q}\delta(\vec{q}+\vec{s}_0-\vec{k})\delta_{nl}+\delta(\vec{q}+\vec{s}_0-\vec{k})\int_{cell} u_n^*(\vec{k},\vec{r})\nabla u_l(\vec{q},\vec{r})d^3r\}$$

$$=[\frac{(2\pi)^3}{\Omega}]\{\delta(\vec{q}+\vec{s}_0-\vec{k})\int_{cell} u_n^*(\vec{k},\vec{r})\nabla u_l(\vec{q},\vec{r})d^3r\} \tag{9}$$

where, Ω is the volume of the unit cell and $\int u_n^*(\vec{k},\vec{r})u_l(\vec{q},\vec{r})d^3r=\delta(\vec{q}-\vec{k})\delta_{nl}=0$, since $n\neq l$.

The delta function expresses the conservation of wave vector in the absorption of light wave and $\vec{s}_0$ is small compared to the dimension of a typical Brillouin zone and we set $\vec{q}=\vec{k}$.

From equations (8) and (9), we can write,

$$\hat{H}'_{nl}=\frac{|e|A_0}{2m}\vec{\varepsilon}_s.\hat{p}_{nl}(\vec{k})\delta(\vec{q}-\vec{k})\cos(\omega t) \tag{10}$$

where, $\hat{p}_{nl}(\vec{k})=-i\hbar\int u_n^*\nabla u_l d^3r=\int u_n^*(\vec{k},\vec{r})\hat{p}u_l(\vec{k},\vec{r})d^3r$

Therefore, we can write

$$\hat{H}'_{nl}=\frac{|e|A_0}{2m}\vec{\varepsilon}\bullet\hat{p}_{nl}(\vec{k}) \tag{11}$$

where, $\vec{\varepsilon}=\vec{\varepsilon}_s\cos\omega t$

When a photon interacts with a semiconductor, the carriers (i.e., electrons) are generated in the bands which are followed by the inter band transitions. For example, when the carriers are generated in the valence band, the carriers then make inter band transition to the conduction band.

The transition of the electrons within the same band i.e., $\hat{H}'_{nn} = \langle n\vec{k} \,|\, \hat{H}' \,|\, n\vec{k} \rangle$ is neglected. Because, in such a case, i.e., when the carriers are generated within the same bands by photons, are lost by recombination within the aforementioned band resulting zero carriers.

Therefore,

$$< n\vec{k} \left| \hat{H}' \right| n\vec{k} > = 0 \tag{12}$$

with $n = c$ stands for conduction band and $l=v$ stand for valance band, the energy equation for the conduction electron can approximately be written as

$$\gamma(E) = (\frac{\hbar^2 k^2}{2m^*}) + \frac{(\frac{|e| A_0}{2m})^2 \left\langle \left| \vec{\varepsilon} \bullet \hat{p}_{cv}(\vec{k}) \right|^2 \right\rangle_{av}}{E_c(\vec{k}) - E_v(\vec{k})} \tag{13}$$

where, $\gamma(E) \equiv E(aE+1)(bE+1)/(cE+1), a \equiv 1/E_{g_0}$,

E_{g_0} is the un-perturbed band-gap, $b \equiv 1/(E_{g_0} + \Delta), c \equiv 1/(E_{g_0} + 2\Delta/3)$ *and* $\left\langle \left| \vec{\varepsilon} \bullet \hat{p}_{cv}(\vec{k}) \right|^2 \right\rangle_{av}$ represents the average of the square of the optical matrix element (OME).

For the three-band model of Kane, we can write,

$$\xi_{1k} = E_c(\vec{k}) - E_v(\vec{k}) = (E_{g_0}^2 + E_{g_0} \hbar^2 k^2 / m_r)^{1/2} \tag{14}$$

where, m_r is the reduced mass and is given by $m_r^{-1} = (m^*)^{-1} + m_v^{-1}$, and m_v is the effective mass of the heavy hole at the top of the valance band in the absence of any field.

The doubly degenerate wave functions $u_1(\vec{k},\vec{r})$ and $u_2(\vec{k},\vec{r})$ can be expressed as [6]

$$u_1(\vec{k},\vec{r}) = a_{k+}[(is)\downarrow'] + b_{k+}[\frac{X' - iY'}{\sqrt{2}}\uparrow'] + c_{k+}[Z'\downarrow_{'}] \tag{15}$$

and

$$u_2(\vec{k},\vec{r}) = a_{k-}[(is)\uparrow'] - b_{k-}[\frac{X' + iY'}{\sqrt{2}}\downarrow'] + c_{k-}[Z'\uparrow'] \tag{16}$$

s is the s-type atomic orbital in both unprimed and primed coordinates, $\downarrow'$ indicates the spin down function in the primed coordinates,

$$a_{k\pm} \equiv \beta[E_{g_0} - (\gamma_{0k\pm})^2(E_{g_0} - \delta')]^{1/2}(E_{g_0} + \delta')^{-1/2}], \beta \equiv [(6(E_{g_0} + 2\Delta/3)(E_{g_0} + \Delta))/\chi]^{1/2},$$
$$\chi \equiv (6E_{g_0}^2 + 9E_{g_0}\Delta + 4\Delta^2), \gamma_{0k\pm} \equiv [\frac{(\xi_{1k} \mp E_{g_0})}{2(\xi_{1k} + \delta')}]^{1/2}, \xi_{1k} \equiv E_c(\vec{k}) - E_v(\vec{k}) = E_{g_0}[1 + 2(1 + \frac{m_c}{m_v})\frac{\gamma(E)}{E_{g_0}}]^{1/2},$$

$\delta' \equiv (E_{g_0}^2\Delta)(\chi)^{-1}$, X', Y', and Z' are the p-type atomic orbitals in the primed coordinates, $\uparrow'$ indicates the spin-up function in the primed coordinates, $b_{k\pm} \equiv \rho\gamma_{0k\pm}$, $\rho \equiv (4\Delta^2/3\chi)^{1/2}$, $c_{k\pm} \equiv t\gamma_{0k\pm}$ and

$$t \equiv [6(E_{g_0} + 2\Delta/3)^2/\chi]^{1/2}.$$

We can, therefore, write the expression for the optical matrix element (OME) as

$$OME = \hat{p}_{cv}(\vec{k}) = \langle u_1(\vec{k},\vec{r})|\hat{p}|u_2(\vec{k},\vec{r})\rangle \tag{17}$$

Since the photon vector has no interaction in the same band for the study of inter-band optical transition, we can therefore write

$$\langle S|\hat{p}|S\rangle = \langle X|\hat{p}|X\rangle = \langle Y|\hat{p}|Y\rangle = \langle Z|\hat{p}|Z\rangle = 0$$

and

$$\langle X|\hat{p}|Y\rangle = \langle Y|\hat{p}|Z\rangle = \langle Z|\hat{p}|X\rangle = 0$$

There are finite interactions between the conduction band (CB) and the valance band (VB) and we can obtain

$$\langle S\,|\,\hat{P}\,|\,X\rangle = \hat{i}.\hat{P} = \hat{i}.\hat{P}_x$$
$$\langle S\,|\,\hat{P}\,|\,Y\rangle = \hat{j}.\hat{P} = \hat{j}.\hat{P}_y$$
$$\langle S\,|\,\hat{P}\,|\,Z\rangle = \hat{k}.\hat{P} = \hat{k}.\hat{P}_z$$

where, $\hat{i}, \hat{j}$ and $\hat{k}$ are the unit vectors along x, y and z axes respectively.

It is well known that

$$\begin{bmatrix} \uparrow' \\ \downarrow' \end{bmatrix} \begin{bmatrix} e^{-i\phi/2}\cos(\theta/2) & e^{i\phi/2}\sin(\theta/2) \\ -e^{-i\phi/2}\cos(\theta/2) & e^{i\phi/2}\sin(\theta/2) \end{bmatrix} \begin{bmatrix} \uparrow \\ \downarrow \end{bmatrix}$$

and $$\begin{bmatrix} X' \\ Y' \\ Z' \end{bmatrix} = \begin{bmatrix} \cos\theta\cos\phi & \cos\theta\sin\phi & -\sin\theta \\ -\sin\phi & \cos\phi & 0 \\ \sin\theta\cos\phi & \sin\theta\sin\phi & \cos\theta \end{bmatrix} \begin{bmatrix} X \\ Y \\ Z \end{bmatrix}$$

Besides, the spin vector can be written as

$\vec{S} = \frac{\hbar}{2}\vec{\sigma}$, where, $\sigma_x = \begin{bmatrix} 0 & 1 \\ 1 & 0 \end{bmatrix}$, $\sigma_y = \begin{bmatrix} 0 & -i \\ i & 0 \end{bmatrix}$ and $\sigma_z = \begin{bmatrix} 1 & 0 \\ 0 & -1 \end{bmatrix}$.

From above, we can write

$$\hat{p}_{CV}(\vec{k}) = \left\langle u_1(\vec{k},\vec{r}) \mid \hat{P} \mid u_2(\vec{k},\vec{r}) \right\rangle \Big\langle \{a_{k_+}[(iS)\downarrow'] + b_{k_+}[(\frac{X'-iY'}{\sqrt{2}})\uparrow'] + c_{k_+}[Z'\downarrow']\} \mid \hat{P} \mid \{[a_{k_-}(iS)\uparrow']$$
$$-b_{k_-}[(\frac{X'+iY'}{\sqrt{2}})\downarrow' + c_{k_-}[Z'\uparrow']]\} \Big\rangle$$

Using above relations, we get

$$\hat{p}_{CV}(\vec{k}) = \left\langle u_1(\vec{k},\vec{r}) \mid \hat{P} \mid u_2(\vec{k},\vec{r}) \right\rangle$$
$$= \frac{b_{k_+} a_{k_-}}{\sqrt{2}} \left\{ \left\langle (X'-iY') \mid \hat{P} \mid iS \right\rangle \left\langle \uparrow' \mid \uparrow' \right\rangle \right\} + c_{k_+} a_{k_-} \left\{ \left\langle Z' \mid \hat{P} \mid iS \right\rangle \left\langle \downarrow' \mid \uparrow' \right\rangle \right\} - \frac{a_{k_+} b_{k_-}}{\sqrt{2}} \left\{ \left\langle iS \mid \hat{P} \mid (X'+iY') \right\rangle \left\langle \downarrow' \mid \downarrow' \right\rangle \right\}$$
$$+ a_{k_+} c_{k_-} \left\{ \left\langle iS \mid \hat{P} \mid Z' \right\rangle \left\langle \downarrow' \mid \uparrow' \right\rangle \right\} \tag{18}$$

From Eqn. (18), we can write

$$\left\langle (X'-iY') \mid \hat{P} \mid iS \right\rangle = \left\langle (X') \mid \hat{P} \mid iS \right\rangle - \left\langle (iY') \mid \hat{P} \mid iS \right\rangle = i\int u_{X'}^* \hat{P} S - \int -iu_{Y'}^* \hat{P} i u_X = i\left\langle X' \mid \hat{P} \mid S \right\rangle - \left\langle Y' \mid \hat{P} \mid S \right\rangle$$

From the above relations, for X', Y' and Z', we get

$$\left| X' \right\rangle = \cos\theta\cos\phi \left| X \right\rangle + \cos\theta\sin\phi \left| Y \right\rangle - \sin\theta \left| Z \right\rangle$$

Thus,

$$\left\langle X' \mid \hat{P} \mid S \right\rangle = \cos\theta\cos\phi \left\langle X \mid \hat{P} \mid S \right\rangle + \cos\theta\sin\phi \left\langle Y \mid \hat{P} \mid S \right\rangle - \sin\theta \left\langle Z \mid \hat{P} \mid S \right\rangle = \hat{P}\hat{r}_1$$

where, $\hat{r}_1 = \hat{i}\cos\theta\cos\phi + \hat{j}\cos\theta\sin\phi - \hat{k}\sin\theta$

$$\left| Y' \right\rangle = -\sin\phi \left| X \right\rangle + \cos\phi \left| Y \right\rangle + 0 \left| Z \right\rangle$$

Thus,

$$\left\langle Y' \mid \hat{P} \mid S \right\rangle = -\sin\phi \left\langle X \mid \hat{P} \mid S \right\rangle + \cos\phi \left\langle Y \mid \hat{P} \mid S \right\rangle + 0 \left\langle Z \mid \hat{P} \mid S \right\rangle = \hat{P}\hat{r}_2$$

where $\hat{r}_2 = -\hat{i}\sin\phi + \hat{j}\cos\phi$

so that $\left\langle (X' - iY') \,|\, \hat{P} \,|\, S \right\rangle = \hat{P}(i\hat{r}_1 - \hat{r}_2)$

Thus,

$$\frac{a_{k_-} b_{k_+}}{\sqrt{2}} \left\langle (X' - iY') \,|\, \hat{P} \,|\, S \right\rangle \left\langle \uparrow' | \uparrow' \right\rangle = \frac{a_{k_-} b_{k_+}}{\sqrt{2}} \hat{P}\left(i\hat{r}_1 - \hat{r}_2\right) \left\langle \uparrow' | \uparrow' \right\rangle \tag{19}$$

Now since,

$$\left\langle iS \,|\, \hat{P} \,|\, \left(X' + iY'\right) \right\rangle = i\left\langle S \,|\, \hat{P} \,|\, X' \right\rangle - \left\langle S \,|\, \hat{P} \,|\, Y' \right\rangle = \hat{P}\left(i\hat{r}_1 - \hat{r}_2\right)$$

We can write,

$$-\left[\frac{a_{k_+} b_{k_-}}{\sqrt{2}} \left\{ \left\langle iS \,|\, \hat{P} \,|\, \left(X' + iY'\right) \right\rangle \left\langle \downarrow' | \downarrow' \right\rangle \right\} \right] = -\left[\frac{a_{k_+} b_{k_-}}{\sqrt{2}} \hat{P}\left(i\hat{r}_1 - \hat{r}_2\right) \left\langle \downarrow' | \downarrow' \right\rangle \right] \tag{20}$$

Similarly, we get

$$\left| Z' \right\rangle = \sin\theta\cos\phi \left| X \right\rangle + \sin\theta\sin\phi \left| Y \right\rangle + \cos\theta \left| Z \right\rangle$$

So that, $\left\langle Z' \,|\, \hat{P} \,|\, iS \right\rangle = i\left\langle Z' \,|\, \hat{P} \,|\, S \right\rangle = i\hat{P}\left\{ \sin\theta\cos\phi\hat{i} + \sin\theta\sin\phi\hat{j} + \cos\theta\hat{k} \right\} = i\hat{P}\hat{r}_3$

where, $\hat{r}_3 = \hat{i}\sin\theta\cos\phi + \hat{j}\sin\theta\sin\phi + \hat{k}\cos\theta$

Thus,

$$c_{k_+} a_{k_-} \left\langle Z' \,|\, \hat{P} \,|\, iS \right\rangle \left\langle \downarrow' | \uparrow' \right\rangle = c_{k_+} a_{k_-} i\hat{P}\hat{r}_3 \left\langle \downarrow' | \uparrow' \right\rangle \tag{21}$$

Similarly, we can write,

$$c_{k_-} a_{k_+} \left\langle iS \mid \hat{P} \mid Z' \right\rangle \left\langle \downarrow' \middle| \uparrow' \right\rangle = c_{k_-} a_{k_+} i\hat{P}\hat{r}_3 \left\langle \downarrow' \middle| \uparrow' \right\rangle \tag{22}$$

Therefore, we obtain

$$\begin{aligned} &\frac{a_{k_-} b_{k_+}}{\sqrt{2}} \{ \left\langle (X' - iY') \mid \hat{P} \mid S \right\rangle \left\langle \uparrow' \middle| \uparrow' \right\rangle \} - \frac{a_{k_+} b_{k_-}}{\sqrt{2}} \{ \left\langle iS \mid \hat{P} \mid (X' + iY') \right\rangle \left\langle \downarrow' \middle| \downarrow' \right\rangle \} \\ &= \frac{\hat{P}}{\sqrt{2}} (-a_{k_+} b_{k_-} \left\langle \downarrow' \middle| \downarrow' \right\rangle + a_{k_-} b_{k_+} \left\langle \uparrow \middle| \uparrow' \right\rangle)(i\hat{r}_1 - \hat{r}_2) \end{aligned} \tag{23}$$

Also, we can write,

$$c_{k_+} a_{k_-} \left\langle Z' \mid \hat{P} \mid iS \right\rangle \left\langle \downarrow' \middle| \uparrow' \right\rangle + c_{k_-} a_{k_+} \left\langle iS \mid \hat{P} \mid Z' \right\rangle \left\langle \downarrow' \middle| \uparrow' \right\rangle = i\hat{P}(c_{k_+} a_{k_-} + c_{k_-} a_{k_+})\hat{r}_3 [\left\langle \downarrow' \middle| \downarrow' \right\rangle] \tag{24}$$

Combining equations (23) and (24), we find

$$\hat{p}_{CV}(\vec{k}) = \frac{\hat{P}}{\sqrt{2}} (i\hat{r}_1 - \hat{r}_2) \{ (b_{k_+} a_{k_-}) \left\langle \uparrow' \middle| \uparrow' \right\rangle - (b_{k_-} a_{k_+}) \left\langle \downarrow' \middle| \downarrow' \right\rangle \} + i\hat{P}\hat{r}_3 (c_{k_+} a_{k_-} - c_{k_-} a_{k_+}) \left\langle \downarrow' \middle| \uparrow' \right\rangle \tag{25}$$

From the above relations, we obtain,

$$\left.\begin{aligned} \uparrow' &= e^{-i\phi/2} \cos(\theta/2) \uparrow + e^{i\phi/2} \sin(\theta/2) \downarrow \\ \downarrow' &= -e^{-i\phi/2} \sin(\theta/2) \uparrow + e^{i\phi/2} \cos(\theta/2) \downarrow \end{aligned}\right\} \tag{26}$$

Therefore,

$$\begin{aligned} \left\langle \downarrow' \middle| \uparrow' \right\rangle_x &= -\sin(\theta/2)\cos(\theta/2) \left\langle \uparrow \middle| \uparrow \right\rangle_x + e^{-i\phi} \cos^2(\theta/2) \left\langle \downarrow \middle| \uparrow \right\rangle_x \\ &- e^{i\phi} \sin^2(\theta/2) \left\langle \uparrow \middle| \downarrow \right\rangle_x + \sin(\theta/2)\cos(\theta/2) \left\langle \downarrow \middle| \downarrow \right\rangle_x \end{aligned} \tag{27}$$

But we know from above that

$\langle\uparrow|\uparrow\rangle_x = 0, \langle\downarrow|\uparrow\rangle = \frac{1}{2}, \langle\downarrow|\uparrow\rangle_x = \frac{1}{2}$ and $\langle\downarrow|\downarrow\rangle_x = 0$

Thus, from equation (27), we get

$$\langle\downarrow'|\uparrow'\rangle_x = \frac{1}{2}[e^{-i\phi}\cos^2(\theta/2) - e^{i\phi}\sin^2(\theta/2)]$$
$$\frac{1}{2}[(\cos\phi - i\sin\phi)\cos^2(\theta/2) - (\cos\phi - i\sin\phi)\sin^2(\theta/2)] = \frac{1}{2}[\cos\phi\cos\theta - i\sin\phi] \tag{28}$$

Similarly, we obtain

$\langle\downarrow'|\uparrow'\rangle_y = \frac{1}{2}[i\cos\phi + \sin\phi\cos\theta]$ and $\langle\downarrow'|\uparrow'\rangle_z = \frac{1}{2}[-\sin\theta]$

Therefore,

$$\frac{1}{2}\{(\cos\theta\cos\phi - i\sin\phi)\hat{i} + (i\cos\phi + \sin\phi\cos\theta)\hat{j} - \sin\theta\hat{k}\}$$

$$\frac{1}{2}[\{(\cos\theta\cos\phi)\hat{i} + (\sin\phi\cos\theta)\hat{j} - \sin\theta\hat{k}\} + i\{-\hat{i}\sin\phi + \hat{j}\cos\phi\}] = \frac{1}{2}[\hat{r}_1 + i\hat{r}_2] = -\frac{1}{2}i[i\hat{r}_1 - \hat{r}_2]$$

Similarly, we can write

$\langle\uparrow'|\uparrow'\rangle = \frac{1}{2}[\hat{i}\sin\theta\cos\phi + \hat{j}\sin\theta\sin\phi + \hat{k}\cos\theta] = \frac{1}{2}\hat{r}_3$ and $\langle\downarrow'|\downarrow'\rangle = -\frac{1}{2}\hat{r}_3$

Using the above results and following equation (25) we can write

$$\hat{p}_{CV}(\vec{k}) = \frac{\hat{P}}{\sqrt{2}}(i\hat{r}_1 - \hat{r}_2)\{(a_{k_-}b_{k_+})\langle\uparrow'|\uparrow'\rangle - (b_{k_-}a_{k_+})\langle\downarrow'|\downarrow'\rangle\} + i\hat{P}\hat{r}_3\{(c_{k_+}a_{k_-} - c_{k_-}a_{k_+})\langle\downarrow'|\uparrow'\rangle\}$$
$$= \frac{\hat{P}}{2}\hat{r}(i\hat{r}_1 - \hat{r}_2)\{(\frac{a_{k_-}b_{k_+}}{\sqrt{2}} + \frac{b_{k_-}a_{k_+}}{\sqrt{2}})\} + \frac{\hat{P}}{2}\hat{r}_3(i\hat{r}_1 - \hat{r}_2)\{(c_{k_+}a_{k_-} + c_{k_-}a_{k_+})\}$$

Thus,

$$\hat{p}_{CV}(\vec{k}) = \frac{\hat{P}}{2}\hat{r}_3(i\hat{r}_1 - \hat{r}_2)\{a_{k_+}(\frac{b_{k_-}}{\sqrt{2}} + c_{k_-}) + a_{k_-}(\frac{b_{k_+}}{\sqrt{2}} + c_{k_+})\} \tag{29}$$

We can write that,

$|\hat{r}_1| = |\hat{r}_2| = |\hat{r}_3| = 1$, also, $\hat{P}\hat{r}_3 = \hat{P}_x \sin\theta\cos\phi\hat{i} + \hat{P}_y \sin\theta\sin\phi\hat{j} + \hat{P}_z \cos\theta\hat{k}$

where, $\hat{P} = \langle S|\hat{P}|X\rangle = \langle S|\hat{P}|Y\rangle = \langle S|\hat{P}|Z\rangle$,

$\langle S|\hat{P}|X\rangle = \int u_C^*(0,\vec{r})\hat{P}u_{VX}(0,\vec{r})d^3r = \hat{P}_{CVX}(0)$, $\langle S|\hat{P}|Y\rangle = \hat{P}_{CVY}(0)$ and $\langle S|\hat{P}|Z\rangle = \hat{P}_{CVZ}(0)$

Thus, $\hat{P} = \hat{P}_{CVX}(0) = \hat{P}_{CVY}(0) = \hat{P}_{CVZ}(0) = \hat{P}_{CV}(0)$

where, $\hat{P}_{CV}(0) \equiv \int u_c^*(0,\vec{r})\hat{P}u_V(0,\vec{r})d^3r \equiv \hat{P}$

For a plane polarized light wave, we have the polarization vector $\vec{\varepsilon}_s = \hat{k}$, when the light wave vector is traveling along the z-axis. Therefore, for a plane polarized light-wave, we have considered $\vec{\varepsilon}_s = \hat{k}$.

Then, from equation (29)

$$(\vec{\varepsilon} \bullet \hat{p}_{CV}(\vec{k})) = \vec{k} \bullet \frac{\hat{P}}{2}\hat{r}_3(i\hat{r}_1 - \hat{r}_2)[A(\vec{k}) + B(\vec{k})]\cos\omega t \tag{30}$$

and

$$\left.\begin{aligned} A(\vec{k}) &= a_{k_-}(\frac{b_{k_+}}{\sqrt{2}} + c_{k_+}) \\ B(\vec{k}) &= a_{k_+}(\frac{b_{k_+}}{\sqrt{2}} + c_{k_-}) \end{aligned}\right\} \tag{31}$$

Thus,

$$\left|\vec{\varepsilon}\bullet\hat{p}_{cv}(\vec{k})\right|^2 = \left|\hat{k}\bullet\frac{\hat{P}}{2}\hat{r}_3\right|^2 \left|i\hat{r}_1 - \hat{r}_2\right|^2 [A(\vec{k}) + B(\vec{k})]^2 \cos^2\omega t$$
$$= \frac{1}{4}\left|\hat{P}_z\cos\theta\right|^2 [A(\vec{k}) + B(\vec{k})]^2 \cos^2\omega t \tag{32}$$

So, the average value of $\left|\vec{\varepsilon}\bullet\hat{p}_{cv}(\vec{k})\right|^2$ for a plane polarized light-wave is given by

$$\left\langle\left|\vec{\varepsilon}\bullet\hat{p}_{cv}(\vec{k})\right|^2\right\rangle_{av} = \frac{2}{4}\left|\hat{P}_z\right|^2 [A(\vec{k}) + B(\vec{k})]^2 (\int_0^{2\pi} d\phi \int_0^{\pi} \cos^2\theta \sin\theta d\theta)(\frac{1}{2}) = \frac{2\pi}{3}\left|\hat{P}_z\right|^2 [A(\vec{k}) + B(\vec{k})]^2 \tag{33}$$

where, $\left|\hat{P}_z\right|^2 = (\frac{1}{2})\left|\vec{k}\bullet\hat{p}_{cv}(0)\right|^2$

and

$$\left|\vec{k}\bullet\hat{p}_{cv}(0)\right|^2 = \frac{m^2}{4m_r}\frac{E_{g_0}(E_{g_0}+\Delta)}{(E_{g_0}+\frac{2}{3}\Delta)} \tag{34}$$

We shall express $A(\vec{k})$ and $B(\vec{k})$ in terms of constants of the energy spectra in the following way:

Substituting $a_{k_\pm}$, $b_{k_\pm}$, $c_{k_\pm}$ and $\gamma_{0k_\pm}$ in $A(\vec{k})$ and $B(\vec{k})$ in equation (31) we get

$$A(\vec{k}) = \beta(t + \frac{\rho}{\sqrt{2}})\{(\frac{E_{g_0}}{E_{g_0}+\delta'})\gamma_{0k_+}^2 - \gamma_{0k_+}^2\gamma_{0k_-}^2(\frac{E_{g_0}-\delta'}{E_{g_0}+\delta'})\}^{1/2} \tag{35}$$

$$B(\vec{k}) = \beta(t + \frac{\rho}{\sqrt{2}})\{(\frac{E_{g_0}}{E_{g_0} + \delta'})\gamma_{0k_-}^2 - \gamma_{0k_+}^2 \gamma_{0k_-}^2 (\frac{E_{g_0} - \delta'}{E_{g_0} + \delta'})\}^{1/2} \tag{36}$$

in which, $\gamma_{0k_+}^2 \equiv \frac{\xi_{1k} - E_{g_0}}{2(\xi_{1k} + \delta')} \equiv \frac{1}{2}[1 - (\frac{E_{g_0} + \delta'}{\xi_{1k} + \delta'})]$

and $\gamma_{0k_+}^2 \equiv \frac{\xi_{1k} - E_{g_0}}{2(\xi_{1k} + \delta')} \equiv \frac{1}{2}[1 + (\frac{E_{g_0} + \delta'}{\xi_{1k} + \delta'})]$

Substituting $x \equiv \xi_{1k} + \delta'$ in $\gamma_{0k_\pm}^2$, we can write,

$$A(\vec{k}) = \beta(t + \frac{\rho}{\sqrt{2}})\{(\frac{E_{g_0}}{E_{g_0} + \delta'})\frac{1}{2}(1 - \frac{E_{g_0} + \delta'}{x}) - \frac{1}{4}(\frac{E_{g_0} - \delta'}{E_{g_0} + \delta'})(1 - \frac{E_{g_0} + \delta'}{x})(1 + \frac{E_{g_0} - \delta'}{x})\}^{1/2}$$

Thus, $A(\vec{k}) = \frac{\beta}{2}(t + \frac{\rho}{\sqrt{2}})\{1 - \frac{2a_0}{x} + \frac{a_1}{x^2}\}^{1/2}$

where, $a_0 \equiv (E_{g_0}^2 + \delta'^2)(E_{g_0} + \delta')^{-1}$ and $a_1 \equiv (E_{g_0} - \delta')^2$.

After tedious algebra, one can show that

$$A(\vec{k}) = \frac{\beta}{2}(t + \frac{\rho}{\sqrt{2}})(E_{g_0} - \delta')[\frac{1}{\xi_{1k} + \delta'} - \frac{1}{E_{g_0} + \delta'}]^{1/2}[\frac{1}{\xi_{1k} + \delta'} - \frac{(E_{g_0} + \delta')}{(E_{g_0} - \delta')^2}]^{1/2} \tag{37}$$

Similarly, from equation (36), we can write,

$$B(\vec{k}) = \beta(t + \frac{\rho}{\sqrt{2}})\{(\frac{E_{g_0}}{E_{g_0} + \delta'})\frac{1}{2}(1 + \frac{E_{g_0} - \delta'}{x}) - \frac{1}{4}(\frac{E_{g_0} - \delta'}{E_{g_0} + \delta'})(1 - \frac{E_{g_0} + \delta'}{x})(1 + \frac{E_{g_0} - \delta'}{x})\}^{1/2}$$

So that, finally we get,

$$B(\vec{k}) = \frac{\beta}{2}(t + \frac{\rho}{\sqrt{2}})(1 + \frac{E_{g_0} - \delta'}{\xi_{1k} + \delta'}) \tag{38}$$

Using equations (33), (34), (37) and (38), we can write

$$(\frac{|e|A_0}{2m})^2 \frac{\left\langle |\vec{\varepsilon}\bullet\hat{p}_{cv}(\vec{k})|^2 \right\rangle_{av}}{E_c(\vec{k})-E_v(\vec{k})} = (\frac{|e|A_0}{2m})^2 \frac{2\pi}{3} |\vec{k}\bullet\hat{p}_{cv}(0)|^2 \frac{\beta^2}{4}(t+\frac{\rho}{\sqrt{2}})^2$$
$$\frac{1}{\xi_{1k}}\{(1+\frac{E_{g_0}-\delta'}{\xi_{1k}+\delta'})+(E_{g_0}-\delta')[\frac{1}{\xi_{1k}+\delta'}-\frac{1}{E_{g_0}+\delta'}]^{1/2}[\frac{1}{\xi_{1k}+\delta'}-\frac{E_{g_0}+\delta'}{(E_{g_0}-\delta')^2}]^{1/2}\}^2 \tag{39}$$

Following Nag [7], it can be shown that

$$A_0^2 = \frac{I\lambda^2}{2\pi^2 c^3\sqrt{\varepsilon_{sc}\varepsilon_0}} \tag{40}$$

where, I is the light intensity of wavelength λ, ε_0 is the permittivity of free space and c is the velocity of light.

Thus, the simplified electron energy spectrum in III-V, ternary and quaternary materials up to the second order in the presence of light waves can approximately be written as

$$\frac{\hbar^2 k^2}{2m^*} = \beta_0(E,\lambda) \tag{41}$$

where, $\beta_0(E,\lambda) \equiv [\gamma(E)-\theta_0(E,\lambda)]$,

$$\theta_0(E,\lambda) \equiv \frac{|e|^2}{96 m_r \pi c^3}\frac{I\lambda^2}{\sqrt{\varepsilon_{sc}\varepsilon_0}}\frac{E_{g_0}(E_{g_0}+\Delta)}{(E_{g_0}+\frac{2}{3}\Delta)}\frac{\beta^2}{4}(t+\frac{\rho}{\sqrt{2}})^2\frac{1}{\phi_0(E)},$$

$$\{(1+\frac{E_{g_0}-\delta'}{\phi_0(E)+\delta'})+(E_{g_0}-\delta')[\frac{1}{\phi_0(E)+\delta'}-\frac{1}{E_{g_0}+\delta'}]^{1/2}[\frac{1}{\phi_0(E)+\delta'}-\frac{E_{g_0}+\delta'}{(E_{g_0}-\delta')^2}]^{1/2}\}^2$$

and $\phi_0(E) \equiv E_{g_0}(1+2(1+\frac{m^*}{m_v})\frac{\gamma(E)}{E_{g_0}})^{1/2}$

Thus, under the limiting condition $\vec{k}\to 0$, from Eqn. (41), we observe that $E \neq 0$ and is positive. Therefore, in the presence of external light waves, the energy of the electron does not tend to zero when $\vec{k}\to 0$, whereas for the un-perturbed three band model of Kane, $I(E)=[\hbar^2k^2/(2m^*)]$ in which $E\to 0$ for $\vec{k}\to 0$. As the conduction band is taken as the reference level of energy, therefore the lowest positive value of E for $\vec{k}\to 0$ provides the increased band gap (ΔE_g) of the semiconductor due to photon excitation. The values of the increased band gap can be obtained by computer iteration processes for various values of I and λ respectively.

2.2. Special Cases

For the two-band model of Kane, we have $\Delta\to 0$. Under this condition,

$\gamma(E)\to E(1+aE)=\dfrac{\hbar^2k^2}{2m^*}$. Since, $\beta\to 1$, $t\to 1$, $\rho\to 0$, $\delta'\to 0$ for $\Delta\to 0$, from equation (41), we can write the energy spectrum of III-V, ternary and quaternary materials in the presence of external photo-excitation whose unperturbed conduction electrons obey the two band model of Kane as

$$\frac{\hbar^2k^2}{2m^*}=\tau_0(E,\lambda) \tag{42}$$

where, $\tau_0(E,\lambda)\equiv E(1+aE)-B_0(E,\lambda)$,

$$B_0(E,\lambda)\equiv\frac{|e|^2 I\lambda^2 E_{g_0}}{384\pi c^3 m_r\sqrt{\varepsilon_{sc}\varepsilon_0}}\frac{1}{\phi_1(E)}\{(1+\frac{E_{g_0}}{\phi_1(E)})+E_{g_0}[\frac{1}{\phi_1(E)}-\frac{1}{E_{g_0}}]\}^2,$$

$$\phi_1(E)\equiv E_{g_0}\{1+\frac{2m^*}{m_r}aE(1+aE)\}^{1/2}$$

For relatively wide band gap semiconductors, one can write, $a\to 0$, $b\to 0$, $c\to 0$ and $\gamma(E)\to E$.

Thus, from equation (42), we get,

$$\frac{\hbar^2 k^2}{2m^*} = \rho_0(E,\lambda)$$
$$\rho_0(E,\lambda) \equiv E - \frac{|e|^2 I\lambda^2}{96\pi c^3 m_r\sqrt{\varepsilon_{sc}\varepsilon_0}}[1+(\frac{2m^*}{m_r})aE]^{-3/2} \qquad (43)$$

$$\rho_0(E,\lambda) \equiv E - \frac{|e|^2 I\lambda^2}{96\pi c^3 m_r\sqrt{\varepsilon_{sc}\varepsilon_0}}[1+(\frac{2m^*}{m_r})aE]^{-3/2} \qquad (44)$$

The dispersion relations of the 1D electron in QWs of optoelectronic materials in the presence of light waves can be expressed from (1), (2) and (3) as

$$k_y^2 = \frac{2m^*\beta_0(E,\lambda)}{\hbar^2} - (\frac{\pi n_{z81}}{d_z})^2 - (\frac{\pi n_{x81}}{d_x})^2 \qquad (45)$$

$$k_y^2 = \frac{2m^*\tau_0(E,\lambda)}{\hbar^2} - (\frac{\pi n_{z82}}{d_z})^2 - (\frac{\pi n_{x82}}{d_x})^2 \qquad (46)$$

$$k_y^2 = \frac{2m^*\rho_0(E,\lambda)}{\hbar^2} - (\frac{\pi n_{z83}}{d_z})^2 - (\frac{\pi n_{x83}}{d_x})^2 \qquad (47)$$

where $n_{x8J}(J=1,2,3)$ is the size quantum number.

The electron concentration per unit length in the presence of light waves in this case are given by

$$n_0 = \frac{2g_v\sqrt{2m^*}}{\pi\hbar}\sum_{n_{x81}=1}^{n_{x81_{max}}}\sum_{n_{z81}=1}^{n_{z81_{max}}}[\phi_{87}(E_{F1DL},n_{x81},n_{z81})+\phi_{88}(E_{F1DL},n_{x81},n_{z81})] \qquad (48)$$

$$n_0 = \frac{2g_v\sqrt{2m^*}}{\pi\hbar}\sum_{n_{x82}=1}^{n_{x82_{max}}}\sum_{n_{z82}=1}^{n_{z82_{max}}}[\phi_{89}(E_{F1DL},n_{x82},n_{z82})+\phi_{810}(E_{F1DL},n_{x82},n_{z82})] \qquad (49)$$

$$n_0 = \frac{2g_v\sqrt{2m^*}}{\pi\hbar}\sum_{n_{x83}=1}^{n_{x83_{\max}}}\sum_{n_{z83}=1}^{n_{z83_{\max}}}[\phi_{811}(E_{F1DL},n_{x83},n_{z83})+\phi_{812}(E_{F1DL},n_{x83},n_{z83})] \tag{50}$$

where E_{F1DL} is the Fermi energy in QWs in the presence of light waves as measured from the edge of the conduction band in the vertically upward direction in the absence of any quantization,

$$\phi_{87}(E_{F1DL},n_{x81},n_{z81}) = [\beta_0(E_{F1DL},\lambda) - G_{81}(n_{x81},n_{z81})]^{1/2}, G_{8i}(n_{x8i},n_{z8i}) = \frac{\hbar^2}{2m^*}[(\frac{\pi n_{x8i}}{d_x})^2 + (\frac{\pi n_{z8i}}{d_z})^2],$$

$$\phi_{88}(E_{F1DL},n_{x81},n_{z81}) = \sum_{r=1}^{s_0} Z_{r,Y}[\phi_{87}(E_{F1DL},n_{x81},n_{z81})], Y = 1DL,$$

$$\phi_{89}(E_{F1DL},n_{x82},n_{z82}) = [\tau_0(E_{F1DL},\lambda) - G_{82}(n_{x82},n_{z82})]^{1/2},$$

$$\phi_{810}(E_{F1DL},n_{x82},n_{z82}) = \sum_{r=1}^{s} Z_{r,Y}[\phi_{89}(E_{F1DL},n_{x82},n_{z82})], \phi_{811}(E_{F1DL},n_{x83},n_{z83}) = [\rho_0(E_{F1DL},\lambda) - G_{83}(n_{x83},n_{z83})]^{1/2}$$

and

$$\phi_{812}(E_{F1DL},n_{x83},n_{z83}) = \sum_{r=1}^{s} Z_{r,Y}[\phi_{811}(E_{F1DL},n_{x83},n_{z83})]$$

The MTP (G_0) can, in general, be expressed as [5]

$$G_0 = \frac{1}{e}(\frac{\partial\overline{S_0}}{\partial n_0})_{E_F,T} \tag{51}$$

where E_F is the Fermi energy (which changes as the dispersion relation of the carriers changes from one system to another system introducing new physical phenomena) corresponding to the electron statistics n_0 per unit length and S_0 is the entropy per unit length which can be written as

$$\overline{S_0} = -\left.\frac{\partial\Omega}{\partial T}\right|_{E=E_F} \tag{52}$$

in which Ω is the thermodynamic potential which, in turn, can be expressed in accordance with the Fermi-Dirac statistics as

$$\Omega = -k_B T \sum \ln \left| 1 + \exp[\frac{E_F - E_{\delta_0}}{k_B T}] \right| \tag{53}$$

where the summation is carried out over all the possible δ_0 states.

Thus, combining (52-54) and extending [5], the magnitude of the MTP can be written in a simplified form as

$$G_0 = (\pi^2 k_B^2 T / 3en_0)(\frac{\partial n_0}{\partial E_F}) \tag{54}$$

It should be noted that being a thermodynamic relation and temperature induced phenomena, the MTP as expressed by (54), in general, is valid for electronic materials having arbitrary dispersion relations and their nanostructures. In addition to bulk materials in the presence of strong magnetic field, (54) is valid under one, two and three dimensional quantum confinement of the charge carriers (such as quantum wells in ultrathin films, nipi structures, inversion and accumulation layers, quantum well superlattices, carbon nanotubes, quantum wires, quantum wire superlattices, quantum dots, magneto inversion and accumulation layers, quantum dot superlattices, magneto nipis, quantum well superlattices under magnetic quantization, ultrathin films under magnetic quantization, etc.). The formulation of G_0 requires the relation between electron statistics and the corresponding Fermi energy which is basically the band structure dependent quantity and changes under different physical conditions. It is worth noting to note that the number $(\pi^2/3)$ has occurred as a consequence of mathematical analysis and is not connected with the well-known Lorenz number. For quantum wells in ultrathin films, nipi structures, inversion and accumulation layers, quantum well superlattices, magneto inversion and accumulation layers, magneto nipis, quantum well superlattices under magnetic quantization and magneto size quantization, the carrier concentration is measured per unit area whereas, for quantum wires, quantum wires under magnetic field, quantum wire superlattices and such allied systems, the same can be measured per unit length [55-60]. Besides, for bulk materials under strong magnetic field, quantum dots, quantum dots under magnetic field, quantum dot superlattices and quantum dot superlattices under magnetic field, the carrier concentration is expressed per unit volume.

Combining the appropriate equations the MTP in the presence of microwave radiation in accordance with perturbed three and two band models of Kane together with perturbed parabolic energy bands can, respectively, be written as

$$G_0 = \left(\frac{\pi^2 k_B^2 T}{3e}\right)\left[\sum_{n_{x81}=1}^{n_{x81_{\max}}} \sum_{n_{z81}=1}^{n_{z81_{\max}}} [\phi_{87}(E_{F1DL}, n_{x81}, n_{z81}) + \phi_{88}(E_{F1DL}, n_{x81}, n_{z81})]\right]^{-1}.$$

$$\left[\sum_{n_{x81}=1}^{n_{x81_{\max}}} \sum_{n_{z81}=1}^{n_{z81_{\max}}} [\phi_{87}'(E_{F1DL}, n_{x81}, n_{z81}) + \phi_{88}'(E_{F1DL}, n_{x81}, n_{z81})]\right]$$

$$G_0 = \left(\frac{\pi^2 k_B^2 T}{3e}\right)\left[\sum_{n_{x82}=1}^{n_{x82_{\max}}} \sum_{n_{z82}=1}^{n_{z82_{\max}}} [\phi_{89}(E_{F1DL}, n_{x82}, n_{z82}) + \phi_{810}(E_{F1DL}, n_{x82}, n_{z82})]\right]^{-1}. \quad (55)$$

$$\left[\sum_{n_{x82}=1}^{n_{x82_{\max}}} \sum_{n_{z82}=1}^{n_{z82_{\max}}} [\phi_{89}'(E_{F1DL}, n_{x82}, n_{z82}) + \phi_{810}'(E_{F1DL}, n_{x82}, n_{z82})]\right] \quad (56)$$

and

$$G_0 = \left(\frac{\pi^2 k_B^2 T}{3e}\right)\left[\sum_{n_{x83}=1}^{n_{x83_{\max}}} \sum_{n_{z83}=1}^{n_{z83_{\max}}} [\phi_{811}(E_{F1DL}, n_{x83}, n_{z83}) + \phi_{812}(E_{F1DL}, n_{x83}, n_{z83})]\right]^{-1}.$$

$$\left[\sum_{n_{x83}=1}^{n_{x83_{\max}}} \sum_{n_{z83}=1}^{n_{z83_{\max}}} [\phi_{811}(E_{F1DL}, n_{x83}, n_{z83}) + \phi_{812}(E_{F1DL}, n_{x83}, n_{z83})]\right] \quad (57)$$

3. RESULT AND DISCUSSIONS

Using the appropriate equations we have plotted the MTP from Figures 1 to 15 for quantum wires of $InSb, GaAs, Hg_{1-x}Cd_xTe$ $and\ In_{1-x}Ga_xAs_yP_{1-x}$ as functions of film thickness, electron concentration, light intensity and wavelength. The Figure 1 shows the variation of MTP for QWs of InSb with film thickness for $d_x = 30nm, n_0 = 20\times10^8 m^{-1}, I_0 = 0.1Wm^{-2}$ and $\lambda = 400nm$. The Figures 5, 9 and 13 exhibit the said variation for the other materials. It appears from the said figures that for QWs, the MTP exhibits rectangular variations with the film thickness and exhibits composite oscillations due to the intermingling between the two quantum numbers along two directions. In Figures 2, 6, 10 and 14, we have presented the variation of the MTP as function of electron concentration for $n_x = 1$, $d_x = 50nm$, $I_0 = 0.1Wm^{-2}$ and $\lambda = 400nm$. From Figures 2, 6, 10 and 14, it appears that MTP decreases with increasing electron concentration in oscillatory manners. For large values of electron concentration per unit length, the MTP exhibits very quick oscillatory decrement whereas for small values of the carrier degeneracy, the MTP shows the converging tendencies. The Figures 3, 7, 11 and 15 exhibits the variation of the MTP for quantum wires of such materials with the light intensity for $n_x = n_y = 1$, $d_x = 10nm$, $d_z = 30nm$, $n_{2DL} = 1\times10^8 m^{-1}$ and $I_0 = 0.1Wm^{-2}$ respectively. It may be noted that the MTP increases with increasing light intensity and due to both quantum limits, the variation is again non-oscillatory. The Figures 4, 8, 12 and 16 show the variation of the MTP of the quantum wires of the aforementioned materials with wavelength for $n_x = n_y = 1$, $d_x = 10nm, d_z = 30nm,$ $n_0 = 20\times10^8 m^{-1}$, $I_0 = 0.1Wm^{-2} and \lambda = 400nm.$ Because of the quantum limits, the MTP changes slowly with wavelength in a non-oscillatory manner although the wide difference in the MTP versus λ plot for the perturbed two and three band models of Kane together with parabolic energy bands exhibit the influence of energy band models.

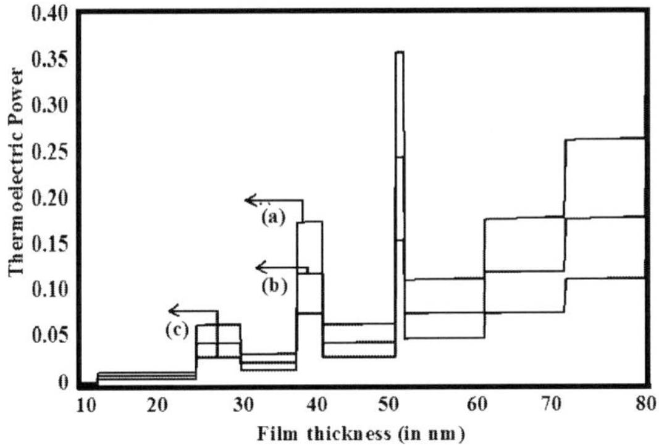

Figure 1. Plot of the MTP as a function of film thickness for the QWs of InSb for all cases of Figure 1 in the presence of external photo-excitation.

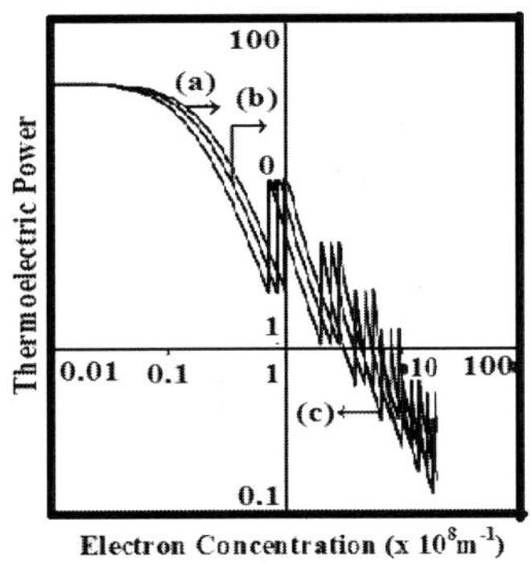

Figure 2. Plot of the MTP as a function of electron concentration for the quantum wires of InSb for all the cases of Figure 1 in the presence of external photo-excitation.

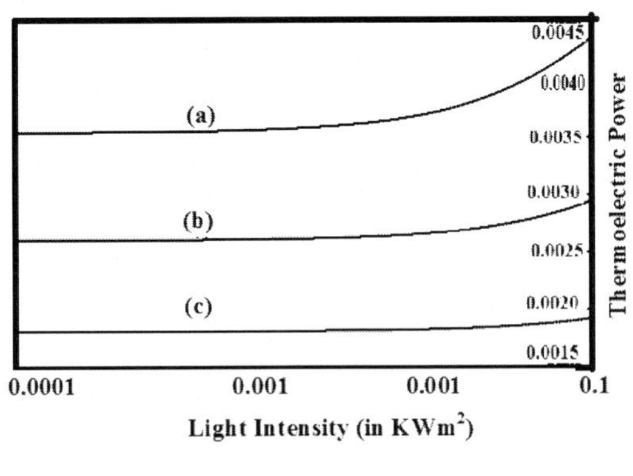

Figure 3. Plot of the MTP as a function of light intensity for the quantum wires of InSb for all the cases of Figure 1 in the presence of external photo-excitation.

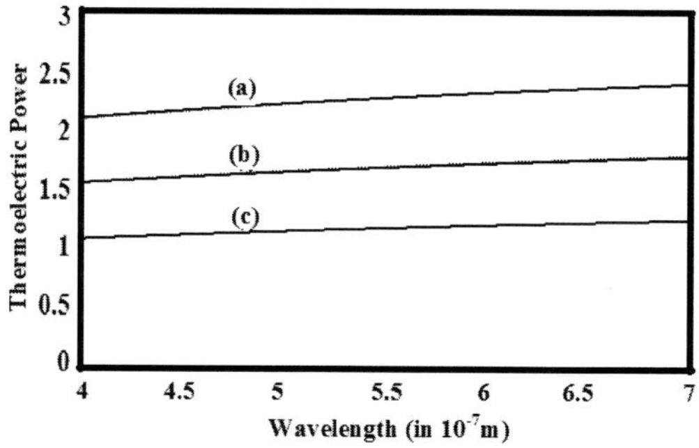

Figure 4. Plot of the MTP as a function of wavelength for the quantum wires of InSb for all the cases of Figure 17 in the presence of external photo-excitation.

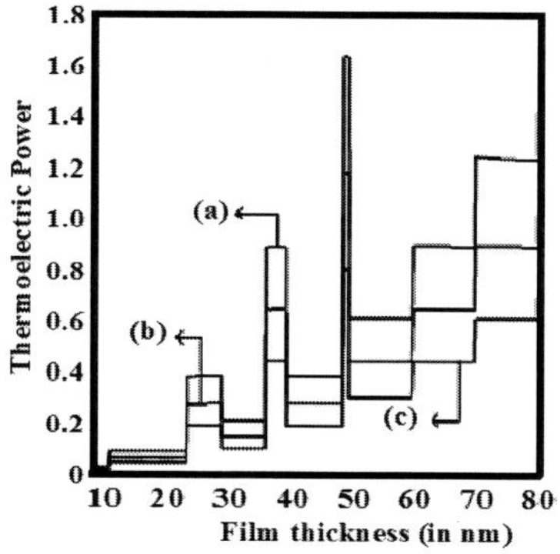

Figure 5. Plot of the MTP as a function of film thickness for the quantum wires of GaAs for all cases of Figure 1 in the presence of external photo-excitation.

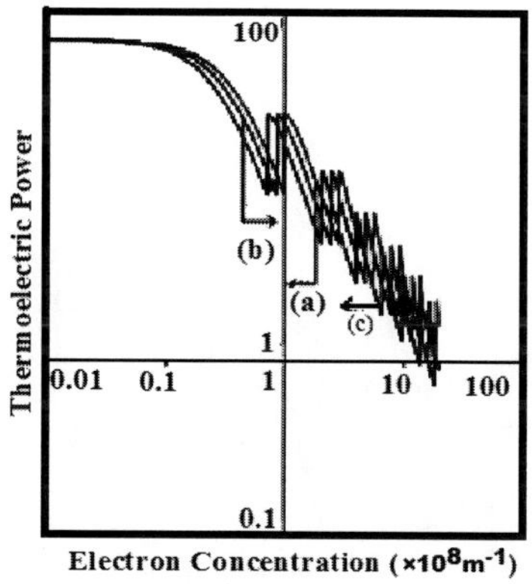

Figure 6. Plot of the MTP as a function of electron concentration for the quantum wires of GaAs for all the cases of Figure 4 in the presence of external photo-excitation.

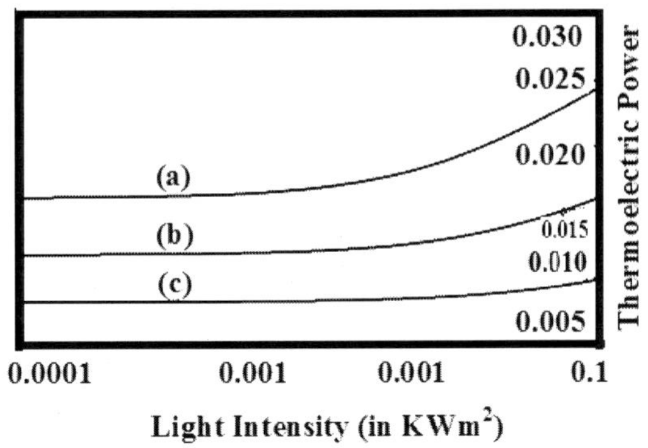

Figure 7. Plot of the MTP as a function of light intensity for the quantum wires of GaAs for all the cases of Figure 4 in the presence of external photo-excitation.

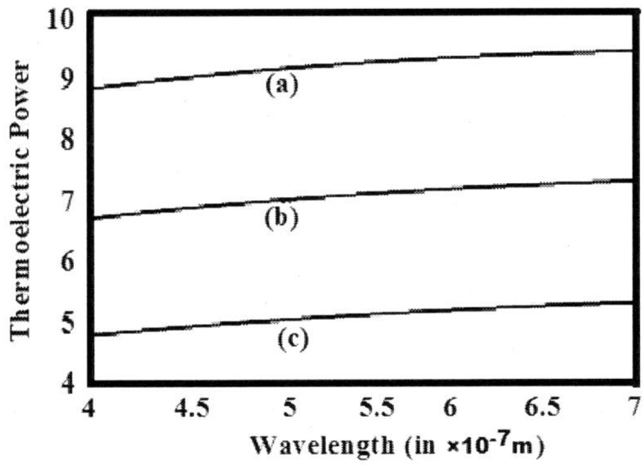

Figure 8. Plot of the MTP as a function of wavelength for the quantum wires of GaAs for all the cases of Figure 4 in the presence of external photo-excitation.

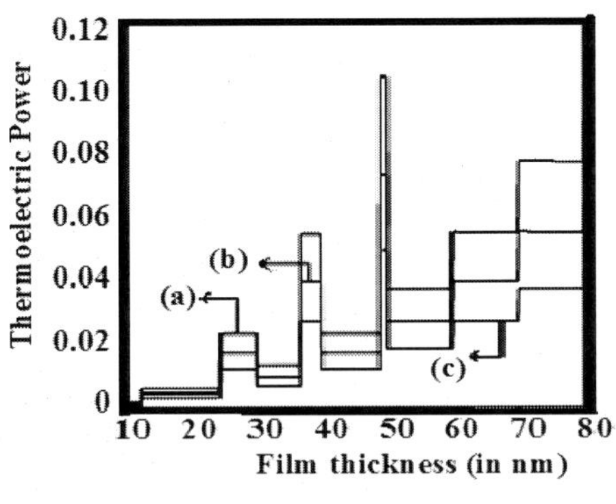

Figure 9. Plot of the MTP as a function of film thickness for the quantum wires of $Hg_{1-x}Cd_xTe$ all cases of Figure 1 in the presence of external photo-excitation.

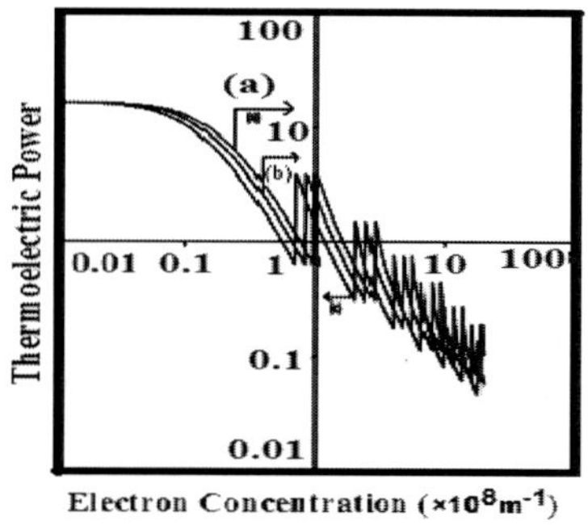

Figure 10. Plot of the MTP as a function of electron concentration for all the cases of Figure 25 in the presence of external photo-excitation.

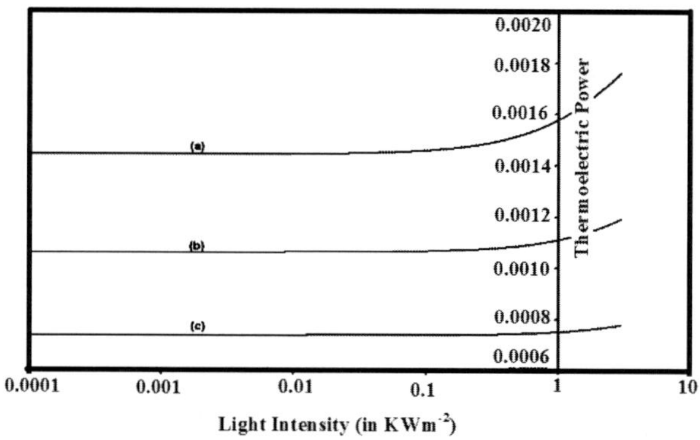

Figure 11. Plot of the MTP as a function of light intensity for the quantum wires of $Hg_{1-x}Cd_xTe$ for all the cases of Figure 9 in the presence of external photo-excitation.

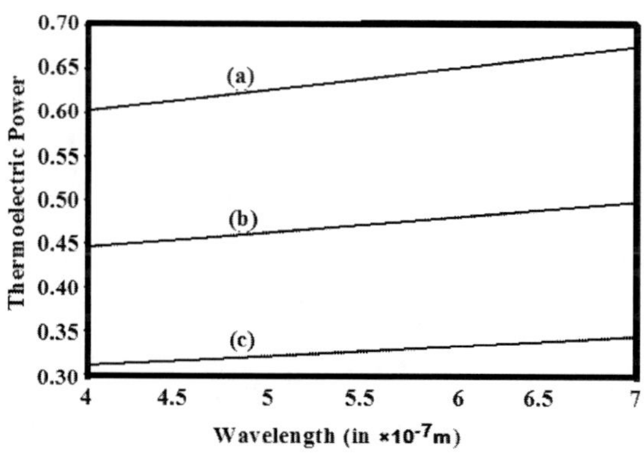

Figure 12. Plot of the MTP as a function of wavelength for the quantum wires of $Hg_{1-x}Cd_xTe$ for all the cases of Figure 9 in the presence of external photo-excitation.

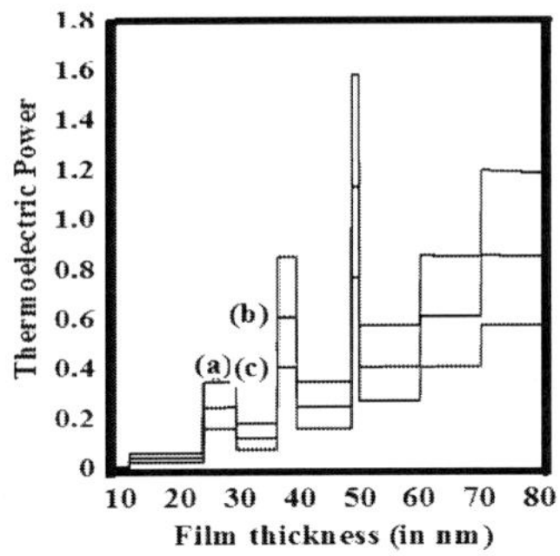

Figure 13. Plot of the MTP as a function of film thickness for the quantum wires of $In_{1-x}Ga_xAs_{1-y}P_y$ for all cases of Figure 1 in the presence of external photo-excitation.

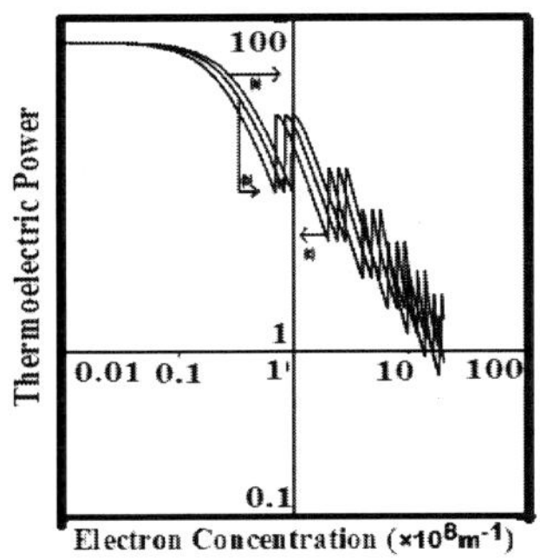

Figure 14. Plot of the MTP as a function of electron concentration for the quantum wires of $In_{1-x}Ga_xAs_{1-y}P_y$ for all the cases of Figure 12 in the presence of external photo-excitation.

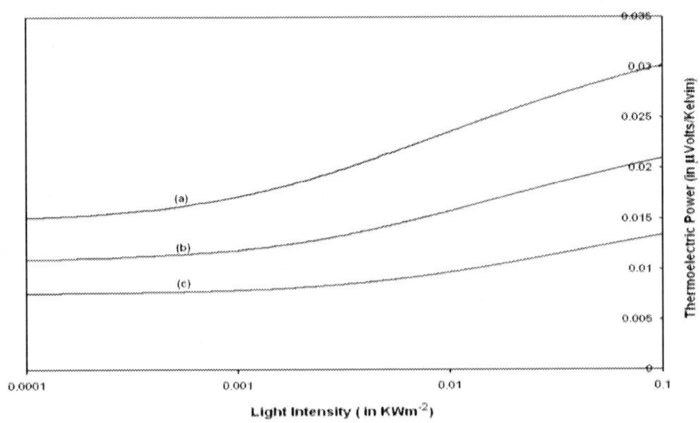

Figure 15. Plot of the MTP as a function of light intensity for the quantum wires of $In_{1-x}Ga_xAs_{1-y}P_y$ for all the cases of Figure 12 in the presence of external photo-excitation.

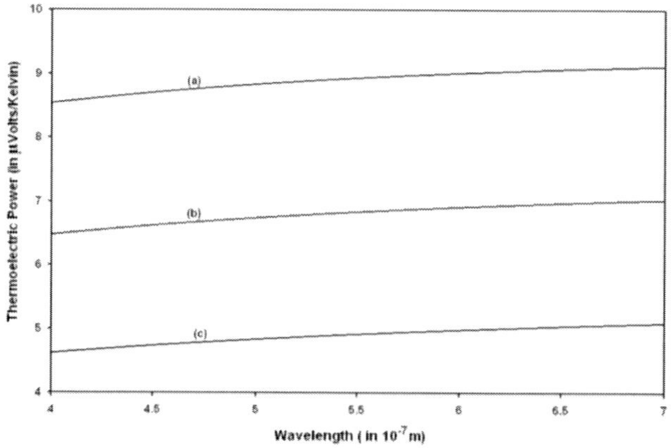

Figure 16. Plot of the MTP as a function of wavelength for the quantum wires of $In_{1-x}Ga_xAs_{1-y}P_y$ for all the cases of Figure 13 in the presence of external photo-excitation.

References

[1] Hajdu, J., G. Landwehr, In: *Strong and Ultrastrong Magnetic Fields and Their Applications,* Topics in Applied Physics, F. Herlach (ed.) Vol. 57, p. 97 (Springer-Verlag, Germany, 1985).

[2] Tsidilkovskii, M., *Band Structures of Semiconductors* (Pergamon Press, London, 1982).

[3] Ghatak, K. P., S. Bhattacharya, S. Bhowmik, R. Benedictus, S. Choudhury, *J. Appl. Phys.* 103, 034303 (2008).

[4] Zawadzki, W., in *Two-Dimensional Systems, Heterostructures and Superlattices*, G. Bauer, F. Kuchar, H. Heinrich (ed.) Springer Ser. Solid-State Sci., Vol. 53 (Springer-Verlag, Germany, 1984); W. Zawadzki, *11th Internet Conf. Phys. of Semicond.* Vol. 1 (Elsevier Publishing Company, Netherlands, 1972); S. P. Zelenin, A. S. Kondrat'ev, A. E. Kuchma, *Sov. Phys. Semicond.* 16, 355 (1982).

[5] Klitzing, K. v., G. Dorda, M. Pepper, *Phys. Rev. Lett.* 45, 494 (1980); K. v. Klitzing, *Rev. Mod. Phys.* 58, 519 (1986).

[6] Arushanov, E. A., A. F. Knyazev, A. N. Natepov, S. I. Radautsan, *Sov. Phys. Semi.* 15, 828 (1981).

[7] Zelenim, S. P., A. S. Kondratev, A. E. Kuchma, *Sov. Phys. Semi.* 16, 355 (1982).
[8] Peeters, F. M., P. Vasilopoulos, *Phys. Rev. B* 46, 4667 (1992).
[9] B. M. Askerov, N. F. Gashimzcede, M. M. Panakhov, *Sov. Phys. Sol. State* 29, 465 (1987).
[10] Chuiko, G. P., *Sov. Phys. Semi.* 19, 1279 (1985).
[11] Pahari, S., S. Bhattacharya, K. P. Ghatak, *J. Comput. Theor. Nanosci.*, 6, 2088, (2009).
[12] Ghatak, K. P., S. Bhattacharya, S. Pahari, D. De, S. Ghosh, M. Mitra, *Annalen der Physik* 17, 195 (2008).
[13] Ghatak, K. P., S. N. Biswas, *J. Appl. Phys.* 70, 299 (1991).
[14] Ghatak, K. P., S. N. Biswas, *J. Low Temp. Phys.* 78, 219 (1990).
[15] Ghatak, K. P., M. Mondal, *J. Appl. Phys.* 65, 3480 (1989).
[16] Ghatak, K. P., B. Nag, *Nanostructured Materials* 5, 769 (1995).
[17] Ghatak, K. P., M. Mondal, *Phys. Stat. Sol. (b)* 185, K5 (1994).
[18] Ghatak, K. P., B. Mitra, *Il Nuovo Cimento D* 15, 97 (1993).
[19] Ghatak, K. P., S. N. Biswas, *Phys. Stat. Solidi (b)* 140, K107 (1987).
[20] Ghatak, K. P., *Il Nuovo Cimento D* 13, 1321 (1991).
[21] Ghatak, K. P., A. Ghoshal, *Phys. Stat. Sol. (b)* 170, K27 (1992).
[22] Ghatak, K. P., B. De, B. Nag, P. K. Chakraborty, *Molecular Crystals and Liquid Crystals Science and Technology Section B: Nonlinear Optics* 16, 221 (1996).
[23] Ghatak, K. P., M. Mitra, B. Goswami, B. Nag, *Molecular Crystals and Liquid Crystals Science and Technology Section B: Nonlinear Optics* 16, 167 (1996).
[24] Ghatak, K. P., D. K. Basu, D. Basu, B. Nag, *Nuovo Cimento della Societa Italiana di Fisica D - Condensed Matter, Atomic, Molecular and Chemical Physics, Biophysics* 18, 947 (1996).
[25] Mitra, B., K. P. Ghatak, *Phys. Letts. A* 141, 81 (1989); S. K. Biswas, A. R. Ghatak, A. Neogi, A. Sharma, S. Bhattacharya, K. P. Ghatak, *Physica E: Low-Dimensional Systems and Nanostructures* 36, 163 (2007).
[26] Mondal, M., A. Ghoshal, K. P. Ghatak, *Il Nuovo Cimento D* 14, 63 (1992); K. P. Ghatak, M. Mondal *Phys. Stat. Sol. (b)*, 135, 819

(1986); L. J. Singh, S. Choudhury, D. Baruah, S. K. Biswas, S. Pahari, K. P. Ghatak, *Phys. B* 368, 188 (2005).
[27] Ghatak, K. P., B. De, M. Mondal, S. N. Biswas, *Materials Research Society Symposium – Proceedings* 198, 327 (1990); K. P. Ghatak, *Proceedings of SPIE - The International Society for Optical Engineering* 1584, 435 (1992).
[28] Ghatak, K. P., B. De, *Materials Research Society Symposium – Proceedings* 234, 55 and 59 (1991).
[29] Ghatak, K. P., B. De, M. Mondal, S. N. Biswas, *Materials Research Society Symposium – Proceedings* 184, 261 (1990).
[30] Ghatak, K. P., *Influence of band structure on some quantum processes in tetragonal semiconductors*, D. Eng. Thesis, Jadavpur University, Kolkata, India, 1991.
[31] Ghatak, K. P., S. N. Biswas, *Materials Research Society Symposium – Proceedings* 216, 465 (1990).
[32] Basu, P. K., *Theory of Optical Processes in Semiconductors, Bulk and Microstructures* (Oxford University Press, Oxford, 1997).
[33] Ghatak, K. P., S. Bhattacharya, *J. Appl. Phys.* 102, 073704, (2007).
[34] Ghatak, K. P., S. Bhattacharya, S. K. Biswas, A. De, A. K. Dasgupta, *Phys. Scr.* 75, 820, (2007).
[35] Ghatak, K. P., S. Bhattacharya, D. De, S. Pahari, A. Dey, A. K. Dasgupta, S. N. Biswas, *J. Comput. Theo. Nanosci.* 5, 1345 (2008).
[36] Lu, P. Y., C. H. Wung, C. M. Williams, S. N. G. Chu, C. M. Stiles, *Appl. Phys. Letts.* 49, 1372 (1986).
[37] Taskar, N. R., I. B. Bhat, K. K. Prat, D. Terry, H. Ehasani, S. K. Ghandhi, *J. Vac. Sci. Tech.* 7A, 281 (1989).
[38] Koch, F., *Springer Series in Solid States Sciences*, Vol. 53, pp. 20 (Springer-Verlag, Germany, 1984).
[39] Tomasetta, L. R., H. D. Law, R. C. Eden, I. Reyhimy, K. Nakano, *IEEE, J. Quant. Electron.* 14, 800 (1978).
[40] Yamato, T., K. Sakai, S. Akiba, Y. Suematsu, *IEEE J. Quant. Electron.* 14, 95 (1978).
[41] Pearsall, T. P., B. I. Miller, R. J. Capik, *Appl. Phys. Letts.* 28, 499 (1976).

[42] Washington, M. A., R. E. Nahory, M. A. Pollack, E. D. Beeke, *Appl. Phys. Letts.* 33, 854 (1978).

[43] Timmons, M. I., S. M. Bedair, R. J. Markunas, J. A. Hutchby, *Proceedings of the 16th IEEE Photovoltaic Specialist Conference* (IEEE, San Diego, California 666 (1982)).

[44] Ghatak, K. P., S. Bhattacharya, D. De, *Einstein Relation in Compound Semiconductors and Their Nanostructures*, Springer Series in Materials Science, Vol. 116 (Springer-Verlag, Germany, 2008).

[45] Ghatak, K. P., S. Bhattacharya, D. De, *Photoemission from Optoelectronic Materials and Their Nanostructures*, Springer Series in Nanostructure Science and Technology (Springer-New York, USA, 2009).

[46] Ghatak, K. P., S. N. Biswas, *Journal of applied physics* 70, 4309-4316, 1991; KP Ghatak, Il Nuovo Cimento D 13, 1321-1324; KP Ghatak, SN Biswas, Orlando'91, Orlando, FL, 206-214.

[47] Ghatak, K. P., S. N. Biswas, Orlando'91, Orlando, FL, 149-166; KP Ghatak, *International journal of electronics* 71, 239 (1991); KP Ghatak, A. Ghoshal, physica status solidi (a) 126, K53-K58 (1991); Ghatak, K. P., A. Ghoshal, B. Mitra, Il Nuovo Cimento D, 13, 867(1991); K. P. Ghatak, S. N. Biswas, *Journal of applied physics* 70, 299(1991); K. P. Ghatak, B. Mitra, *Physics Letters A* 156, 233(1991); K. P. Ghatak, *Acta Physica Hungarica* 70, 91(1991).

[48] Ghatak, K. P., S. Bhattacharyya, S. N. Biswas, *Acta Physica Hungarica* 70, 83 (1991); K. P. Ghatak, *Acta Physica Hungarica* 70), 77(1991); K. P. Ghatak, B. De, M. Mondal, *physica status solidi* (b) 165, K53(1991); KP Ghatak, M. Mondal, *physica status solidi* (b) 164, 477(1991).

[49] Mitra, B., K. P. Ghatak, *physica status solidi* (b) 164, K13(1991); K. P. Ghatak, M. Mondal, *Acta Physica Hungarica*, 69, 113(1991); K. P. Ghatak, *Acta Physica Hungarica* 69, 121(1991); K. P. Ghatak, B. Mitra, *International Journal of Electronics Theoretical and Experimental*, 70, 343(1991).

[50] Ghatak, K. P., M. Mondal, *Journal of applied physics* 69, 1666(1991); SN Biswas, KP Ghatak, *International Journal of Electronics Theoretical and Experimental,* 70, 125(1991); KP Ghatak, B De, *MRS Proceedings* 234, 59(1991).

[51] Ghatak, K. P., A. Ghoshal, S. Bhattacharyya, M. Mondal, San Diego'90, 8-13 July, 471-489(1991); K. P. Ghatak, SN Biswas, SPIE 1484, 149(1991); K. P. Ghatak, B. De, *MRS Proceedings* 234, 55(1991); K. P. Ghatak, B. De, *MRS Proceedings* 228, 237(1991).

[52] Ghatak, K. P., *Acta Physica Hungarica* 68, 253-256, (1990); KP Ghatak, A. Ghoshal, B. De, San Dieg-DL Tentative, 111-133(1990); K. P. Ghatak, A. Ghoshal, S. Bhattacharyya, M. Mondal, Orlando'90, 16-20 April, 414-424(1990).

[53] Ghatak, K. P., M. Mondal, S. Bhattacharyya, Semi-DL tentative, 113-126(1990); K. P. Ghatak, S. Bhattacharyya, M. Mondal, Orlando'90, 16-20 April, 586-599(1990); K. P. Ghatak, M. Mondal, S. N. Biswas, *Journal of Applied Physics* 68 3032(1990); K. P. Ghatak, The Hague'90, 12-16 April, 53-67; KP Ghatak, M Mondal, *physica status solidi* (b) 160, 673-681(1990).

[54] Ghoshal, A., B. Mitra, K. P. Ghatak, Il Nuovo Cimento D 12, 891(1990); B. Mitra, K. P. Ghatak, *Physics Letters A* 146 (6), 357(1990); S. N. Biswas, K. P. Ghatak, *International Journal of Electronics Theoretical and Experimental*, 68, 957(1990).

[55] Ghatak, K. P., A. Ghoshal, S. Bhattacharyya, OE/LASE'90, 14-19 Jan., Los Angeles, CA, 282-290(1990); A. Ghoshal, M. Mondal, K. P. Ghatak, *physica status solidi* (b) 158, K123(1990); N Chattopadhyay, KP Ghatak, M Mondal, *Acta Physica Hungarica* 67, 163-170(1990).

[56] Ghatak, K. P., A. Ghoshal, S. Bhattacharyya, M. Mondal, Orlando'90, 16-20 April, 225-237(1990); KP Ghatak, SN Biswas, *Journal of Low Temperature Physics*;(USA) 78(1990); KP Ghatak, SN Biswas, *Journal of low temperature physics* 78, 219(1990).

[57] Ghatak, K. P., S. Bhattacharyya, B. Mitra, *Fizika* 22, 391(1990); K. P. Ghatak, *MRS Proceedings* 216, 469(1990); K. P. Ghatak, S. N. Biswas, *MRS Proceedings* 216, 465(1990); KP Ghatak, A. Ghoshal,

S. Bhattacharyya, M. Mondal, Ultrahigh-and High-speed Photography, Videography, Photonics, and Velocimetry'90: Eighth in a Series: 10-13 July (1990), San Diego, California, 8, 471.

[58] Ghatak, K. P., B. De, M. Mondal, S. N. Biswas, *MRS Proceedings* 184, 261(1990); K. P. Ghatak, B. De, M. Mondal, S. N. Biswas, *MRS Proceedings* 198, 333(1990); KP Ghatak, B Mitra, *Physica Scripta* 42, 103, (1990).

[59] Ghatak, K. P., A. Ghoshal, S. N. Biswas, *33rd Annual Techincal Symposium*, 229-242, (1989); B Mitra, KP Ghatak, *Physics Letters A* 142, 401(1989); B Mitra, KP Ghatak, *Physics Letters A* 141, 81(1989).

[60] Ghatak, K. P., M. Mondal, *Journal of applied physics* 66, 3056(1989); B. Mitra, K. P. Ghatak, *Solid-state electronics* 32, 810(1989); B. Mitra, A. Ghoshal, K. P. Ghatak, physica status solidi (b) 155, K23(1989); B Mitra, A Ghoshal, KP Ghatak, *physica status solidi* (b) 154, K147(1989).

In: Quantum Wires: An Overview
Editor: Kamakhya Prasad Ghatak
ISBN: 978-1-53617-676-6

Chapter 6

ON THE SIMPLIFIED EXPRESSION OF THE DENSITY-OF-STATES FUNCTION IN HEAVILY DOPED SUPERLATTICES IN THE PRESENCE OF ELECTRIC FIELD

***N. Debbarma*[1], *M. Mondal*[2] *and M. Debbarma*[3]**

[1]Department of Computer Science and Engineering, National Institute of Technology, Agartala,Tripura, India
[2]Department of Physics, Palpara, Purba Medinipur, West Bengal, India
[3]Department of Physics, Women's College, Agartala, Tripura, India

ABSTRACT

In recent years with the advent of low dimensional electronic devices, there has been considerable interest in investigating the different electronic properties in the presence of strong electric field. The Density-of-States function (DOS) affects nearly all the physical properties of different semiconductors and their nano structures. In this chapter we study the DOS in the presence of intense electric field in Heavily Doped

(HD) Quantum Wires (QWs) of III-V, ternary and quaternary semiconductors. We also derive the expressions of the DOS in QWHD superlattices of Kane type semiconductors with graded interfaces in the presence of intense electric field together with QWHD superlattices with graded interfaces under strong electric field.

With the advent of modern nano devices, there has been considerable interest in studying the electric field induced processes in semiconductors having different band structures. It appears from the literature that the studies have been made on the assumption that the carrier dispersion laws are invariant quantities in the presence of intense electric field, which is not fundamentally true [1, 2]. In this chapter, we shall study the Density-of-states function (DOS) in quantum confined optoelectronic semiconductors under strong electric field. The importance of DOS function is already well-known in the whole field of Nano Technology in general [3-30]. In what follows in section 2.1, and attempt is made to investigate the DOS in the presence of intense electric field in Heavily Doped (HD) Quantum Wires (QWs) of III-V, ternary and quaternary semiconductors. In section 2.2 the DOS in QWHD superlattices of Kane type semiconductors with graded interfaces in the presence of intense electric field has been studied. In section 2.3 the DOS in QWHD superlattices of Kane type semiconductors with graded interfaces in the presence of intense electric field has been investigated.

THE DOS IN THE PRESENCE OF INTENSE ELECTRIC FIELD IN HD III-V, TERNARY AND QUATERNARY SEMICONDUCTORS

The expression of the inter-band transition matrix element (X_{12}) in this case can be written as

$$X_{12} = i\int u^{*}_{\bar{k}1}(\bar{r}).\frac{\partial}{\partial k_x} u_{\bar{k}2}(\bar{r}) d^3r \tag{1}$$

where $u_{\bar{k}1}(\bar{r}) \equiv u_1(\bar{k},\bar{r})$ and $u_{\bar{k}2}(\bar{r}) \equiv u_2(\bar{k},\bar{r})$ in which $u_1(\bar{k},\bar{r})$ and $u_2(\bar{k},\bar{r})$ are given by (1) and (2) respectively.

In the case of the presence of an external electric field, F_s along x axis, the inter-band transition matrix-element, X_{12}, has finite interaction band same band, e.g.,

$$\langle S|S\rangle = \langle X|X\rangle = \langle Y|Y\rangle = \langle Z|Z\rangle = 1$$
$$\langle X|Y\rangle = \langle Y|Z\rangle = \langle Z|X\rangle = 0$$
$$\langle S|X\rangle = \langle X|S\rangle = 0; \langle S|Y\rangle = \langle Y|S\rangle = 0 \; and \; \langle S|Z\rangle = \langle Z|S\rangle = 0$$

Using the appropriate equations, we can write

$$X_{12} = i\int d^3r\{a_{k_+}[(iS)\downarrow'] + b_{k_+}[(\frac{X'-iY'}{\sqrt{2}})\uparrow'] + c_{k_+}[Z'\downarrow']\}^* \frac{\partial}{\partial k_x}\{a_{k_-}[(iS)\uparrow'] - b_{k_-}[(\frac{X'+iY'}{\sqrt{2}}\downarrow')] + c_{k_-}[Z'\uparrow']\}$$
$$= i\int d^3r[\{(a_{k_+}\frac{\partial}{\partial k_x}a_{k_-}).([(iS)\downarrow']^*.[(iS)\uparrow']) + (\frac{b_{k_+}}{\sqrt{2}}\frac{\partial}{\partial k_x}a_{k_-})[(X'-iY')\uparrow']^*.[(iS)\uparrow']$$
$$+(c_{k_+}\frac{\partial}{\partial k_x}a_{k_-})[(Z'\downarrow')]^*.[(iS)\uparrow']\}]$$
$$+i\int d^3r[\{(-\frac{a_{k_+}}{\sqrt{2}}\frac{\partial}{\partial k_x}b_{k_-}).[(iS)\downarrow']^*.[(X'+iY')\downarrow'] - (\frac{b_{k_+}\frac{\partial}{\partial k_x}b_{k_-}}{2})[(X'-iY')\uparrow']^*.[(X'+iY')\downarrow']$$
$$-(\frac{c_{k_+}\frac{\partial}{\partial k_x}b_{k_-}}{\sqrt{2}})[(Z'\downarrow')]^*.[(X'+iY')\downarrow']\}]$$
$$= i\int d^3r[\{(a_{k_+}\frac{\partial}{\partial k_x}c_{k_-}).([(iS)\downarrow']^*.[Z'\uparrow']) + (\frac{b_{k_+}}{\sqrt{2}}\frac{\partial}{\partial k_x}c_{k_-})[(X'-iY')\uparrow']^*.[Z'\uparrow']$$
$$+(c_{k_+}\frac{\partial}{\partial k_x}c_{k_-})[(Z'\downarrow')]^*.[Z'\uparrow']\}] \qquad (2)$$

Therefore,

$$\frac{X_{12}}{i} = [\{[(a_{k_+}\frac{\partial}{\partial k_x}a_{k_-}).\langle iS|iS\rangle\langle\downarrow'|\uparrow'\rangle] + [(\frac{b_{k_+}}{\sqrt{2}}\frac{\partial}{\partial k_x}a_{k_-})\langle(X'-iY')|iS\rangle\langle\uparrow'|\downarrow'\rangle]$$
$$+[(c_{k_+}\frac{\partial}{\partial k_x}a_{k_-})\langle Z'|iS\rangle\langle\downarrow'|\uparrow'\rangle]\}]$$
$$-[\{[(\frac{a_{k_+}}{\sqrt{2}}\frac{\partial}{\partial k_x}b_{k_-}).\langle iS|(X'+iY')\rangle\langle\downarrow'|\downarrow'\rangle] + [(\frac{b_{k_+}}{2}\frac{\partial}{\partial k_x}b_{k_+})\langle(X'-iY')|(X'+iY')\rangle\langle\uparrow'|\downarrow'\rangle]$$
$$+[(\frac{c_k}{\sqrt{2}}\frac{\partial}{\partial k_x}b_{k_-})\langle Z'|(X'+iY')\rangle\langle\downarrow'|\downarrow'\rangle]\}]$$
$$+\{[(a_k\frac{\partial}{\partial k_x}c_{k_-}).\langle iS|Z'\rangle\langle\downarrow'|\uparrow'\rangle] + [(\frac{b_{k_+}}{\sqrt{2}}\frac{\partial}{\partial k_x}c_{k_-}).\langle(X'-iY')|Z'\rangle\langle\uparrow'|\uparrow'\rangle] + [(c_k\frac{\partial}{\partial k_x}c_{k_-}).\langle Z'|Z'\rangle\langle\downarrow'|\uparrow'\rangle]\}(3)$$

Therefore we can write,

$$X_{12} = i\{-(a_{k_+} \frac{\partial}{\partial k_x} a_{k_-})\langle\downarrow'|\uparrow'\rangle + (c_{k_+} \frac{\partial}{\partial k_x} c_{k_-})\langle\downarrow'|\uparrow'\rangle\} \quad (4)$$

We can prove that

$$\langle\downarrow'|\uparrow'\rangle = \frac{1}{2}(\hat{r}_1 + i\hat{r}_2) \quad (5)$$

Therefore (4) and (5) we get

$$X_{12} = i\{-(a_{k+} \frac{\partial}{\partial k_x} a_{k-}) + (c_{k+} \frac{\partial}{\partial k_x} c_{k-})\}\langle\downarrow'|\uparrow'\rangle = -i\{(a_{k+} \frac{\partial}{\partial k_x} a_{k-}) - (c_{k+} \frac{\partial}{\partial k_x} c_{k-})\}.\frac{1}{2}(\hat{r}_1 + i\hat{r}_2)$$

$$= \frac{-iA(k)}{2}(\hat{r}_1 + i\hat{r}_2) \quad (6)$$

where,

$$A(k) = (a_{k+} \frac{\partial}{\partial k_x} a_{k-}) - (c_{k+} \frac{\partial}{dk_x} c_{k-}) \quad (7)$$

From (6), we find,

$$|X_{12}|^2 = \frac{1}{4}A^2(k)(1+1) = [since, |\hat{r}_1| = |\hat{r}_2| = 1] \quad (8)$$

Considering spin-up and spin-down, we have to multiply by 2

$$|X_{12}|^2 = 2 \times \frac{1}{2}A^2(k) = A^2(k) \quad (9)$$

We can evaluate X11 and X22. in the following way:

$$X_{11} = i\int u^*_{\bar{k}1}(\bar{r}).\frac{\partial}{\partial k_x} u_{\bar{k}1}(\bar{r}).d^3r$$

$$= i\int d^3r\{(a_{k_+} \frac{\partial}{\partial k_x} a_{k_+}) + (b_{k_+} \frac{\partial}{\partial k_x} b_{k_+}) + (c_{k_+} \frac{\partial}{\partial k_x} c_{k_+})\}$$

$$= \frac{1}{2}i\int d^3r\{\frac{\partial}{\partial k_x}(a_k^2 + b_k^2 + c_k^2)\} = \frac{1}{2}i\int d^3r\{\frac{\partial}{\partial k_x}(1)\} = 0$$

since $a_{k+}^2 + b_{k+}^2 + c_{k+}^2 = 1$.

Therefore $X_{11} = 0$, and similarly we can prove $X_{22} = 0$. Thus, we conclude that intraband momentum matrix element due to external electric field (X_{CC}) is zero. From the expression of $a_{k\pm}$ we can write

$$a^2_{k_+} = r_0^2\left[\frac{E_g - \gamma_{k_+}^2 (E_g - \delta')}{E_g + \delta'}\right]^2 \textit{and } a_{k-}^2 = r_0^2\left[\frac{E_g - \gamma_{k-}^2 (E_g - \delta')}{E_g + \delta'}\right]^2.$$

Therefore,

$$2a_{k-}\frac{\partial}{\partial k_x}a_{k-} = r_0^2\left[-\left(\frac{E_g - \delta'}{E_g + \delta'}\right)\right]\frac{\partial \gamma_{k-}^2}{\partial k_x} \textit{ and } \frac{\partial a_{k-}}{\partial k_x} = -\frac{r_0^2}{2}\left(\frac{E_g - \delta'}{E_g + \delta'}\right)\frac{1}{a_{k-}}\frac{\partial \gamma_{k-}^2}{\partial k_x}$$

Combining we can write $a_{k+}\dfrac{\partial a_{k-}}{\partial k_x} = -\dfrac{r_0^2}{2}\left(\dfrac{E_g - \delta'}{E_g + \delta'}\right)\dfrac{a_{k+}}{a_{k-}}\dfrac{\partial \gamma_{k-}^2}{\partial k_x}$

Similarly, $c_{k_+} = t\gamma_{k_+}$ and $c_{k\_} = t\gamma_{k\_}$.

Therefore, $c_{k+}\dfrac{\partial}{\partial k_x}c_{k-} = \dfrac{t^2}{2}\dfrac{c_{k+}}{c_{k-}}\dfrac{\partial \gamma_{k-}^2}{\partial k_x}$

$$A(k) = \left\{-\frac{r_0^2}{2}\left[\frac{E_g - \delta'}{E_g + \delta'}\right]\frac{a_{k+}}{a_{k-}} - \frac{t^2}{2}\frac{c_{k+}}{c_{k-}}\right\}\frac{\partial \gamma_{k-}^2}{\partial k_x}$$

Now,

$$\left(\frac{a_{k+}}{a_{k-}}\right)^2 = \frac{E_g - \gamma_{k+}^2 (E_g - \delta')}{E_g - \gamma_{k-}^2 (E_g - \delta')} = \frac{E_g - \dfrac{\eta - E_g}{2(\eta + \delta')}(E_g - \delta')}{E_g - \dfrac{\eta + E_g}{2(\eta + \delta')}(E_g - \delta')}$$

$$= \frac{2E_g - (\eta + \delta') - (\eta - E_g)(E_g - \delta')}{2E_g(\eta + \delta') - (\eta + E_g)(E_g - \delta')} = \frac{\eta(E_g + \delta') + E_g(E_g + \delta')}{\eta(E_g + \delta') - E_g(E_g - 3\delta')}$$

Therefore,

$$\left(\frac{a_{k+}}{a_{k-}}\right)^2 = \frac{\eta + E_g}{\eta - E_g\left(\frac{E_g - 3\delta'}{E_g + \delta'}\right)} = \frac{\eta + E_g}{\eta - E_g'} \quad \text{where,} \quad E_g' = \frac{E_g(E_g - 3\delta')}{E_g + \delta'}.$$

Thus, $\frac{a_{k+}}{a_{k-}} = \sqrt{\frac{\eta + E_g}{\eta - E_g'}}$ similarly, $\frac{c_{k+}}{c_{k-}} = \frac{\gamma_{k+}}{\gamma_{k-}} = \sqrt{\frac{\eta - E_g}{\eta + E_g'}}$ and thus,

$$A(k) = -\{P(\frac{\eta + E_g}{\eta - E_g'})^{1/2} + Q(\frac{\eta - E_g}{\eta + E_g})^{1/2}\}\frac{\partial \gamma_{k-}^2}{\partial k_x} \tag{10}$$

where, $P = \frac{r_0^2}{2}\left(\frac{E_g - \delta'}{E_g + \delta'}\right)$ *and* $Q = t^2/2$

Now $\quad \gamma_{k-}^2 = \frac{\eta + E_g}{2(\eta + \delta')}$ so that

$$\frac{\partial \gamma_{k-}^2}{\partial k_x} = \frac{1}{2}\left[\frac{\partial \eta / \partial k_x}{(\eta + \delta')} - \frac{\eta + E_g}{(\eta + \delta')^2}\frac{\partial \eta}{\partial k_x}\right]$$

Thus,

$$\frac{\partial \gamma_{k-}^2}{\partial k_x} = \frac{1}{2}\left[\frac{\eta + \delta' - \eta - E_g}{(\eta + \delta')}\right]\frac{\partial \eta}{\partial k_x} = -\frac{1}{2}\frac{(E_g - \delta')}{(\eta + \delta')^2}\cdot\frac{\partial \eta}{\partial k_x} \tag{11}$$

From (10) and (11), we get

$$A(k) = \frac{1}{2}\frac{(E_g - \delta')}{(\eta + \delta')^2}\cdot\frac{\partial \eta}{\partial k_u}\cdot\{P(\frac{\eta + E_g}{\eta - E_g'})^{1/2} + Q(\frac{\eta - E_g}{\eta + E_g})^{1/2}\} \tag{12}$$

This implies

$$\frac{\partial \eta}{\partial k_x} = \frac{E_g \hbar^2}{m_r} . \frac{k_x}{\eta} \tag{13}$$

From (12) and (13), one can write

$$A(k) = \frac{E_g \hbar^2}{2m_r} . \frac{k_u}{\eta} . \frac{(E_g - \delta')}{(\eta + \delta')^2} . \{P(\frac{\eta + E_g}{\eta - E_g^{'}})^{1/2} + Q(\frac{\eta - E_g}{\eta + E_g})^{1/2}\} \tag{14}$$

Thus,

$$|A(k)|^2 = \frac{E_g^2 (E_g - \delta')\hbar^2}{4m_r} \frac{\hbar^2 k_x^2}{m_r} \frac{1}{\eta^2} \frac{1}{(\eta + \delta')^4} \{P(\frac{\eta + E_g}{\eta - E_g^{'}})^{1/2} + Q(\frac{\eta - E_g}{\eta + E_g})^{1/2}\}^2 \tag{15}$$

$$and\ |X_{12}|^2 = |A(\bar{k})|^2 \tag{16}$$

From (15) and (16) we can write the square of the magnitude of the inter-band transition matrix element due to external electric field ($|X_{CV}|^2$)

It is well-known that the energy Eigen value, $E_n^{(2)}(\bar{k})$ in the presence of a perturbed Hamiltonian, H', is given by [1]

$$E_n^{(2)}(\bar{k}) = E_n(\bar{k}) + \langle n\bar{k} | H' | n\bar{k} \rangle + \{| \langle n\bar{k} | H' | n\bar{k} \rangle |^2 / [E_n(\bar{k}) - E_m(\bar{k})]\} \tag{17}$$

where

$$H\psi_n(\bar{k}, \bar{r}) = E\psi(\bar{k}, \bar{r}) \tag{18}$$

$$H = H_0 + H' \tag{19}$$

$$H_0 u_n(\bar{k}, \bar{r}) = E_n(\bar{k}) u_n(\bar{k}, \bar{r}) \tag{20}$$

in which, H is the total Hamiltonian, $\psi(\bar{k}, \bar{r})$ is the wave function, where $u_n(\bar{k}, \bar{r})$ is the periodic function of it, H_0 is the unperturbed Hamiltonian, n is the band index, and $E_n(\bar{k})$ is the energy of an electron in the periodic lattice.

For an external electric field (F_S) applied along the x-axis, the perturbed Hamiltonian (H') can be written as

$$H' = -F.x \tag{21}$$
$$where\, F = (eF_s)$$

Therefore we get

$$E_n^{(2)}(\bar{k}) = E_n(\bar{k}) - F\langle n\bar{k} \mid H' \mid n\bar{k}\rangle + F^2\{|\langle n\bar{k} \mid H' \mid n\bar{k}\rangle|^2 /[E_n(\bar{k}) - E_m(\bar{k})]\} \tag{22}$$

In (22), the second and the third terms are due to the perturbation factor.

For

$$X_{nm}(\bar{k}) = \langle n\bar{k} \mid x \mid m\bar{k}\rangle \tag{23}$$

we find

$$X_{nm}(\bar{k}) = i\int u_n^{*}(\bar{k},\bar{r})(\partial/\partial u)[u_m(\bar{k},\bar{r})]d^3r \tag{24}$$

where k_x is the x component of the $\bar{k}$ and the integration in (24) extends over the unit cell. From (22), (23) and (24) with the *n* corresponds to the conduction band (C) and *m* corresponds to the valance band (V), we get

$$E_c^{(2)}(\bar{k}) = E_c(\bar{k}) - FX_{cc} + \{F^2 \mid X_{CV} \mid^2 /[E_C(\bar{k}) - E_V(\bar{k})]\} \tag{25}$$

Thus combining the appropriate equations, the dispersion relation of the conduction electrons in the presence of electric field along x-axis can be written as

$$I_{11}(E) = \frac{\hbar^2 k_x^2}{2m_c} + \frac{\hbar^2 k_y^2}{2m_c} + \frac{\hbar^2 k_z^2}{2m_c} + \frac{F^2 \mid X_{12} \mid^2}{\eta} = [\frac{\hbar^2 k_x^2}{2m_c} + \frac{\hbar^2 k_y^2}{2m_c} + \frac{\hbar^2 k_z^2}{2m_c}$$
$$+\{\frac{\hbar^2 k_x^2}{2m_c}.\frac{2m_c}{m_r}.\frac{F^2\hbar^2 E_g^2(E_g - \delta')^2}{4m_r}\frac{1}{\eta^3}\frac{1}{(\eta+\delta')^4}[P(\frac{\eta+E_g}{\eta-E_g'})^{1/2} + Q(\frac{\eta-E_g}{\eta+E_g})^{1/2}]^2\}] \tag{26}$$

When $F \rightarrow 0$, we have from (26), $k^2 \rightarrow \frac{2m_c}{\hbar^2} I_{11}(E)$ and $\eta_1^2 = [E_g^2 + E_g \frac{2m_c}{m_r} I_{11}(E)]$. Using the method of successive approximation one can write

$$1 = \frac{\hbar^2 k_x^2}{2m_c I_{11}(E)} + \frac{\hbar^2 k_y^2}{2m_c I_{11}(E)} + \frac{\hbar^2 k_z^2}{2m_c I_{11}(E)} + \frac{\hbar^2 k_x^2}{2m_c I_{11}(E)} .\Phi(E,F) \tag{27}$$

$$where,\ \Phi(E,F) = \frac{2m_c}{m_r} \frac{F^2 \hbar^2 E_g^2 (E_g - \delta')^2}{4m_r} \frac{1}{\eta_1^3} \frac{1}{(\eta_1 + \delta')^4} [P(\frac{\eta_1 + E_g}{\eta_1 - E_g'})^{1/2} + Q(\frac{\eta_1 - E_g}{\eta_1 + E_g})^{1/2}]^2$$

Therefore, the *E-k* dispersion relation in the presence of an external electric field for III-V, ternary and quaternary materials whose unperturbed energy band structures are defined by the three band model of Kane can be expressed as

$$\frac{k_x^2}{\frac{2m_c}{\hbar^2}[\frac{I_{11}(E)}{1+\Phi(E,F)}]} + \frac{k_y^2}{\frac{2m_c}{\hbar^2} I_{11}(E)} + \frac{k_z^2}{\frac{2m_c}{\hbar^2} I_{11}(E)} = 1 \tag{28}$$

In (28), the coefficients of k_x, k_y and k_z are not same and for this reason, this basic equation is “anisotropic” in nature together with the fact that the anisotropic dispersion relation is the ellipsoid of revolution in the k-space.

From (28) the expressions of the effective electron masses along *x*, *y*, and *z* directions can, respectively, be written as

$$m_x^*(E,F) = \hbar^2 k_x \left.\frac{\partial k_x}{\partial E}\right|_{k_y=0,k_z=0} = m_c[1+\Phi(E,F)]^{-2}[[1+\Phi(E,F)]I_{11}'(E) - I_{11}(E)\Phi'(E,F)] \tag{29}$$

$$m_y^*(E,F) = \hbar^2 k_y \left.\frac{\partial k_y}{\partial E}\right|_{k_x=0,k_z=0} = m_0 I_{11}'(E) \tag{30}$$

$$m_z^*(E,F) = \hbar^2 k_z \left.\frac{\partial k_z}{\partial E}\right|_{k_x=0,k_y=0} = m_0 I_{11}'(E) \tag{31}$$

where $I'_{11}(E)=\frac{\partial}{\partial E}(I_{11}(E))$ and $\Phi'(E,F)=\frac{\partial}{\partial E}[\Phi(E,F)]$. It may be noted from (29) that the effective mass along x- direction is a function of both electron energy and electric field respectively whereas from (30) and (31) we can infer the expressions of the effective masses along y and z directions are same and they depend on the electron energy only. Thus in the presence of an electric field, the mass anisotropy for Kane type semiconductors depends both on electron energy and electric field respectively.

The use of the usual approximation [2]

$$k_x^2 \approx \frac{1}{3}k^2 \tag{32}$$

in (28), leads to the simplified expression of the electron energy spectrum in the present case as

$$I_{11}(E)=\frac{\hbar^2k^2}{2m_c}+\frac{F^2\hbar^2E_g^2(E_g-\delta')^2}{12m_r}\frac{2m_c}{m_r}I_{11}(E)\frac{1}{\eta_1}\frac{1}{(\eta_1+\delta')^4}\{P(\frac{\eta_1+E_g}{\eta_1-E'_g})^{1/2}+Q(\frac{\eta_1-E_g}{\eta_1+E_g})^{1/2}\}^2 \tag{33}$$

The (33) can approximately be written as

$$where,\ \frac{\hbar^2k^2}{2m_c}=[e_1E^4+e_2E^3+e_3E^2+e_4E+e_5-\frac{e_6}{1+CE}+e_7(1+CE)^{-2}] \tag{34a}$$

$e_1=Q_f.\omega_1,\ Q_f=\frac{m_c}{m_r}E_g^{-4}[5e_fE_g^{-2}-6G_f+7h_fE_g^{-4}],\ e_f=A_fP_f,\ A_f=[F\hbar E_g(E_g-\delta')]^2m_c(6m_r^2(\delta')^4)^{-1},$

$F=eF_s,\ \ G_f=e_f(4\delta'+C_f),\ \ C_f=(2E_gQ^2+PQ(E_g-E'_g)-2P^2E_g),$

$E'_g=\frac{E_g(E_g-3\delta')}{E_g+\delta'},\ \ P=\frac{r_0^2}{2}(\frac{E_g-\delta'}{E_g+\delta'}),\ \ r_0=[\frac{6}{\chi}(E_g+\Delta)(E_g+\frac{2}{3}\Delta)]^{1/2},$

$Q=\frac{t^2}{2},t=[\frac{6}{\chi}(E_g+\frac{2}{3}\Delta)]^{1/2},\ \ h_f=(4\delta'e_fC_f)(B_f)^{-1},\ \ B_f=(P+Q)^2$

$P_f=E_g^{-3}(e_fE_g^{-2}-G_f+h_fE_g^{-4}),\ \ \omega_1=a_1^2,\ \ a_1=\frac{ab}{c},\ \ a=\frac{1}{E_g},\ \ b=\frac{1}{E_g+\Delta},\ \ c=(E_g+\frac{2}{3}\Delta)^{-1},$

$e_2=Q_f.\omega_2,\ \ \omega_2=2a_1b_1,\ \ b_1=(c)^{-2}(ac+bc-ab),\ e_3=(1-P_f)a_1+Q_f.\omega_3,\ \ \omega_3=(b_1^2+2a_1c_1),$

$c_1=[\frac{1}{c}(1-\frac{a}{c})(1-\frac{b}{c})],\ \ e_4=[(1-P_f)b_1+Q_f\omega_4],\ \ \omega_4=2b_1c_1,\ e_5=[(1-P_f)c_1+\omega_5Q_f]$

$\omega_5=(c_1^2-2c_1b_1),e_7=Q_f\omega_7,\omega_7=c_1^2,\ e_6=[(1-P_f)c_1-Q_f\omega_6]\ and\ \omega_6=\frac{2c_1b_1}{c}(1-\frac{cc_1}{b_1})]$

The Gaussian distribution $F(V)$ of the impurity potential is given by [121]

$$F(V) = (\pi\eta_g^2)^{-1/2} \exp(-V^2 / \eta_g^2) \tag{34b}$$

where, η_g is the impurity scattering potential. It appears from (34b) that the variance parameter η_g is not equal to zero, but the mean value is zero. Further, the impurities are assumed to be uncorrelated and the band mixing effect has been neglected in this simplified theoretical formalism.

We have to average the kinetic energy in the order to obtain the E-k dispersion relation in nonlinear optical materials in the presence of band tails.

Using (1.26b) and (34) we get

$$\frac{\hbar^2 k^2}{2m_c}\int_{-\infty}^{E} F(V)dV = e_1\int_{-\infty}^{E}(E-V)^4 F(V)dV + e_2\int_{-\infty}^{E}(E-V)^3 F(V)dV + e_3\int_{-\infty}^{E}(E-V)^2 F(V)dV +$$
$$e_4\int_{-\infty}^{E}(E-V)F(V)dV + e_5\int_{-\infty}^{E}F(V)dV - e_6\int_{-\infty}^{E}\frac{F(V)dV}{[1+c(E-V)]} + e_7\int_{-\infty}^{E}\frac{F(V)dV}{[1+c(E-V)]^2} \tag{35}$$

We can prove that

$$\int_{-\infty}^{E}(E-V)^4 F(V)dV = \frac{E^4}{2}[1+Erf(E/\eta_g)] + \frac{3\eta_g^4}{8\pi}[1+Erf(E/\eta_g) - \frac{2\eta_g}{3E}\exp(\frac{-E^2}{\eta_g^2})]$$
$$+\frac{3}{2}(E\eta_g)^2[1+Erf(E/\eta_g)] + \frac{E^3\eta_g}{\sqrt{\pi}}\exp(\frac{-E^2}{\eta_g^2}) + \frac{2}{\sqrt{\pi}}E\eta_g^3\exp(\frac{-E^2}{\eta_g^2})[1+\frac{E^2}{\eta_g^2}] = \psi_0(\mathrm{E},\eta_g) \tag{36}$$

We know that

$$\int_{-\infty}^{E}\frac{F(V)dV}{[1+c(E-V)]} = c_1(c,E,\eta_g) - ic_2(c,E,\eta_g) \tag{37}$$

where

$$c_1(c,E,\eta_g) = [\frac{2}{c\eta_g\sqrt{\pi}}\exp(-u^2)[\sum_{p=1}^{\infty}\exp(\frac{-p^2}{4})(p)^{-1}Sinh(pu)]],\ c_2(c,E,\eta_g) = \frac{\sqrt{\pi}}{c\eta_g}\exp(-u^2)$$

and $u = \dfrac{1+cE}{c\eta_g}$

We know that

$$\frac{\partial}{\partial x}\int_{A(x)}^{B(x)} F(x,\zeta)d\zeta = \int_{A(x)}^{B(x)} \frac{\partial}{\partial x}[F(x,\zeta)]d\zeta + F(x,B(x))\frac{\partial B(x)}{\partial x} - F(x,A(x))\frac{\partial A(x)}{\partial x} \tag{38}$$

Using (37) and (38), we get

$$\int_{-\infty}^{E} \frac{F(V)dV}{[1+c(E-V)]^2} = c_3(c,E,\eta_g) - ic_4(c,E,\eta_g) \tag{39}$$

$$where\, c_3(c,E,\eta_g) = [\frac{-4u\exp(-u^2)}{c^2\eta_g^2\sqrt{\pi}}[\sum_{p=1}^{\infty}\exp(\frac{-p^2}{4})(p)^{-1} Sinh(pu)] + \frac{1}{\pi c\eta_g}\exp(\frac{-E^2}{\eta_g^2})$$
$$-\frac{2\exp(-u^2)}{c^2\eta_g^2\sqrt{\pi}}\sum_{p=1}^{\infty}\exp(\frac{-p^2}{4})\mathrm{Cos}\,h(pu)] \;\; and \;\; D_3(c,E,\eta_g) = \frac{2u}{c^2\eta_g^2}\exp(-u^2)$$

Therefore the Dispersion Relation in HD Kane type semiconductors can be written as

$$\frac{\hbar^2 k^2}{2m_c} = J_1(E,c,\eta_g,F) + i\, J_2(E,c,\eta_g,F) \tag{40}$$

where,

$$J_1(E,c,\eta_g,F) = 2[1+Erf(E/\eta_g)]^{-1}[e_1\psi_0(E,\eta_g) + e_2\psi_1(E,\eta_g) + e_3\theta_0(E,\eta_g) + e_4\gamma_0(E,\eta_g)$$
$$+e_5\frac{1}{2}[1+Erf(E/\eta_g)] - e_6c_1(E,c,\eta_g) + e_7c_3(E,c,\eta_g)], \psi_1(E,\eta_g) = [\frac{E}{2}[1+Erf(E/\eta_g)][E^2+\frac{3}{2}\eta_g^2]$$
$$+\frac{\eta_g}{2\sqrt{\pi}}\exp(\frac{-E^2}{\eta_g^2})(4E^2+\eta_g^2)]$$

and

$$J_2(E,c,\eta_g,F) = 2[1+Erf(E/\eta_g)]^{-1}[e_6c_2(E,c,\eta_g) + e_7D_3(E,c,\eta_g)]$$

The DR in HDQWs of III-V, ternary and quaternary materials this case is given by

$$\frac{\hbar^2}{2m_c}(\frac{n_z\pi}{d_z})^2 + \frac{\hbar^2}{2m_c}(\frac{n_y\pi}{d_y})^2 + \frac{\hbar^2 k_x^2}{2m_c} = [J_1(E,c,\eta_g,F) + i\, J_2(E,c,\eta_g,F)] \tag{41}$$

The DOS function can be expressed as

$$N_{1D}(E,c,\eta_g,F)=\frac{g_v\sqrt{2m_c}}{\pi\hbar}\sum_{n_y=1}^{n_{y\max}}\sum_{n_z=1}^{n_{z\max}}\frac{[J_1'(E,c,\eta_g,F)+i\,J_2'(E,c,\eta_g,F)]H(E-E_{n_{z18,2HD}})}{\sqrt{[J_1(E,c,\eta_g,F)+i\,J_2(E,c,\eta_g,F)]-G_2(n_y,n_z)}} \quad (42)$$

where, $E_{n_{z18,2HD}}$ is the sub-band energy in this case and can be expressed as

$$\frac{\hbar^2}{2m_c}(\frac{n_z\pi}{d_z})^2+\frac{\hbar^2}{2m_c}(\frac{n_y\pi}{d_y})^2=[J_1(E_{n_{z18,2HD}},c,\eta_g,F)+i\,J_2(E_{n_{z18,2HD}},c,\eta_g,F)] \quad (43)$$

where, $G_2(n_y,n_z)=\frac{\hbar^2}{2m_c}[(\frac{n_y\pi}{d_y})^2+(\frac{n_z\pi}{d_z})^2]$

The EEM in this case is given by

$$m^*(E,\eta_g,F)=m_cJ_1'(E,c,\eta_g,F) \quad (44)$$

The electron concentration can be given as

$$n_0=\frac{2g_v\sqrt{2m_c}}{\pi\hbar}\text{Real part of}\sum_{n_z=1}^{n_{\max}}[\Theta_{20,7}(E_{f20,4},\eta_g,F)+\Theta_{20,8}(E_{f20,4},\eta_g,F)] \quad (45)$$

where $\Theta_{20,7}(E_{f20,4},\eta_g,F)=[J_1(E_{f20,4},c,\eta_g,F)+jJ_2(E_{f20,4},c,\eta_g,F)-(\frac{\pi n_z}{d_z})^2-(\frac{\pi n_y}{d_y})^2]^{\frac{1}{2}}$,

$\Theta_{20,8}(E_{f20,4},\eta_g,F)=\sum_{r=1}^{s}L(r)[\Theta_{20,7}(E_{f20,4},\eta_g,F)]$ *and* $E_{f20,4}$ *is the Fermi energy in this case.*

THE DOS IN QWHD SUPERLATTICES OF KANE TYPE SEMICONDUCTORS WITH GRADED INTERFACES IN THE PRESENCE OF INTENSE ELECTRIC FIELD

The DR in NWHD III-V SLs with graded interfaces in the presence of intense electric field can be expressed as [41]

$$k_z^2 = G_{8,20} + iH_{8,20} \tag{46}$$

where,

$$G_{8,20} = [\frac{C_{7,20}^2 - D_{7,20}^2}{L_0^2} - k_s^2], C_{7,20} = \cos^{-1}(\bar{\omega}_{7,20}), \bar{\omega}_{7,20} = (2)^{\frac{-1}{2}}[(1 - G_{7,20}^2 - H_{7,20}^2) - \sqrt{(1 - G_{7,20}^2 - H_{7,20}^2)^2 + 4G_{7,20}^2}]^{\frac{1}{2}}$$

$$G_{7,20} = [G_{1,20} + (\rho_{5,20}G_{2,20}/2) - (\rho_{6,20}H_{2,20}/2) + (\Delta_0/2)\{\rho_{6,20}H_{2,20} - \rho_{8,20}H_{3,20} + \rho_{9,20}H_{4,20} - \rho_{10,20}H_{4,20} +$$
$$\rho_{11,20}H_{5,20} - \rho_{12,20}H_{5,20} + (1/12)(\rho_{12,20}G_{6,20} - \rho_{14,20}H_{6,20})\}],$$

$$G_{1,20} = [(\cos(h_{1,20}))(\cosh(h_{2,20}))(\cosh(g_{1,20}))(\cos(g_{2,20})) + (\sin(h_{1,20}))(\sinh(h_{2,20}))(\sinh(g_{1,20}))(\sin(g_{2,20}))],$$

$$h_{1,20} = e_{1,20}(b_0 - \Delta_0), e_{1,20} = 2^{\frac{-1}{2}}(\sqrt{t_{1,20}^2 + t_{2,20}^2} + t_{1,20})^{\frac{1}{2}},$$

$$t_{1,20} = [(2m_{c1}/\hbar^2).J_{11}(E, \Delta_1, E_{g01}, F) - G_2(n_y, n_z)], \quad t_{2,20} = [(2m_{c1}/\hbar^2)J_{21}(E, \Delta_1, E_{g01}, F)],$$

$$h_{2,20} = e_{2,20}(b_0 - \Delta_0), e_{2,20} = 2^{\frac{-1}{2}}(\sqrt{t_{1,20}^2 + t_{2,20}^2} - t_{1,20})^{\frac{1}{2}}, g_{1,20} = d_{1,20}(a_0 - \Delta_0), d_{1,20} = 2^{\frac{-1}{2}}(\sqrt{x_{1,20}^2 + y_{1,20}^2} + x_{1,20})^{\frac{1}{2}},$$

$$x_{1,20} = [-(2m_{c2}/\hbar^2).J_{11}(E - V_0, E_{g2}, \Delta_2, \eta_{g2}, F) + G_2(n_y, n_z)], y_1 = [(2m_{c2}/\hbar^2).J_{22}(E - V_0, E_{g2}, \Delta_2, \eta_{g2}, F)],$$

$$g_{2,20} = d_{2,20}(a_0 - \Delta_0), d_{2,20} = 2^{\frac{-1}{2}}(\sqrt{x_{1,20}^2 + y_{1,20}^2} - x_{1,20})^{\frac{1}{2}}, \rho_{5,20} = (\rho_{3,20}^2 + \rho_{4,20}^2)^{-1}[\rho_{1,20}\rho_{3,20} - \rho_{2,20}\rho_{4,20}],$$

$$\rho_{1,20} = [d_{1,20}^2 + e_{2,20}^2 - d_{2,20}^2 - e_{1,20}^2], \rho_{3,20} = [d_{1,20}e_{1,20} + d_{2,20}e_{2,20}], \rho_{2,20} = 2[d_{1,20}d_{2,20} + e_{1,20}e_{2,20}], \rho_{4,20} = [d_{1,20}e_{2,20} - e_{1,20}d_{2,20}],$$

$$G_{2,20} = [(\sin(h_{1,20}))(\cosh(h_{2,20}))(\sinh(g_{1,20}))(\cos(g_{2,20})) + (\cos(h_{1,20}))(\sinh(h_{2,20}))(\cosh(g_{1,20}))(\sin(g_{2,20}))],$$

$$\rho_{6,20} = (\rho_{3,20}^2 + \rho_{4,20}^2)^{-1}[\rho_{1,20}\rho_{4,20} + \rho_{2,20}\rho_{3,20}],$$

$$H_{2,20} = [(\sin(h_{1,20}))(\cos(h_{2,20}))(\sin(g_{2,20}))(\cosh(g_{1,20})) - (\cos(h_{1,20}))(\sinh(h_{2,20}))(\sinh(g_{1,20}))(\cos(g_{2,20}))],$$

$$\rho_{7,20} = [(e_{1,20}^2 + e_{2,20}^2)^{-1}[e_{1,20}(d_{1,20}^2 - d_{2,20}^2) - 2d_{1,20}d_{2,20}e_{2,20}] - 3e_{1,20}],$$

$$G_{3,20} = [(\sin(h_{1,20}))(\cosh(h_{2,20}))(\cosh(g_{1,20}))(\cos(g_{2,20})) + (\cos(h_{1,20}))(\sinh(h_{2,20}))(\sinh(g_{1,20}))(\sin(g_{2,20}))],$$

$$\rho_{8,20} = [(e_{1,20}^2 + e_{2,20}^2)^{-1}[e_{2,20}(d_{1,20}^2 - d_{2,20}^2) + 2d_{1,20}d_{2,20}e_{1,20}] + 3e_{2,20}],$$

$$H_{3,20} = [(\sin(h_{1,20}))(\cosh(h_{2,20}))(\sin(g_{2,20}))(\sinh(g_{1,20})) - (\cos(h_{1,20}))(\sinh(h_{2,20}))(\cosh(g_{1,20}))(\cos(g_{2,20}))],$$

$$\rho_{9,20} = [(d_{1,20}^2 + d_{2,20}^2)^{-1}[d_{1,20}(e_{2,20}^2 - e_{1,20}^2) + 2e_{2,20}d_{2,20}e_{1,20}] + 3d_{1,20}],$$

$$G_{4,20} = [(\cos(h_{1,20}))(\cosh(h_{2,20}))(\cos(g_{2,20}))(\sinh(g_{1,20})) - (\sin(h_{1,20}))(\sinh(h_{2,20}))(\cosh(g_{1,20}))(\sin(g_{2,20}))],$$

$$\rho_{10,20} = [-(d_{1,20}^2 + d_{2,20}^2)^{-1}[d_{2,20}(-e_{2,20}^2 + e_{1,20}^2) + 2e_{2,20}d_{2,20}e_{1,20}] + 3d_{2,20}],$$

$$H_{4,20} = [(\cos(h_{1,20}))(\cosh(h_{2,20}))(\cosh(g_{1,20}))(\sin(g_{2,20})) + (\sin(h_{1,20}))(\sinh(h_{2,20}))(\sinh(g_{1,20}))(\cos(g_{2,20}))],$$

$$\rho_{11,20} = 2[d_{1,20}^2 + e_{2,20}^2 - d_{2,20}^2 - e_{1,20}^2],$$

$$G_{5,20} = [(\cos(h_{1,20}))(\cosh(h_{2,20}))(\cos(g_{2,20}))(\cosh(g_{1,20})) - (\sin(h_{1,20}))(\sinh(h_{2,20}))(\sinh(g_{1,20}))(\sin(g_{2,20}))],$$

$$\rho_{12,20} = 4[d_{1,20}d_{2,20} + e_{1,20}e_{2,20}],$$

$$H_{5,20} = [(\cos(h_{1,20}))(\cosh(h_{2,20}))(\sinh(g_{1,20}))(\sin(g_{2,20})) + (\sin(h_{1,20}))(\sinh(h_{2,20}))(\cosh(g_{1,20}))(\cos(g_{2,20}))],$$

$$\rho_{13,20} = [\{5(d_{1,20}e_{1,20}^3 - 3e_{1,20}e_{2,20}^2d_{1,20}) + 5d_{2,20}(e_{1,20}^3 - 3e_{1,20}^2e_{2,20})\}(d_{1,20}^2 + d_{2,20}^2)^{-1} +$$
$$(e_{1,20}^2 + e_{2,20}^2)^{-1}\{5(e_{1,20}\,d_{1,20}^3 - 3d_{2,20}e_{1,20}^2d_{1,20}) + 5(d_{2,20}^3e_{2,20} - 3d_{1,20}^2d_{2,20}e_{2,20})\} - 34(d_{1,20}e_{1,20} + d_{2,20}e_{2,20})],$$

$$G_{6,20} = [(\sin(h_{1,20}))(\cosh(h_{2,20}))(\sinh(g_{1,20}))(\cos(g_{2,20})) + (\cos(h_{1,20}))(\sinh(h_{2,20}))(\cosh(g_{1,20}))(\sin(g_{2,20}))],$$

$$\rho_{14,20} = [\{5(d_{1,20}e_{2,20}^3 - 3e_{2,20}e_{1,20}^2d_{1,20}) + 5d_{2,20}(-e_{1,20}^3 + 3e_{2,20}^2e_{1,20})\}(d_{1,20}^2 + d_{2,20}^2)^{-1} + (e_{1,20}^2 + e_{2,20}^2)^{-1}$$
$$\{5(-e_{1,20}\,d_{2,20}^3 + 3d_{1,20}^2d_{2,20}e_{1,20}) + 5(-d_{1,20}^3e_{2,20} + 3d_{2,20}^2d_{1,20}e_{2,20})\} + 34(d_{1,20}e_{2,20} - d_{2,20}e_{1,20})],$$

$$H_{6,20} = [(\sin(h_{1,20}))(\cosh(h_{2,20}))(\cosh(g_{1,20}))(\sin(g_{2,20})) - (\cos(h_{1,20}))(\sinh(h_{2,20}))(\sinh(g_{1,20}))(\cos(g_{2,20}))],$$

$$H_{7,20} = [H_{1,20} + (\rho_{5,20}H_{2,20}/2) + (\rho_{6,20}G_{2,20}/2) + (\Delta_0/2)\{\rho_{8,20}G_{3,20} + \rho_{7,20}H_{3,20} + \rho_{10,20}G_{4,20} + \rho_{9,20}H_{4,20} +$$
$$\rho_{12,20}G_{5,20} + \rho_{11,20}H_{5,20} + (1/12)(\rho_{14,20}G_{6,20} + \rho_{13,20}H_{6,20})\}],$$

$$H_{1,20} = [(\sin(h_{1,20}))(\sinh(h_{2,20}))(\cosh(g_{1,20}))(\cos(g_{2,20})) + (\cos(h_{1,20}))(\cosh(h_{2,20}))(\sinh(g_{1,20}))(\sin(g_{2,20}))],$$

$$D_{7,20} = \sinh^{-1}(\bar{\omega}_{7,20}), H_{8,20} = (2C_{7,19}D_{7,20}/L_0^2)$$

The sub-band equation in this case can be expressed as

$$0 = [G_{8,20} + iH_{8,20}]\big|_{E=E_{610}} \tag{47}$$

where E_{610} is the sub-band energy in this case.

At low temperatures where the quantum effects become prominent, the DOS function for the lowest SL mini-band is given by

$$N_{HDSL}(\mathrm{E},\eta_g,\mathrm{F}) = \frac{g_v}{\pi}\sum_{n_x=1}^{n_{x\max}}\sum_{n_y=1}^{n_{y\max}}\frac{[\mathrm{G}'_{8,20}+i\,\mathrm{H}'_{8,20}]}{\sqrt{\mathrm{G}_{8,20}+i\,\mathrm{H}_{8,20}}}H(\mathrm{E}-\mathrm{E}_{610}) \tag{48}$$

The EEM can be written as

$$m^*(\mathrm{E},\mathrm{n}_y,\mathrm{n}_z,\eta_g,\mathrm{F}) = \frac{\hbar^2}{2}(\mathrm{G}'_{8,20}) \tag{49}$$

The electron concentration is given by

$$n_0 = \frac{2g_v}{\pi}\,\mathrm{Re}\,al\ Part\ of \sum_{n_x=1}^{n_{x\max}}\sum_{n_y=1}^{n_{y\max}}[[\mathrm{G}'_{8,20}+i\,\mathrm{H}'_{8,20}]+\sum_{r=1}^{s}L(r)[\mathrm{G}'_{8,20}+i\,\mathrm{H}'_{8,20}]]\big|_{E=E_{f20,26}} \tag{50}$$

where $E_{f20,26}$ is the Fermi energy in this case.

The DOS in QWHD Effective Mass Superlattices of Kane Type

Semiconductors in the Presence of Intense Electric Field

Following Sasaki [8], the magneto DR in HD III-V effective mass superlattices (EMSLs) in the presence of intense electric field, the dispersion relations of whose constituent materials in the absence of any perturbation are defined by the three band model of Kane can be written

$$k_x^2 = [\frac{1}{L_0^2}\{\cos^{-1}(\mathrm{f}_{18HD2})\}^2 - G_2(n_y,n_z)] \tag{51}$$

In which,

$$f_{18HD2} = a_{1HD1,18}\cos[a_0C_{1HD2,18}+b_0D_{1HD2,18}]-a_{2HD1,18}\cos[a_0C_{1HD2,18}+b_0D_{1HD2,18}]$$

$$a_{1HD1,18}=[\sqrt{\frac{m_{c2}\,\mathrm{Re}\,al\ part\ of[J_3(0,c,\eta_{g2},F)]}{m_{c1}\,\mathrm{Re}\,al\ part\ of[J_3(0,c,\eta_{g1},F)]}}+1]^2[4(\frac{m_{c2}\,\mathrm{Re}\,al\ part\ of[J_3(0,c,\eta_{g2},F)]}{m_{c1}\,\mathrm{Re}\,al\ part\ of[J_3(0,c,\eta_{g1},F)]})^{\frac{1}{2}}]^{-1},$$

$$a_{2HD1,18}=[\sqrt{\frac{m_{c2}\,\mathrm{Re}\,al\ part\ of[J_3(0,c,\eta_{g2},F)]}{m_{c1}\,\mathrm{Re}\,al\ part\ of[J_3(0,c,\eta_{g1},F)]}}-1]^2[4(\frac{m_{c2}\,\mathrm{Re}\,al\ part\ of[J_3(0,c,\eta_{g2},F)]}{m_{c1}\,\mathrm{Re}\,al\ part\ of[J_3(0,c,\eta_{g1},F)]})^{\frac{1}{2}}]^{-1},$$

$$J_1(E,c,\eta_g,F)+i\,J_2(E,c,\eta_g,F)=J_3(E,c,\eta_g,F)$$

$$C_{1HD2,18}\equiv[(\frac{2m_{c1}}{\hbar^2})J_3(E,c,\eta_{g1},F)-G_2(n_y,n_z)]^{\frac{1}{2}}$$

$$and\ D_{1HD2,18}\equiv[(\frac{2m_{c1}}{\hbar^2})J_3(E,c,\eta_{g2},F)-G_2(n_y,n_z)]^{\frac{1}{2}}$$

The DOS function in this case is given by

$$N_{100}(E,F)=\frac{g_v}{\pi}\sum_{n_y=1}^{n_{y\max}}\sum_{n_z=1}^{n_{z\max}}\frac{\delta'_{99}}{\sqrt{\delta_{99}}}H(E-E_{99}) \tag{52}$$

$$where,\ \delta_{99}=[\frac{1}{L_0^2}\{\cos^{-1}(f_{18HD2})\}^2-G_2(n_y,n_z)],$$

and H_{99} is the real positive root of (19.82) with $k_x=0$

REFERENCES

[1] Ghatak, KP, S Chakrabarti, B Chatterjee, PK Das, P Dutta, A Halder, *Journal of Material Focus*, 7, 390 (2018).

[2] Chatterjee, B, N Debbarma, M Mitra, T Datta, KP Ghatak, *Journal of Nanoscience and Nanotechnology*, 17, 3352 (2017).

[3] Bhattacharjee, R, KP Ghatak, *Journal of nanoscience and nanotechnology* 17, 640 (2017).

[4] Mitra, M, TN Sen, T Datta, R Bhattacharjee, LS Singh, KP Ghatak, *Journal of nanoscience and nanotechnology* 17, 256 (2017).

[5] Mitra, M, M Chakraborty, S Debbarma, S Chakraborty, SK Sen, B Chatterjee, KP Ghatak, *Journal of Quantum Matter*, 5, 58 (2016).

[6] Schiff, L, *Quantum Mechanics*, McGraw-Hill, USA (1968).

[7] Nag, B. R, *Electron Transport in compound Semiconductors* (Springer, Heidelberg, 1980).

[8] Sasaki, H, (1984) *Phys. Rev.* B 30, 7016.

[9] Jiang, HX, Lin, JY, (1987) *J. Appl. Phys.* 61, 624.

[10] Ghatak, KP, B Mitra, A Ghoshal, *physica status solidi* (b) 154, K121(1989); B Mitra, KP Ghatak, *physica status solidi* (b) 154, K35(1989); KP Ghatak, *physica status solidi* (b) 154, K29(1989).

[11] Mitra, B, KP Ghatak, *Solid-state electronics* 32, 515(1989).

[12] Mitra, B, A Ghoshal, KP Ghatak, *physica status solidi* (b) 153, K209(1989).

[13] Mitra, B, KP Ghatak, *Physics Letters* A 137, 413(1989).

[14] Ghatak, KP, M Mondal, *Journal of applied physics* 65, 3480(1989).

[15] Ghatak, KP, N Chattopadhyay, M Mondal, *Applied Physics* A 48, 365(1989).

[16] Mitra, B, KP Ghatak, *Physics Letters* A 135, 397(1989).

[17] Mondal, M, KP Ghatak, AK Das, *Acta Physica Hungarica* 65, 43(1989).

[18] Mondal, M, S Banik, KP Ghatak, *Journal of low temperature physics* 74, 423(1989).

[19] Ghatak, KP, A Ghosal, *physica status solidi* (b) 151, K135(1989).

[20] Mitra, B, KP Ghatak, *Solid-state electronics* 32, 177(1989).

[21] Ghatak, KP, SN Biswas, *MRS Proceedings* 161, 371 (1990).

[22] Ghatak, KP, S Biswas, *Journal of Vacuum Science & Technology B* 7, 104(1989).

[23] Mondal, M, SN Banik, KP Ghatak, *Canadian journal of physics* 67, 72(1989).

[24] Mondal, M, KP Ghatak, *Annalen der Physik* 501, 502(1989).

[25] Mitra, B, KP Ghatak, *Physica Scripta* 40, 776, (1989).

[26] Ghatak, KP, Mondal, M, *Zeitschrift für Physik B Condensed Matter* 69, 471, (1988).

[27] Mondal, MN, Chattopadhyay, KP Ghatak, *Journal of low temperature physics* 73, 321(1989).

[28] Mitra, B, Ghosal, A, KP Ghatak, *physica status solidi* (b) 150, K67(1988).

[29] Ghatak, KP, M Mondal, *Solid-State Electronics* 31, 1561(1989).

[30] Chakravarti, AN, KP Ghatak, KK Ghosh, S Ghosh, HM Mukherjee, *physica status solidi* (b) 108, 609, (1981).

In: Quantum Wires: An Overview
Editor: Kamakhya Prasad Ghatak
ISBN: 978-1-53617-676-6

Chapter 7

THE QUANTUM CAPACITANCE IN QUANTUM WIRE FIELD EFFECT TRANSISTORS

***J. Pal*[1,*], *R. Paul*[2] *and M. Mitra*[3]**
[1]Department of Physics, Meghnad Saha Institute of Technology, Nazirabad, Anandapur, Kolkata, India
[2]Department of CSE, University of Engineering and Management, Kolkata, India
[3]Department of ETCE, Indian Institute of Engineering Science and Technology, Shibpur, Kolkata, India

ABSTRACT

This chapter explores the quantum capacitance (C_g) in quantum wire field effect transistors (QWFET's) manufactured from completely different technologically vital nonstandard materials by using all types of anisotropies of band structures in addition to splitting of bands due to large fields of the crystals inside the frame work of Kane's matrix methodology that successively generates new 1D dimensional electron energy versus wave vector relation. We derive the (C_g) under very low temperature so that the Fermi function tends to unity for QWFET's of

* Corresponding Author's Email: jayita241@gmail.com.

$Cd_3As_2, CdGeAs_2, InSb, Hg_{1-x}Cd_xTe, InAs, GaAs, In_{1-x}Ga_xAs_yP_{1-y}$ IV-VI, stressed materials, $Te, GaP, PtSb_2, Bi_2Te_3$, *Ge, GaSb* and II-V compounds using the appropriate band models. The (C_g) becomes the functions of the thickness of the quantum-confined transistors. The (C_g) varies with varying film thickness in various quantized steps and saw-tooth manners with different numerical values.

The (C_g) for the QWFET's of all the materials as considered here increases with increasing 1D electron statistics in step like fashion and decreases in an oscillatory way with the increasing gate voltage. All the results for all models gets simplified to the result of (C_g) in QWFET having wide energy band gap materials subjected to various constraints which exhibit the mathematical compatibility of our simplified theory.

INTRODUCTION

A carrier in semiconducting samples, moves on the closed 3D wave vector (k) space wherever the overall electron energy (E) is constant. The system spatiality will manufacture 1D electron motion once; these carriers are forced to move in 1 direction where as other two wave vector components assume quantized values. Hence, the one-dimensional energy wave vector dispersion relations become straight lines whose length decreases with the increase of the size quantum numbers leading to quantization. The physical properties of such 1D system have been investigated in the recent literature extensively [1-40]. Although many researchers had reported their investigations in various forms, even so it seems that the study of other new aspects of such 1D structure is becoming more and more necessary. One such important quantum quantity that in recent year

Generates substantial interest is that the quantum capacitance (C_g) of quantum wire field effect transistors (QWFET) which is biased by a gate voltage creating a large surface field of force [41-44]. The 1D carrier statistics in such 1D layer will rather simply be varied by varying V_g that, in turn, one dimensional potential well is formed at the semiconductor interface. The spatial variation brings an amendment of the electric field of force, the (C_g) depends on the gate-voltage. This variation has been investigated for QWFET's of wide band gap materials shows few necessary periodical characteristics of such devices. In this chapter we shall study the (C_g) in QWFET of all the materials as stated in the abstract.

THEORETICAL BACKGROUND

In QWFET, the (C_g) can be expressed by the following equation:

$$C_g = e\bar{D}n_s \tag{1}$$

where e is the electron charge, $\bar{D} = \frac{\partial}{\partial V_g}$, V_g is the gate voltage and n_s is the 1D electron statistics .

1. The C_g in QWFET of Tetragonal Materials

The derivation of (C_g) depends on n_s and the same can be derived by using Heisenberg's scientific theory at low temperatures with the unit value of Fermi Dirac function without using the difficult DOS technique as used by us in [41] as

$$n_s = \frac{2g_v}{\pi} S[B_{11}(E'_F, n_x, n_y)] \tag{2}$$

where, $S = \sum_{n_x=1}^{n_{z\max}} \sum_{n_y=1}^{n_{y\max}}$, $E'_F = [e\bar{V}_g - E_{fb} - \frac{e^2 n_0}{C_{ins}}]$, E_{fb} is Fermi energy of the bulk material,

$C_{ins} = \frac{2\pi\varepsilon_{ins}}{\ln(\frac{2t_{ins} + t_{wire}}{t_{wire}})}$, ε_{ins} is the permittivity of the insulator layer, t_{ins} is the thickness of the insulator layer and t_{wire} is the diameter of the wire.and the other notations are given in [41].

Using (1) and (2) the C_g can be expressed as

$$C_g = [\frac{2g_v e^2}{\pi} S(\theta_1)][1 + \frac{2g_v e^2}{\pi C_{ins}} S(\theta_1)]^{-1} \tag{3}$$

where, $\theta_1 = \frac{\partial}{\partial E'_F}(B_{11}(E'_F, n_x, n_y)$

2. The (C_g) in QWFET of Kane Type Compounds

2.1. The Three Band Model of Kane

Under certain constraints the equation (2) and (3) get simplified as

$$n_s = \frac{2g_v}{\pi} S[B_{13}(E'_F, n_x, n_y)] \tag{4}$$

$$C_g = [\frac{2g_v e^2}{\pi} S(\theta_2)][1 + \frac{2g_v e^2}{\pi C_{ins}} S(\theta_2)]^{-1} \tag{5}$$

where, $\theta_2 = \frac{\partial}{\partial E'_F}(B_{13}(E'_F, n_x, n_y))$ and $B_{13}(E'_F, n_x, n_y)$ are defined in [41]

2.2. The Two Band Model of Kane

Under certain constraints, the equations (4) and (5) assume the forms

$$n_s = \frac{2g_v}{\pi} S[B_{15}(E'_F, n_x, n_y)] \tag{6a}$$

$$C_g = [\frac{2g_v e^2}{\pi} S(\theta_3)][1 + \frac{2g_v e^2}{\pi C_{ins}} S(\theta_3)]^{-1} \tag{6b}$$

where, $\theta_3 = \frac{\partial}{\partial E'_F}(B_{15}(E'_F, n_x, n_y))$ and $B_{15}(E'_F, n_x, n_y)$ are defined in [41]

2.3. The Isotropic Parabolic Energy Bands

Under the condition of very large band gap, the equations (6a) and (6b) assume the forms

$$n_s = \frac{2g_v}{\pi} S[\bar{B}_{15}(E'_F, n_x, n_y)] \tag{7a}$$

$$C_g = [\frac{2g_v e^2}{\pi} S(\bar{\theta}_3)][1 + \frac{2g_v e^2}{\pi C_{ins}} S(\bar{\theta}_3)]^{-1} \tag{7b}$$

where, $\bar{B}_{15}(E'_F, n_x, n_y) = [\frac{2\bar{m}_c}{\hbar^2}(E'_F) - ((\frac{n_x \pi}{d_x})^2 + (\frac{n_y \pi}{d_y})^2)]^{1/2}$ and $\bar{\theta}_3 = \frac{\partial}{\partial E'_F}(\bar{B}_{15}(E'_F, n_x, n_y))$

2.4. The Model of Stillman et al.

The 1D electron statistics under the condition of unity Fermi function at low temperatures can be written following [41] as

$$n_s = \frac{2g_v}{\pi} S[B_{17}(E'_F, n_x, n_y)] \quad (8)$$

Using (1) and (8) the C_g can be expressed as

$$C_g = [\frac{2g_v e^2}{\pi} S(\theta_4)][1 + \frac{2g_v e^2}{\pi C_{ins}} S(\theta_4)]^{-1} \quad (9)$$

$$where,\ \theta_4 = \frac{\partial}{\partial E'_F}(B_{17}(E'_F, n_x, n_y))$$

2.5. The Model of Newson and Kurobe

The 1D electron statistics under the condition of unity Fermi function at low temperatures can be written following [41] as

$$n_s = \frac{2g_v}{\pi} S[B_{19}(E'_F, n_x, n_y)] \quad (10)$$

Using (1) and (10) the C_g can be expressed as

$$C_g = [\frac{2g_v e^2}{\pi} S(\theta_5)][1 + \frac{2g_v e^2}{\pi C_{ins}} S(\theta_5)]^{-1} \quad (11)$$

$$where,\ \theta_5 = \frac{\partial}{\partial E'_F}(B_{19}(E'_F, n_x, n_y))$$

2.6. The Model of Palik et al.

The 1D electron statistics under the condition of unity Fermi function at low temperatures can be written following [41] as

$$n_s = \frac{2g_v}{\pi} S[B_{21}(E'_F, n_x, n_y)] \quad (12)$$

Using (1) and (12) the C_g can be expressed as

$$C_g = [\frac{2g_v e^2}{\pi} S(\theta_6)][1 + \frac{2g_v e^2}{\pi C_{ins}} S(\theta_6)]^{-1} \quad (13)$$

$$where,\ \theta_6 = \frac{\partial}{\partial E'_F}(B_{21}(E'_F, n_x, n_y))$$

3. The C_g in QWFET of II-VI Materials

The 1D electron statistics under the condition of unity Fermi function at low temperatures can be written following [41] as

$$n_s = \frac{g_v}{\pi\sqrt{B_0}} S[t_7(E'_F, n_x, n_y)] \tag{14}$$

Using (1) and (14) the C_g can be expressed as

$$C_g = [\frac{g_v e^2}{\pi\sqrt{B_0}} S(\theta_7)][1 + \frac{g_v e^2}{C_{ins}\pi\sqrt{B_0}} S(\theta_7)]^{-1} \tag{15}$$

$$where,\ \theta_7 = \frac{\partial}{\partial E'_F}(t_7(E'_F, n_x, n_y))$$

4. The $\bar{C}_G$ in QWFET of IV-VI Materials

The 1D electron statistics under the condition of unity Fermi function at low temperatures can be written following [41] as

$$n_s = \frac{2g_v}{\pi} S[B_{32}(E'_F, n_x, n_y)] \tag{16}$$

Using (1) and (16) the C_g can be expressed as

$$C_g = [\frac{2g_v e^2}{\pi} S(\theta_8)][1 + \frac{2g_v e^2}{\pi C_{ins}} S(\theta_8)]^{-1} \tag{17}$$

$$where,\ \theta_8 = \frac{\partial}{\partial E'_F}(B_{32}(E'_F, n_x, n_y))$$

5. The C_g in QWFET of Stressed Materials

The 1D electron statistics under the condition of unity Fermi function at low temperatures can be written following [41] as

$$n_s = \frac{2g_v}{\pi} S[B_{34}(E'_F, n_x, n_y)] \tag{18}$$

Using (1) and (18) the C_g can be expressed as

$$C_g = [\frac{2g_v e^2}{\pi} S(\theta_9)][1 + \frac{2g_v e^2}{\pi C_{ins}} S(\theta_9)]^{-1} \tag{19}$$

$$where,\ \theta_9 = \frac{\partial}{\partial E'_F}(B_{34}(E'_F, n_x, n_y))$$

6. The C_g in QWFET of GaP Material

The 1D electron statistics under the condition of unity Fermi function at low temperatures can be written following [41] as

$$n_s = \frac{2g_v}{\pi} S[B_{38}(E'_F, n_x, n_y)] \tag{20}$$

Using (1) and (20) the C_g can be expressed as

$$C_g = [\frac{2g_v e^2}{\pi} S(\theta_{10})][1 + \frac{2g_v e^2}{\pi C_{ins}} S(\theta_{10})]^{-1} \tag{21}$$

$$where,\ \theta_{10} = \frac{\partial}{\partial E'_F}(B_{38}(E'_F, n_x, n_y))$$

7. The C_g in QWFET of $PtSb_2$ Material

The 1D electron statistics under the condition of unity Fermi function at low temperatures can be written following [41] as

$$n_s = \frac{2g_v}{\pi} S[B_{40}(E'_F, n_x, n_y)] \tag{22}$$

Using (1) and (22) the C_g can be expressed as

$$C_g = [\frac{2g_v e^2}{\pi} S(\theta_{11})][1 + \frac{2g_v e^2}{\pi C_{ins}} S(\theta_{11})]^{-1} \tag{23}$$

$$where,\ \theta_{11} = \frac{\partial}{\partial E'_F}(B_{40}(E'_F, n_x, n_y))$$

8. The C_g in QWFET of Bi_2Te_3 Material

The 1D electron statistics under the condition of unity Fermi function at low temperatures can be written following [41] as

$$n_s = \frac{2g_v}{\pi} S[B_{42}(E'_F, n_x, n_y)] \tag{24}$$

Using (1) and (24) the C_g can be expressed as

$$C_g = [\frac{2g_v e^2}{\pi} S(\theta_{12})][1 + \frac{2g_v e^2}{\pi C_{ins}} S(\theta_{12})]^{-1} \tag{25}$$

$$where,\ \theta_{12} = \frac{\partial}{\partial E'_F}(B_{42}(E'_F, n_x, n_y))$$

9. The C_g in QWFET of Ge

9.1. Model of Cardona et al.

The 1D electron statistics under the condition of unity Fermi function at low temperatures can be written following [41] as

$$n_s = \frac{2g_v}{\pi} S[B_{44}(E'_F, n_x, n_y)] \tag{26}$$

Using (1) and (26) the C_g can be expressed as

$$C_g = [\frac{2g_v e^2}{\pi} S(\theta_{13})][1 + \frac{2g_v e^2}{\pi C_{ins}} S(\theta_{13})]^{-1} \tag{27}$$

$$where,\ \theta_{13} = \frac{\partial}{\partial E'_F}(B_{44}(E'_F, n_x, n_y))$$

9.2. Model of Wang and Ressler

The 1D electron statistics under the condition of unity Fermi function at low temperatures can be written following [41] as

$$n_s = \frac{2g_v}{\pi} S[B_{46}(E'_F, n_x, n_y)] \tag{28}$$

Using (1) and (28) the C_g can be expressed as

$$C_g = [\frac{2g_v e^2}{\pi} S(\theta_{14})][1 + \frac{2g_v e^2}{\pi C_{ins}} S(\theta_{14})]^{-1} \tag{29}$$

$$where,\ \theta_{14} = \frac{\partial}{\partial E'_F}(B_{46}(E'_F, n_x, n_y))$$

10. The C_g in QWFET of *GaSb*

The 1D electron statistics under the condition of unity Fermi function at low temperatures can be written following [41] as

$$n_s = \frac{2g_v}{\pi} S[B_{48}(E'_F, n_x, n_y)] \tag{30}$$

Using (1) and (30) the C_g can be expressed as

$$C_g = [\frac{2g_v e^2}{\pi} S(\theta_{15})][1 + \frac{2g_v e^2}{\pi C_{ins}} S(\theta_{15})]^{-1} \tag{31}$$

$$where,\ \theta_{15} = \frac{\partial}{\partial E'_F}(B_{48}(E'_F, n_x, n_y))$$

11. The C_g in QWFET of $II - V$ Materials

The 1D electron statistics under the condition of unity Fermi function at low temperatures can be written following [41] as

$$n_s = \frac{2g_v}{\pi} S[B_{49}(E'_F, n_x, n_y)] \tag{32}$$

Using (1) and (32) the C_g can be expressed as

$$C_g = [\frac{2g_v e^2}{\pi} S(\theta_{16})][1 + \frac{2g_v e^2}{\pi C_{ins}} S(\theta_{16})]^{-1} \tag{33}$$

$$where,\ \theta_{16} = \frac{\partial}{\partial E'_F}(B_{49}(E'_F, n_x, n_y))$$

RESULTS AND DISCUSSION

The normalized C_g in QWFETs made of $CdGeAs_2$ with film thickness as the independent variable has been premeditated in Figure 1 by taking the different values of the energy band parameters as given in [41] using (a) equation 3 (orange) (b) equation 3 with $\delta = 0$ (brown), (c) equation 5 (green) (d) equation 6b (red) and (e) equation 7b (blue). From Figure 1 it appears that the normalized C_g enhances in quantized saw-tooth manner with increasing film thickness. The normalized C_g in QWFETs made of $InAs$ with film thickness as the independent variable has been premeditated in Figure 2 by taking the different values of the energy band parameters as given in [41] using (a) equation 5 (green) (d) equation 6b (red) and (e) equation 7b (blue). From Figure 2 it appears that the normalized C_g enhances in quantized distorted saw-tooth manner with increasing film thickness. The normalized C_g in QWFETs made of $InSb$ with film thickness as the independent variable has been premeditated in Figure 3 by taking the different values of the energy band parameters as given in [41] using (a) equation 5 (green) (d) equation 6b (red) and (e) equation 7b (blue). From Figure 3 it appears that the normalized C_g in this case enhances in quantized saw-tooth manner with increasing film thickness. The normalized C_g in QWFETs made of $Hg_{1-x}Cd_xTe$ with film thickness as the independent variable has been premeditated in Figure 4 by taking the different values of the energy band parameters as given in [41] using (a) equation 5 (green) (d) equation 6b (red) and (e) equation 7b (blue). From Figure 4 it appears that the normalized C_g in this case enhances in quantized saw-tooth manner with increasing film thickness. The normalized C_g in QWFETs made of $Ga_{1-x}Al_xAs$ with film thickness as the independent variable has been premeditated in Figure 5 by taking the different values of the energy band parameters as given in [41] using (a) equation 5 (green) (d) equation 6b (red) and (e) equation 7b (blue). From Figure 5 it appears that the normalized C_g in this case enhances in quantized saw-tooth manner with increasing film thickness.

The normalized C_g in QWFETs made of $In_{1-x}Ga_xAs_yP_{1-y}$ with film thickness as the independent variable has been premeditated in Figure 6 by taking the different values of the energy band parameters as given in [41] using (a) equation 5 (green) (d) equation 6b (red) and (e) equation 7b (blue). From Figure 6 it appears that the normalized C_g enhances in quantized saw-tooth manner with increasing film thickness. The normalized C_g in QWFETs made of Cd_3As_2 with 1D electron statistics as the independent variable has been premeditated in Figure 7 by taking the

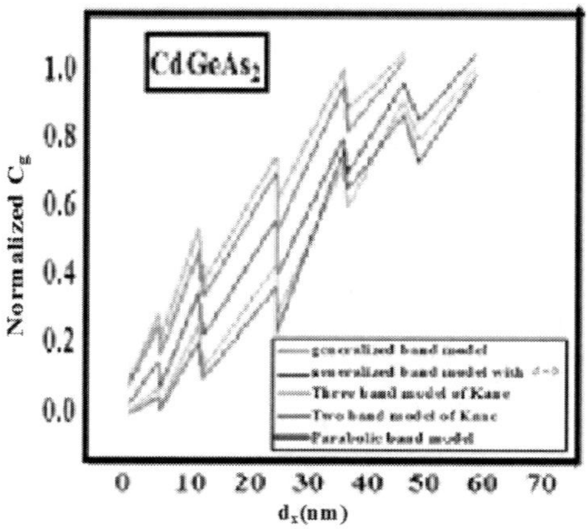

Figure 1. The normalized C_g in QWFETs made of $CdGeAs_2$ with film thickness as the independent variable has been premeditated using (a) equation 3 (orange) (b) equation 3 with $\delta = 0$ (brown), (c) equation 5 (green) (d) equation 6b (red) and (e) equation 7b (blue).

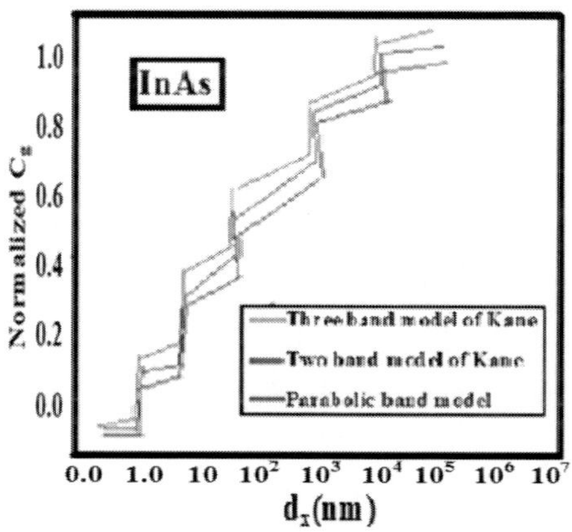

Figure 2. The normalized C_g in QWFETs made of *InAs* with film thickness as the independent variable has been premeditated using (a) equation 5 (green) (d) equation 6b (red) and (e) equation 7b (blue).

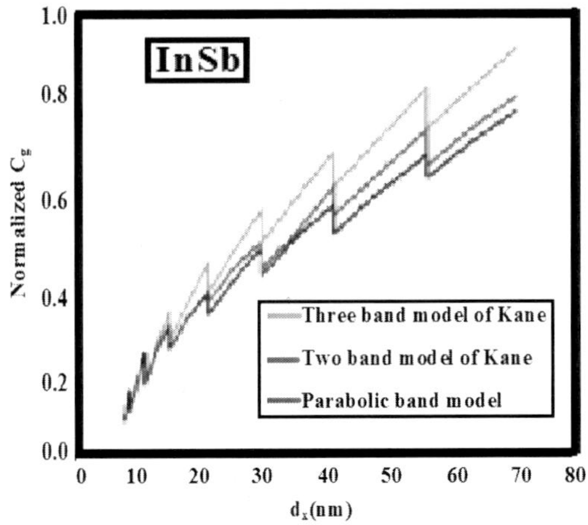

Figure 3. The normalized C_g in QWFETs made of $InSb$ with film thickness as the independent variable has been premeditated using (a) equation 5 (green) (d) equation 6b (red) and (e) equation 7b (blue).

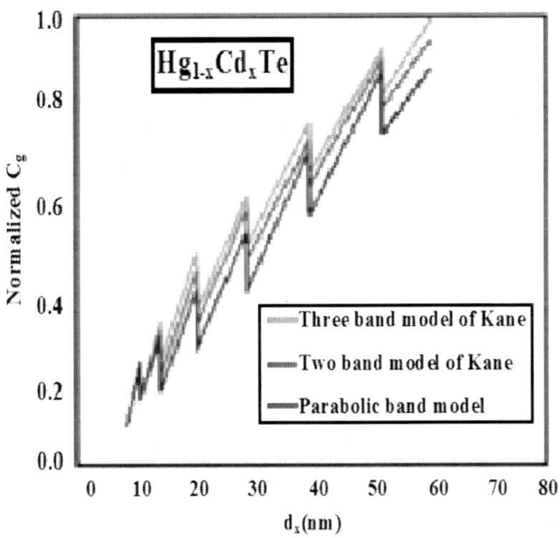

Figure 4. The normalized C_g in QWFETs made of $Hg_{1-x}Cd_xTe$ with film thickness as the independent variable has been premeditated using (a) equation 5 (green) (d) equation 6b (red) and (e) equation 7b (blue).

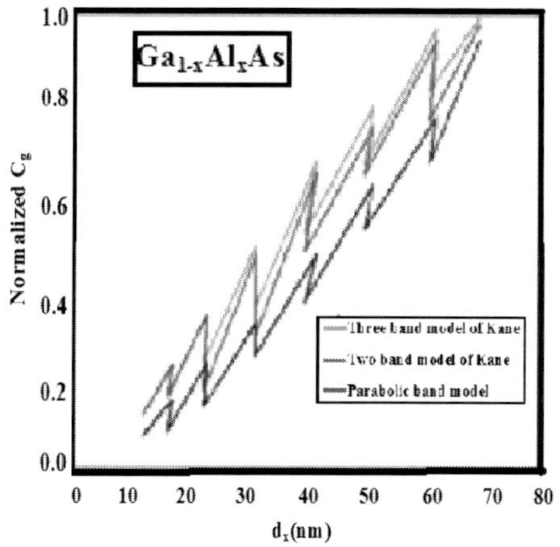

Figure 5. The normalized C_g in QWFETs made of $Ga_{1-x}Al_xAs$ with film thickness as the independent variable has been premeditated using (a) equation 5 (green) (d) equation 6b (red) and (e) equation 7b (blue).

different values of the energy band parameters as given in [41] using (a) equation 3 (brown) (b) equation 3 with $\delta = 0$ (orange), (c) equation 5 (red) (d) equation 6b (green) and (e) equation 7b (blue). From Figure 7 it appears that the normalized C_g enhances in quantized step like manner with increasing 1D electron statistics. The normalized C_g in QWFETs made of $CdGeAs_2$ with 1D electron statistics as the independent variable has been premeditated in Figure 8 by taking the different values of the energy band parameters as given in [41] using (a) equation 3 (brown) (b) equation 3 with $\delta = 0$ (orange), (c) equation 5 (green) (d) equation 6b (red) and (e) equation 7b (blue). From Figure 8 it appears that the normalized C_g in this case enhances in quantized step like manner with increasing 1D electron statistics. The normalized C_g in QWFETs made of $InAs$ with 1D electron statistics as the independent variable has been premeditated in Figure 9 by taking the different values of the energy band parameters as given in [41] using (a) equation 5 (green) (d) equation 6b (red) and (e) equation 7b (blue). From Figure 9 it appears that the normalized C_g enhances in quantized step like manner with increasing 1D electron statistics. The normalized C_g in QWFETs made of $InSb$ with 1D electron statistics as the independent variable has been premeditated in Figure 10 by taking the different values of the energy band parameters as given in [41]using (a) equation 5 (green) (d) equation 6b (red) and (e) equation 7b (blue). From Figure 10 it appears that the normalized Cg enhances in quantized step like manner with increasing 1D electron statistics.

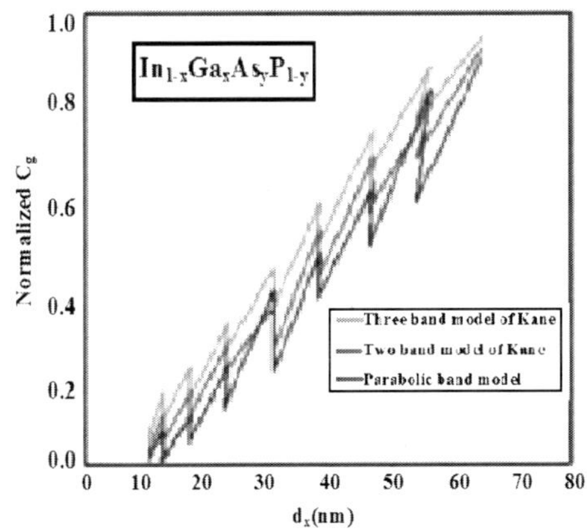

Figure 6. The normalized C_g in QWFETs made of $In_{1-x}Ga_xAs_yP_{1-y}$ with film thickness as the independent variable has been premeditated using (a) equation 5 (green) (d) equation 6b (red) and (e) equation 7b (blue).

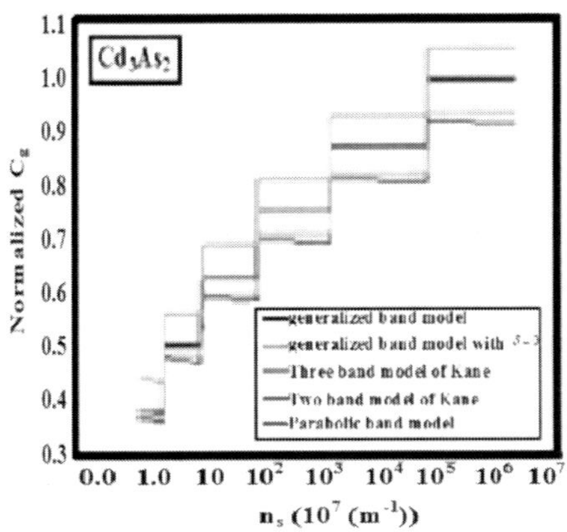

Figure 7. The normalized C_g in QWFETs made of Cd_3As_2 with 1D electron statistics as the independent variable has been premeditated using (a) equation 3 (brown) (b) equation 3 with $\delta = 0$ (orange), (c) equation 5 (red) (d) equation 6b (green) and (e) equation 7b (blue).

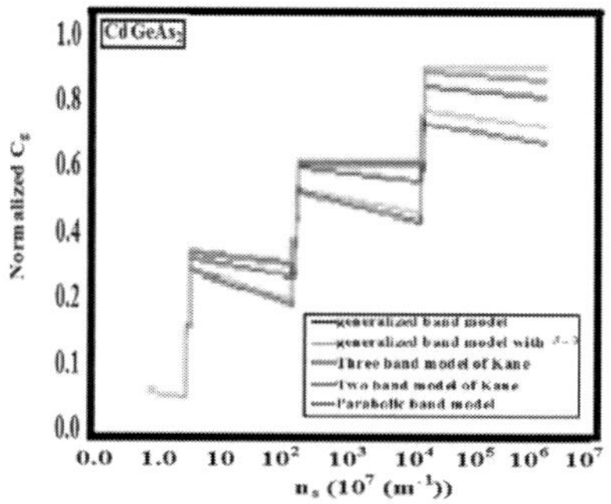

Figure 8. The normalized C_g in QWFETs made of $CdGeAs_2$ with 1D electron statistics as the independent variable has been premeditated using (a) equation 3 (brown) (b) equation 3 with $\delta = 0$ (orange), (c) equation 5 (green) (d) equation 6b (red) and (e) equation 7b (blue).

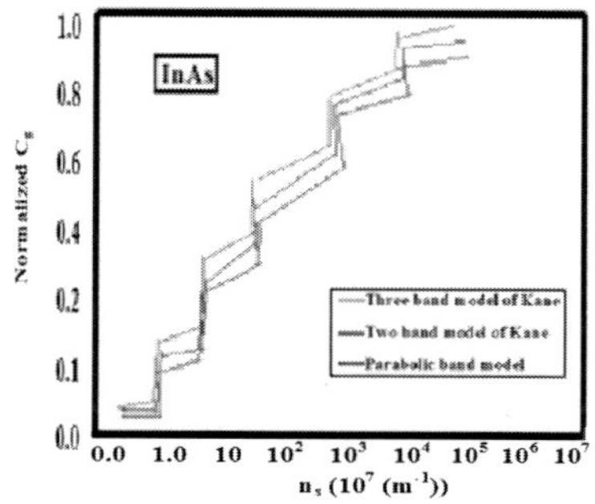

Figure 9. The normalized C_g in QWFETs made of $InAs$ with 1D electron statistics as the independent variable has been premeditated using (a) equation 5 (green) (d) equation 6b (red) and (e) equation 7b (blue).

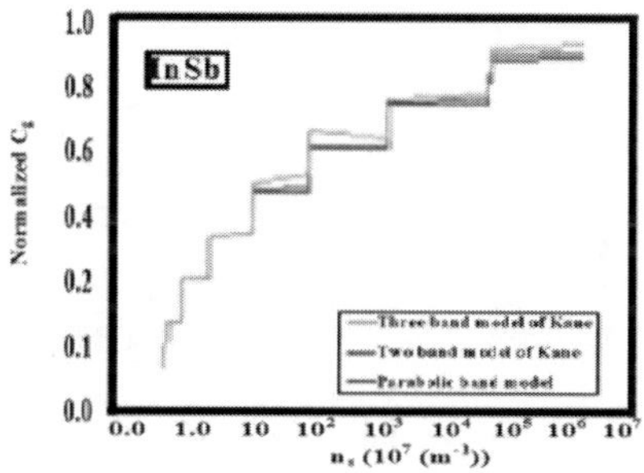

Figure 10. The normalized C_g in QWFETs made of $InSb$ with 1D electron statistics as the independent variable has been premeditated using (a) equation 5 (green) (d) equation 6b (red) and (e) equation 7b (blue).

The normalized C_g in QWFETs made of $CdGeAs_2$ with V_g as the independent variable has been premeditated in Figure 15 by taking the different values of the energy band parameters as given in [41] using (a) equation 3 (brown) (b) equation 3 with $\delta = 0$ (orange), (c) equation 5 (green) (d) equation 6b (red) and (e) equation 7b (blue). From Figure 15 it appears that the normalized C_g decreases with increasing gate voltage in quantized oscillatory fashion for all cases as consider in this case. The rate of oscillatory decrease changes from one type of band model to another type. The normalized C_g in QWFETs made of $InAs$ with V_g as the independent variable has been premeditated in Figure 16 by taking the different values of the energy band parameters as given in [41] using (a) equation 5 (green) (d) equation 6b (red) and (e) equation 7b (blue). From Figure 16 it appears that the normalized C_g decreases with increasing gate voltage in quantized oscillatory fashion for all cases as consider in this case. The rate of oscillatory decrease changes from one type of band model to another type. The normalized C_g in QWFETs made of $InSb$ with V_g as the independent variable has been premeditated in Figure 17 by taking the different values of the energy band parameters as given in [41] using (a) equation 5 (green) (d) equation 6b (red) and (e) equation 7b (blue).

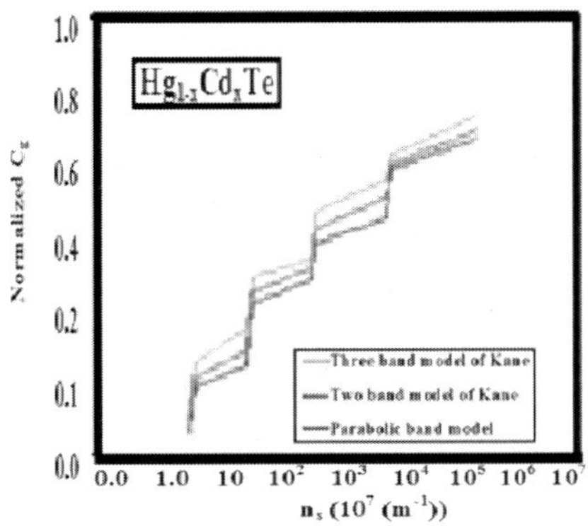

Figure 11. The normalized C_g in QWFETs made of $Hg_{1-x}Cd_xTe$ with 1D electron statistics as the independent variable has been premeditated using (a) equation 5 (green) (d) equation 6b (red) and (e) equation 7b (blue).

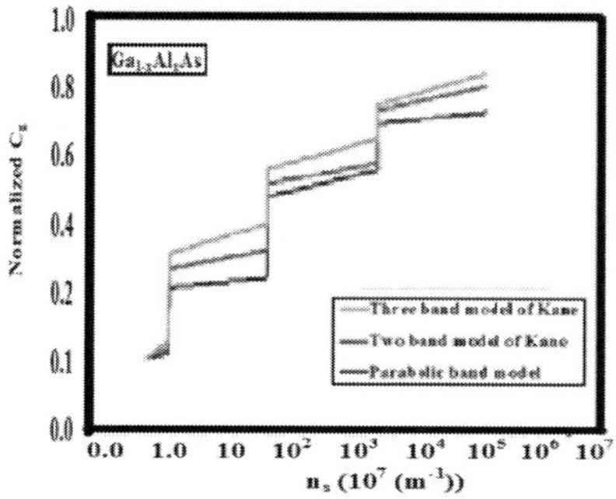

Figure 12. The normalized C_g in QWFETs made of $Ga_{1-x}Al_xAs$ with 1D electron statistics as the independent variable has been premeditated using (a) equation 5 (green) (d) equation 6b (red) and (e) equation 7b (blue).

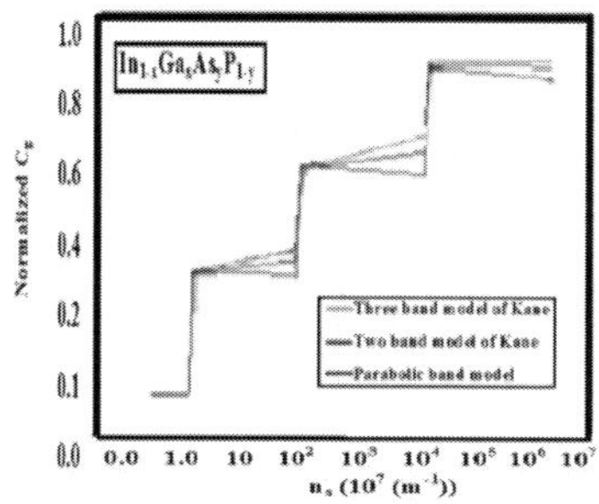

Figure 13. The normalized C_g in QWFETs made of $In_{1-x}Ga_xAs_yp_{1-y}in$ with 1D electron statistics as the independent variable has been premeditated using (a) equation 5 (green) (d) equation 6b (red) and (e) equation 7b (blue).

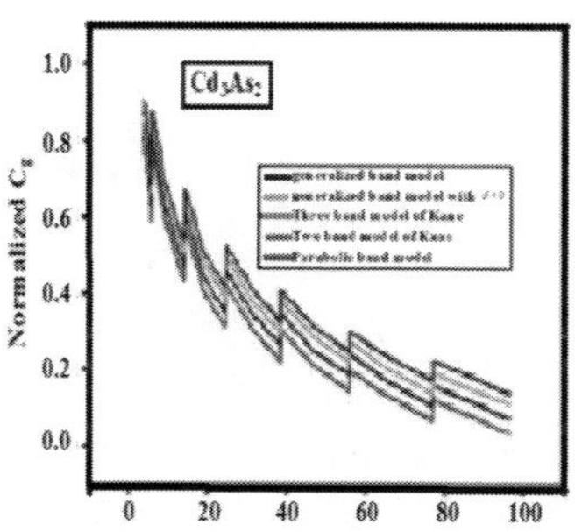

Figure 14. The normalized C_g in QWFETs made of Cd_3As_2 with V_g as the independent variable has been premeditated using (a) equation 3 (Violet) (b) equation 3 with $\delta = 0$ (orange), (c) equation 5 (green) (d) equation 6b (red) and (e) equation 7b (blue).

From Figure 17 it appears that the normalized C_g decreases with increasing gate voltage in quantized oscillatory fashion for all cases as consider in this case. The rate of oscillatory decrease changes from one type of band model to another type. The normalized C_g in QWFETs made of $Hg_{1-x}Cd_xTe$ with V_g as the independent variable has been premeditated in Figure 18 by taking the different values of the energy band parameters as given in [41] using (a) equation 5 (green) (d) equation 6b (red) and (e) equation 7b (blue). From Figure 18 it appears that the normalized C_g decreases with increasing gate voltage in quantized oscillatory fashion for all cases as consider in this case. The rate of oscillatory decrease changes from one type of band model to another type. The normalized C_g C_g in QWFETs made of $Ga_{1-x}Al_xAs$ with V_g as the independent variable has been premeditated in Figure 19 by taking the different values of the energy band parameters as given in [41] using (a) equation 5 (green) (d) equation 6b (red) and (e) equation 7b (blue).

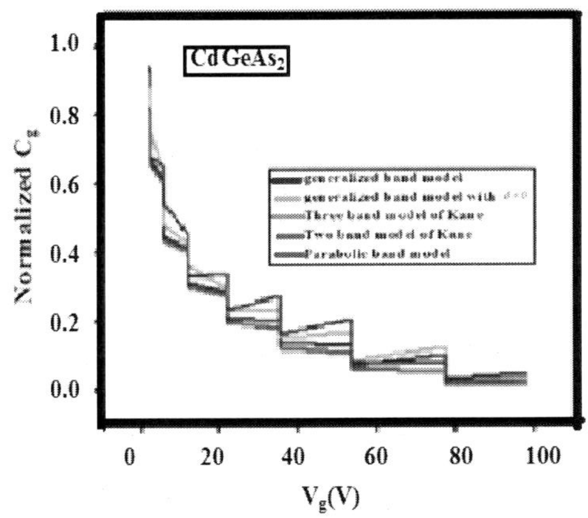

Figure 15. The normalized C_g in QWFETs made of $CdGeAs_2$ with V_g as the independent variable has been premeditated using (a) equation 3 (brown) (b) equation 3 with $\delta = 0$ (orange), (c) equation 5 (green) (d) equation 6b (red) and (e) equation 7b (blue).

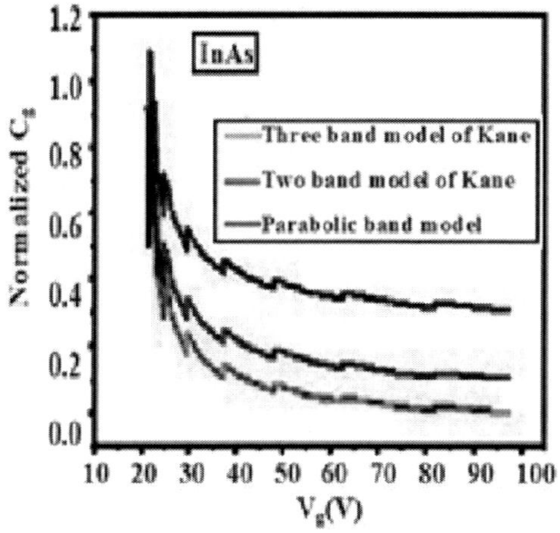

Figure 16. The normalized C_g in QWFETs made of $InAs$ with V_g as the independent variable has been premeditated using (a) equation 5 (green) (d) equation 6b (red) and (e) equation 7b (blue).

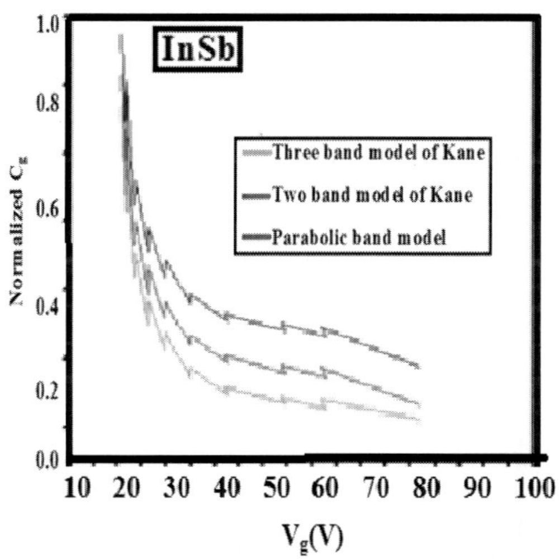

Figure 17. The normalized C_g in QWFETs made of $InSb$ with V_g as the independent variable has been premeditated using (a) equation 5 (green) (d) equation 6b (red) and (e) equation 7b (blue).

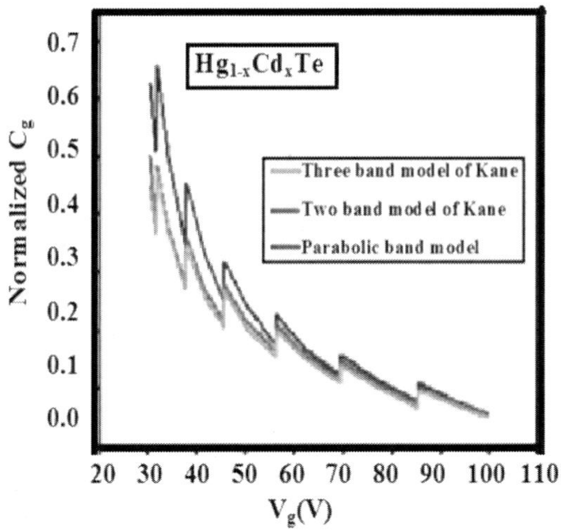

Figure18. The normalized C_g in QWFETs made of $Hg_{1-x}Cd_xTe$ with V_g as the independent variable has been premeditated using (a) equation 5 (green) (d) equation 6b (red) and (e) equation 7b (blue).

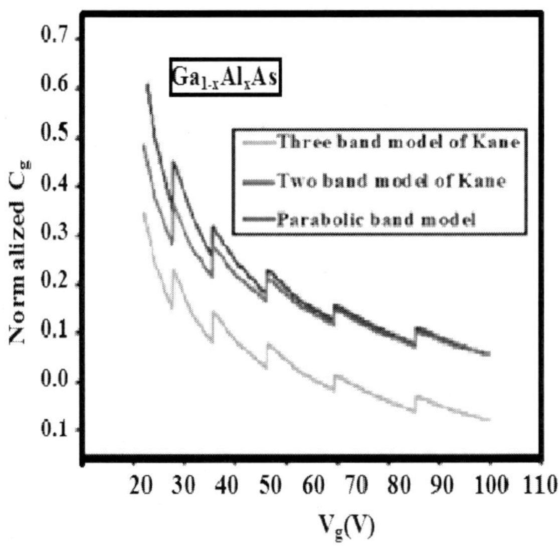

Figure 19. The normalized C_g in QWFETs made of $Ga_{1-x}Al_xAs$ with V_g as the independent variable has been premeditated using (a) equation 5 (green) (d) equation 6b (red) and (e) equation 7b (blue).

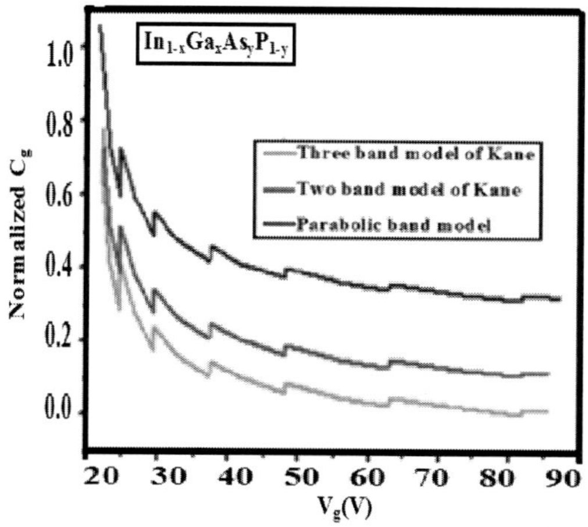

Figure 20. The normalized C_g in QWFETs made of $In_{1-x}Ga_xAs_yP_{1-y}$ with V_g as the independent variable has been premeditated using (a) equation 5 (green) (d) equation 6b (red) and (e) equation 7b (blue).

From Figure 19 it appears that the normalized C_g decreases with increasing gate voltage in quantized oscillatory fashion for all cases as consider in this case. The rate of oscillatory decrease changes from one type of band model to another type. The normalized C_g in QWFETs made of $In_{1-x}Ga_xAs_yP_{1-y}$ with V_g as the independent variable has been premeditated in Figure 20 by taking the different values of the energy band parameters as given in [41] using (a) equation 5 (green) (d) equation 6b (red) and (e) equation 7b (blue). From Figure 20 it appears that the normalized C_g decreases with increasing gate voltage in quantized oscillatory fashion for all cases as consider in this case. The rate of oscillatory decrease changes from one type of band model to another type.

We conclude that the impact of investigation in this context is rather large, since the C_g in QWFET is a very important quantity for the purpose of technical application and experimental studies of physical properties of quantum confined QWFET. It should be noted that for rigorous treatment the multi carrier effects, and therefore the contribution of the non-idealities should be considered, our simplified approach shows the fundamental qualitative behavior of the C_g in QWFETs of small energy gap materials, since these compounds obey the model of Kane, wherever as our study is predicted by extending $\vec{k}.\vec{p}$ theory.

REFERENCES

[1] Ghatak, K.P. and M. Mondal., *Journal of Applied Physics*, 70, 299 (1991).
[2] Ghatak, K.P. and S.N. Biswas, *Journal of Vacuum Science & Technology*, 7B, 104(1989).
[3] Mitra, B. and K.P. Ghatak., *Journal of Solid-State Electronics*, 32, 177(1989).
[4] Ghatak, K.P. and M. Mondal., *Journal of Applied Physics*, 62, 922 (1987).
[5] Mondal, M. and K.P. Ghatak., *Journal of Magnetism and Magnetic Materials*, 62, 115(1986).
[6] Mondal, M. and K.P. Ghatak., *Journal of Physica Scripta*, 31, 613 (1985).
[7] Ghatak, K.P. and M. Mondal., *Journal of Zeitschrift für Physik B Condensed Matter,* 64, 223(1986).
[8] Ghatak, K.P. and S.N. Biswas., 1994, *Journal of Solid-State Electronics*, 37, 1437 (1994).
[9] Ghatak, K.P., N. Chattopadhyay and M. Mondal., *Journal of Applied Physics A,* 48, 365 (1989).
[10] Ghatak, K.P. and M. Mondal, Z. *fur Physik B*, 64, 223 (1986).
[11] Ghatak, K.P. and S.N. Biswas, *Sol. State Electron.,* 37, 1437 (1994).
[12] Choudhury, D.R., A. K. Chowdhury, K.P. Ghatak, A.N. Chakravarti, *Phys. Stat. Sol*. (b) 98, (K141) (1980).
[13] Chakravarti, N.A., K. Chowdhury, K.P. Ghatak, *Phys. Stat. Sol.* (a) 63, K97 (1981).
[14] Mondal, M.K., P. Ghatak, *Acta Phys. Polon.* A 67, 983 (1985); M. Mondal, K.P. Ghatak, *Phys. Stat. Sol.* (b) 128, K21 (1985).
[15] Mondal, M.K., P. Ghatak, *Phys. Stat. So*l. (a) 93, 377 (1986).
[16] Ghatak, K.P., M. Mondal, *Phys. Stat. Sol.* (b) 135, 819 (1986).

[17] Mondal, M.K., P. Ghatak, *Phys. Stat. Sol.* (b) 139, 185 (1987).

[18] Ghatak, K.P. N., Chattopadhyay, S. N. Biswas, *OE/Fibers'* 87, 203 (1987).

[19] Ghatak, K.P., N. Chatterjee, M. Mondal, *Phys. Stat. Sol.* (b) 139, K25 (1987).

[20] Ghatak, K.P. M. Mondal, *Phys. Stat. Sol.* (b) 148, 645 (1988); K.P. Ghatak, A. Ghosal *Phys. Stat. Sol.* (b) 151, K135 (1989); Ghatak, K.P., N. Chattopadhyay, M. Mondal, A, l. *Phys.* A 48, 365 (1989).

[21] Das, P.K., Kamakhya Prasad Ghatak, *Journal of nanoscience and nanotechnology*, 19, 2909 (2019); KP Ghatak, S. Chakrabarti, B. Chatterjee, Materials Focus, 7, 361 (2018); KP Ghatak. Chakrabarti, S., B. Chatterjee, PK Das, P Dutta, A Halder, Materials Focus, 7, 390 (2018); B Chatterjee, N Debbarma, M Mitra, T Datta, KP Ghatak, *Journal of Nanoscience and Nanotechnology*, 17, 3352 (2017).

[22] Ghatak, Kamakhya Prasad, Debashis De, *Materials Focus*, 6, 114 (2017); P.K. Das, P. Dutta, A. Halder, J. Pal, N. Debbarma, S. Debbarma, K.P. Ghatak, *Materials Focus,* 6, 167 (2017); PK Das, P Dutta, A Halder, R Bhattacharjee, KP Ghatak, *Materials Focus*, 6, 133 (2017); R Bhattacharjee, KP Ghatak, *Journal of Nanoscience and Nanotechnology*, 17, 640 (2017).

[23] Mitra, M., T.N. Sen, T. Datta, R. Bhattacharjee, L.S. Singh, K.P. Ghatak, *Journal of Nanoscience and Nanotechnology*, 17, 256 (2017); T.N. Sen, K.P. Ghatak, *Quantum Matter,* 5, 732 (2016);TN Sen, K.P. Ghatak, *Quantum Matter*, 5, 721 (2016); K.P. Ghatak, K. Sarkar, S. Chakrabarti, M. Kumar, M. Debbarma, T.N. Sen, M. Chakraborty, *Reviews in Theoretical Science*, 4, 199 (2016).

[24] Bhattacharya, R., K. Sarkar, M. Kumar, B. Chatterjee, K.P. Ghatak, *Quantum Matter*, 5, 557 (2016); KP Ghatak, K Sarkar, N Debbarma, L Suraj Singh, *Quantum Matter*, 5, 427 (2016); Kamakhya Prasad Ghatak, Debashis De, *Journal of Nanoengineering and Nanomanufacturing*, 6, 1 (2016); SK Biswas, M. Mitra, K.P. Ghatak, *Journal of Nanoengineering and Nanomanufacturing*, 6, 63 (2016). Paitya, N., K., P. Ghatak, *Quantum Matter*, 5, 191 (2016);B

Chatterjee, S Chakrabarti, S.K. Sen, M Mitra, KP Ghatak, *Quantum Matter*, 5, 85 (2016);Kamakhya Prasad Ghatak, Magneto.

[25] *Thermoelectric Power in Heavily Doped Quantized Structures* (Series on the Foundations of Natural Science and Technology, World Scientific Publishing Company, 7, 758 2016); Kamakhya Prasad Ghatak, *Dispersion Relations in Heavily-Doped Nanostructures*, Springer International Publishing, 615 (2016).

[26] Chatterjee, B., K. Sarkar, K.P. Ghatak, *Advances in Optical Science and Engineering*, Springer India, 621 (2015); Kamakhya Prasad Ghatak, Einstein's Photoemission, Springer International Publishing, 295 (2015).

[27] Adhikari,S.M., K.P. Ghatak, *Journal of Advanced Physics*, 2, 130 (2013); S. Bhattacharya, D. De, S. Ghosh, K.P. Ghatak, *Journal of Computational and Theoretical Nanoscience*, 10, 664 (2013); N Paitya, S Bhattacharya, D De, S Ghosh, KP Ghatak, *Journal of Nanoengineering and Nanomanufacturing*, 2, 211 (2012).

[28] Pahari, S., S. Bhattacharya, K.P. Ghatak, *Journal of Computational and Theoretical Nanoscience*, 6, 2088 (2009); L.J. Singh, S. Choudhury, S. Singha Roy, K.P. Ghatak, *Electrical Engineering*, 87, 19 (2005); PK Chakraborty, GC Datta, KP Ghatak, *Physica Scripta*, 68, 368 (2003).

[29] Ghatak, K.P., J. Mukhopadhyay, J.P. Banerjee, *SPIE proceedings series*, *4746*, 1292 (2002); K.P .Ghatak, J. Mukhopadhyay, J.P. Banerjee, *SPIE proceedings series*, *4746*, 1296 (2002); K.P. Ghatak, J. Mukhopadhyay, JP Banerjee, *Proceedings-Spie The International society for Optical Engineering*, 2, 1296 (2002); K.P. Ghatak, J. Mukhopadhyay, J.P. Banerjee, *SPIE proceedings series*, 4746, 347 (2002).

[30] Ghatak, K.P., P.K. Bose, J.P. Banerjee, SPIE proceedings series, 4746, 351 (2002); K.P. Ghatak, SPIE proceedings series, 4746, 292 (2002); K.P. Ghatak, *Journal of Wave Material Interaction*, 14, 157 (1999); KP Ghatak, PK Bose, *Journal of Wave Material Interaction*, 12, 53 (1997); K.P. Ghatak, B. Nag, *J Journal of Wave Material Interaction*, 12, 85 (1997);B Mitra, DK Basu, B Nag, KP Ghatak,

Nonlinear Optics-Reading, 17, 171 (1997);KP Ghatak, PK Bose, *Journal of Wave Material Interaction*, 12, 60 (1997); PK Bose, KP Ghatak, *Journal of Wave Material Interaction*, 12, 67 (1997).

[31] Ghatak, Kamakhya P., P.K. Bose, Gautam Majumder, MRS Proceedings, 494, 157 (1997); KP Ghatak, DK Basu, D Basu, B Nag, Il Nuovo Cimento D, 18, 947 (1996); K.P. Ghatak, S. Dutta, A. Ali, S. Banerjee, B. Nag, *Journal of Wave Material Interaction*, 11, 127 (1996); PK Chakrabarty, B Nag, S Dutta, KP Ghatak, *Journal of Wave Material Interaction*, 11, 55(1996).

[32] Nag, B., P.K. Chakrabarty, K.P. Ghatak, *Journal of Wave Material Interaction*, 11, 211(1996); K.P. Ghatak, PK Chakrabarty, B. Nag, *Journal of Wave Material Interaction*, 11, 159(1996); Kamakhya P. Ghatak, Sambhu N. Banik, *FIZIKA A-ZAGREB-*, 5, 31 (1996).

[33] Ghatak, K.P., J.P. Banerjee, P.K. Chakrabarty, B. Nag, *Journal of Wave Material Interaction*, 11, 166(1996); PK Chakraborty, B Nag, S Dutta, KP Ghatak, *Journal of Wave Material Interaction*, 11, 111 (1996).

[34] Ghatak, K., P.S. Bera, A. Ali, B. Nag, *FIZIKA A-ZAGREB-*, 5, 111 (1996); K.P. Ghatak, J.P. Banerjee, B. Goswami, B. Nag, Nonlinear Optics-Reading-, 16, 241 (1996); K.P. Ghatak, M. Mitra, B. Goswami, B. Nag, Nonlinear Optics-Reading -, 16, 167 (1996); K.P. Ghatak, D. Bhattacharya, D. Basu, B. Nag, physica status solidi (b), 191, 141 (1995).

[35] Ghatak, K.P., B. Nag, M. Mitra, J.P. Bannerjee, *Journal of Wave Material Interaction*, 10, 11 (1995); K.P. Ghatak, D.K. Basu, B. Nag, *Journal of Wave Material Interaction*, 10, 29 (1995); K.P. Ghatak, S. Dutta, D. Basu, B. Nag, *Journal of Wave Material Interaction*, 10, 1 (1995).

[36] Ghatak, K.P., B. Nag, G. Mazumder, MRS Proceedings, 379, 85 (1995); K.P. Ghatak, B. Nag, G. Mazumder, *MRS Proceedings,* 379, 109 (1995); K. P Ghatak, S. N Banik, *FIZIKA A, 3,* 155 (1994); S.N. Banik, K. P Ghatak, S.N. Biswas, *FIZIKA A, 3,* 77 (1994); K.P. Ghatak, Fizika A., 2, 133 (1993); K. P Ghatak, S.N. Biswas, *MRS*

Proceedings, 313, 375 (1993); K. P Ghatak, S. N. Biswas, *MRS Proceedings*, 308, 445 (1993).

[37] Ghatak, K.P., D. Bhattacharyya, *Journal of Wave Material Interaction*, 8, 233 (1993); K.P. Ghatak, S.N. Biswas, *Acta physica Slovaca*, 43, 425 (1993); K.P. Ghatak, S.N. Biswas, *Nanostructured materials*, 2, 91 (1993).

[38] Ghatak, K.P., M. Mondal, Fizika A, 1, 113 (1992); K. P Ghatak, B. De, MRS Proceedings, 262,911 (1992); K. P Ghatak, B. De, MRS Proceedings, 242, 373 (1992); K. P Ghatak, B. De, *MRS Online Proceedings Library Archive*, 228, (1992); K. P Ghatak, Fizika A, 1, 197 (2006).

[39] Ghatak, K.P., S. Bhattacharyya, S.N. Biswas, *Acta Physica Hungarica*, 70, 83 (1991); K.P. Ghatak, Acta Physica Hungarica, 69, 121 (1991); K.P. Ghatak, Badal De, *MRS Proceedings*, 234, 59 (1991); K.P. Ghatak, Badal De, *MRS Proceedings*, 234, 55 (1991); K. P Ghatak, Acta Physica Hungarica, 68, 253 (1990); K. P. Ghatak, Acta Physica Hungarica, 67, 407 (1990); K.P. Ghatak, *MRS Proceedings*, 216, 469 (1990); K.P. Ghatak, *MRS Proceedings*, 216, 465 (1990).

[40] Ghatak, K.P., B. De, M. Mondal, S.N. Biswas, *MRS Proceedings*, 184, 261 (1990); K.P. Ghatak, B. De, M. Mondal, S.N. Biswas, MRS Proceedings, 198, 333 (1990);B Mitra, K. P Ghatak, Solid-state electronics, 32, 810 (1989); K.P. Ghatak, M. Mondal, *Journal of applied physics*, 62, 922 (1987); S.N. Biswas, K.P. Ghatak, Megagauss Technology and Pulsed Power Applications, 8, 219 (1987); M. Mondal, K.P. Ghatak, *Czechoslovak journal of physics*, 36, 1396 (1986); A. Arti, K. P Ghatak, K.K. Ghosh, S. Ghosh, A. Dhar, *Physica Status Solidi B-Basic Research*, 103, K55(1981).

In: Quantum Wires: An Overview
Editor: Kamakhya Prasad Ghatak
ISBN: 978-1-53617-676-6

Chapter 8

EINSTEIN'S PHOTOEMISSION FROM HEAVILY DOPED QUANTUM WIRES

***S. D. Devi*[1], *L. S. Singh*[1], *J. Pal*[2], *N. Debbarma*[3] *and P. K. Das*[4,*]**

[1]Department of Electronics and Communication Engineering, National Institute of Technology, Manipur, Imphal, India
[2]Department of Physics, Meghnad Saha Institute of Technology, Nazirabad, Uchepota, Anandapur, Kolkata, India
[3]Department of Computer Science and Engineering, National Institute of Technology, Agartala, Tripura, India
[4]Department of Basic Science Institute of Engineering & Management, Salt Lake, Kolkata, West Bengal, India

ABSTRACT

In this chapter we will study Einstein's Photoemission (EP) from Heavily Doped (HD) quantum wires (QWs) of non-linear optical, III-V, II-VI, Gallium Phosphide, Germanium, Platinum Antimonide, stressed,

[*] Corresponding Author's Email: prabir.das@iemcal.com.

IV-VI, Lead Germanium Telluride, Tellurium, II-V, Zinc and Cadmium diphosphides and Bismuth Telluride respectively by formulating the respective expressions of EP in each case. The influence of quantum confinement is immediately apparent from figures (1), (4), (7), (10), (13), (16), (19), (22) and (25) since the EP depends strongly on the thickness of the quantum-confined materials in contrast with the corresponding bulk specimens. The EP decreases with increasing film thickness in an oscillatory way with different numerical magnitudes for HD QWs. It appears from the aforementioned figures that the EP exhibits spikes for particular values of film thickness which, in turn, depends on the particular band structure of the specific material. Moreover, the EP from HD QWs of different compounds can become several orders of magnitude larger than of bulk specimens of the same materials, which is also a direct signature of quantum confinement. This oscillatory dependence will be less and less prominent with increasing film thickness. From figures (3), (6), (9), (12), (15), (18), (21) and (24) we conclude that the EP increases with increasing degeneracy and also exhibits spikes. For bulk specimens of the same material, the EP will be found to increase continuously with increasing electron degeneracy in a non-oscillatory manner. The figures (2), (5), (8), (11), (14), (17), (20) and (23) illustrate the dependence of the EP from quantum-confined materials on the normalized incident photon energy. The EP increases with increasing photon energy in a step like manner for all the figures. The appearance of the discrete jumps in all the figures is due to the redistribution of the electrons among the quantized energy levels when the size quantum number corresponding to the highest occupied level changes from one fixed value to the others. With varying electron degeneracy, a change is reflected in the EP through the redistribution of the electrons among the size-quantized levels. It may be noted that at the transition zone from one sub band to another, the height of the peaks between any two sub-bands decreases with the increasing in the degree of quantum confinement and is clearly shown in all the curves. It should be noted that although, the EP varies in various manners with all the variables as evident from all the figures, the rates of variations are totally band-structure dependent.

INTRODUCTION

The importance of Quantum Wires (QWs) is already well-known in the whole field of nano-science and nano-technology. Although many new effects in quantized structures have already been reported, the interest for further research of other aspects of such quantum-confined materials is becoming increasingly important. One such significant property is Einstein's Photoemission (EP) which is a physical phenomenon and occupies a singular position in the whole arena of Modern Physics and related disciplines in general and whose importance has already been established since the inception of Einstein's photoelectric effect (for which Einstein won Nobel Prize in 1921), which in recent years finds extensive applications in modern optoelectronics, characterization and investigation of condensed matter systems, photoemission spectroscopy and related aspects in connection with the investigations of the optical properties of nanostructures [1-5]. Interest in low dimensional silicon nanostructures also grew up and gained momentum, after the discovery of room temperature photoluminescence and electroluminescence of silicon nano-wires in porous silicon [1]. Work on ultrathin layers of $SiSiO_2$ superlattices resulting into visible light emission at room temperature clearly exhibited low dimensional quantum confinement effect [2] and one of the most popular techniques for analyzing the low dimensional structures is to employ photoemission techniques. Recent observation of room temperature photoluminescence and electro luminescence in porous silicon has stimulated vigorous research activities in silicon nanostructures [3].

The EP depends on the density-of-states (DOS) function which, in turn, is significantly affected by the different carrier energy spectra of different semiconductors having various band structures. In recent years, various energy wave vector dispersion relations of the carriers of different materials have been proposed [6] which have created the interest in studying the EP from HD materials and their quantized counterparts. In this chapter we will study the EP from HD quantized wires (QWs) of non-linear optical, III-V, II-VI, Gallium Phosphide, Germanium, Platinum Antimonide, stressed, IV-VI, Lead Germanium Telluride, Tellurium, II-V, Zinc and Cadmium diphosphides and Bismuth Telluride respectively.

It is well-known that the classical equation of the photo-emitted current density is [7] $J=[4\pi\alpha_0 e m_c g_v (k_B T)^2 / h^3]\exp[(hv-\phi)/(k_B T)]$ (where α_0 $e, \mathrm{m_c}, g_v, k_B, T, h, h\upsilon$ and ϕ are the probability of photoemission, electron charge, effective electron mass at the edge of the conduction band, valley degeneracy, the Boltzmann constant, temperature, the Planck constant, incident photon energy along z-axis and work function respectively). The afore-mentioned equation is valid for both the charge carriers and in this conventional form it appears that, the photoemission changes with the effective mass, temperature, work function and the incident photon energy respectively. This relation holds only under the condition of carrier non-degeneracy.

The EP has different values for different materials and varies with doping and with external fields which creates quantization of the wave-vector space of the carriers leading to various types of quantized structures. The nature of these variations has been studied in [1-32] and some of the significant features are as follows:

1. The EP from bulk materials increases with the increase in doping.
2. The EP exhibits oscillatory dependence with inverse quantizing magnetic field because of the Shubnikov de Haas (SdH) effect.
3. The EP changes significantly with the magnitude of the externally applied quantizing electric field in electronic materials.
4. The EP from quantum confined Bismuth, nonlinear optical, III-V, II-VI and IV-VI materials oscillate with nano-thickness in various manners which are totally band structure dependent.
5. The nature of variations is significantly influenced by the energy band constants of various materials having different band structures.
6. The EP has significantly different values in quantum confined semiconductor superlattices and various other quantized structures.

In this chapter in theoretical background we have chronologically investigated the EP from QWs of HD non-linear optical, III-V, II-VI, stressed Kane type, Te, GaP, $PtSb_2$, Bi_2Te_3, Ge and GaAs respectively. The section 2.3 contains the result and discussions pertaining to this chapter.

THEORETICAL BACKGROUND

The EP from Nano Wires of HD Nonlinear Optical Semiconductors

The 1D DOS function per sub-band is given by

$$N_{1D}(E) = \frac{2g_v}{\pi}\frac{\partial k_x}{\partial E} \tag{1}$$

The velocity of the emitted electrons along the x-direction can be written as

$$v_x(E) = \frac{1}{\hbar}\frac{\partial E}{\partial k_x} \tag{2}$$

Therefore the photocurrent is given by

$$I = \frac{\alpha_o e g_v}{2}\sum_{n_y=1}^{n_{y\max}}\sum_{n_z=1}^{n_{z\max}}\int_{\Delta_1}^{\infty}\left(\frac{2}{\pi}\frac{\partial k_x}{\partial E}\right)\left(\frac{1}{\hbar}\frac{\partial E}{\partial k_x}\right)f(E)dE \tag{3}$$

where, $\Delta_1 \equiv E' + W - h\nu$.

Using (3), one can write,

$$I = \frac{\alpha_o e g_v k_B T}{\pi\hbar}\,\mathrm{Re}\,al\ part\ of \sum_{n_y=1}^{n_{y\max}}\sum_{n_z=1}^{n_{z\max}} F_0(\eta_{61HD}) \tag{4}$$

$where,\ \eta_{61HD} \equiv [\frac{E_{F1HDNW} - (E'_{1HDNW} + W - h\nu)}{k_B T}]$ E_{F1HDNW} in the Fermi energy in this case, E'_{1HDNW} is the complex sub-band energy which can be expressed in this case as

$$\frac{\hbar^2 (n_z \pi / d_z)^2}{2m_{\|}^* T_{21}(E'_{1HDNW}, \eta_g)} + \frac{\hbar^2 (n_y \pi / d_y)^2}{2m_{\|}^* T_{22}(E'_{1HDNW}, \eta_g)} = 1 \tag{5}$$

In the absence of band-tails, the EP in this case is given by

$$I = \frac{\alpha_o e g_v k_B T}{\pi \hbar} \sum_{n_y=1}^{n_{y\max}} \sum_{n_z=1}^{n_{z\max}} F_0(\eta_{62}) \tag{6}$$

$where,\ \eta_{62} \equiv [\frac{E_{F1d} - (E'_1 + W - h\nu)}{k_B T}].$

where E_{F1d} is the Fermi energy in this case and (E'_1) can be expressed by the equation

$$\gamma(E_1') = f_1(E_1')(\pi n_y / d_y)^2 + f_2(E_1')(\pi n_z / d_z)^2 \tag{7}$$

The EP from Nano Wires of HD III-V Semiconductors

(i) Three Band Model of Kane

The EP in this case is given by

$$I = \frac{\alpha_o e g_v k_B T}{\pi \hbar}\ Real\ part\ of \sum_{n_y=1}^{n_{y\max}} \sum_{n_z=1}^{n_{z\max}} F_0(\eta_{63HD}) \tag{8}$$

$$where,\ \eta_{63HD} \equiv [\frac{E_{F1HDNW} - (E'_{2HDNW} + W - h\nu)}{k_B T}]$$

and E'_{2HDNW} is the sub-band energy in this case which can be expressed as

$$\frac{\hbar^2 (\mathrm{n}_z \pi / \mathrm{d}_z)^2}{2m_c} + \frac{\hbar^2 (\mathrm{n}_y \pi / \mathrm{d}_y)^2}{2m_c} = T_{31}(E'_{2HDNW}, \eta_g) + iT_{31}(E'_{2HDNW}, \eta_g) \tag{9}$$

(ii) Two Band Model of Kane

The EP in this case is given by

$$I = \frac{\alpha_o e g_v k_B T}{\pi \hbar} \sum_{n_y=1}^{n_{y\max}} \sum_{n_z=1}^{n_{z\max}} F_0(\eta_{64HD}) \tag{10}$$

$$where,\ \eta_{64HD} \equiv [\frac{E_{F1HDNW} - (E'_{3HDNW} + W - h\nu)}{k_B T}]$$

and E'_{3HDNW} is the sub-band energy in this case which can be expressed as

$$\frac{\hbar^2 (n_z \pi / d_z)^2}{2m_c} + \frac{\hbar^2 (n_y \pi / d_y)^2}{2m_c} = \gamma_2(E'_{3HDNW}, \eta_g) \tag{11}$$

(iii) Parabolic Energy Bands

The EP in this case is given by

$$I = \frac{\alpha_o e g_v k_B T}{\pi \hbar} \sum_{n_y=1}^{n_{y\max}} \sum_{n_z=1}^{n_{z\max}} F_0(\eta_{66HD}) \tag{12}$$

$$where,\ \eta_{66HD} \equiv [\frac{E_{F1HDNW} - (E'_{5HDNW} + W - h\nu)}{k_B T}]$$

and E'_{5HDNW} is the sub-band energy in this case which can be expressed as

$$\frac{\hbar^2(\mathrm{n}_z\pi/\mathrm{d}_z)^2}{2m_c}+\frac{\hbar^2(\mathrm{n}_y\pi/\mathrm{d}_y)^2}{2m_c}=\gamma_3(E'_{5HDNW},\eta_g) \quad (13)$$

The expression of the EP in this case is given by

$$I=\frac{\alpha_o e g_v k_B T}{\pi\hbar}\sum_{n_y=1}^{n_{y\max}}\sum_{n_z=1}^{n_{z\max}}F_0(\eta_{67}) \quad (14)$$

$$where,\ \eta_{67}\equiv[\frac{E_{F1d}-(E'_7+W-h\nu)}{k_BT}]$$

Converting the summation over quantum numbers to the corresponding integrations in (2.38), the photocurrent density from semiconductors having isotropic parabolic energy bands with non-degenerate electron concentration gets transformed in to the well known form as given in the preface. Besides, (39) is well-known in the literature.

(iv) The Model of Stillman et al.

The dispersion relation of the 1D electrons in this case can be written as

$$\frac{\hbar^2(n_z\pi/d_z)^2}{2m_c}+\frac{\hbar^2(n_y\pi/d_y)^2}{2m_c}+\frac{\hbar^2 k_x^{\ 2}}{2m_c}=\theta_4(E,\eta_g) \quad (15)$$

$$where,\ \theta_4(E,\eta_g)=I_{12}(E,\eta_g)$$

The EP in this case is given by

$$I=\frac{\alpha_o e g_v k_B T}{\pi\hbar}\sum_{n_y=1}^{n_{y\max}}\sum_{n_z=1}^{n_{z\max}}F_0(\eta_{69HD}) \quad (16)$$

$$where,\ \eta_{69HD}\equiv[\frac{E_{F1HDNW}-(E'_{9HDNW}+W-h\nu)}{k_BT}]$$

and E'_{9HDNW} is the sub-band energy in this case which can be expressed as

$$\frac{\hbar^2(n_z\pi/d_z)^2}{2m_c}+\frac{\hbar^2(n_y\pi/d_y)^2}{2m_c}=\theta_4(E'_{9HDNW},\eta_g) \tag{17}$$

The expression the EP for QWs of III-V materials whose energy band structures are defined by the model of Stillman et al. in the absence of band tailing assumes the form

$$I=\frac{\alpha_o e g_v k_B T}{\pi\hbar}\sum_{n_y=1}^{n_{y\max}}\sum_{n_z=1}^{n_{z\max}}F_0(\eta_{69}) \tag{18}$$

where, $\eta_{69}\equiv[\frac{E_{F1d}-(E'_9+W-h\nu)}{k_B T}]$

(v) The Model of Palik et al.

The dispersion relation of the 1D electrons in this case can be written as

$$\frac{\hbar^2(n_z\pi/d_z)^2}{2m_c}+\frac{\hbar^2(n_y\pi/d_y)^2}{2m_c}+\frac{\hbar^2k_x{}^2}{2m_c}=\theta_5(E,\eta_g) \tag{19}$$

where, $\theta_5(E,\eta_g)=I_{13}(E,\eta_g)$

The EP in this case is given by

$$I=\frac{\alpha_o e g_v k_B T}{\pi\hbar}\sum_{n_y=1}^{n_{y\max}}\sum_{n_z=1}^{n_{z\max}}F_0(\eta_{610HD}), \tag{20}$$

where, $\eta_{610HD}\equiv[\frac{E_{F1HDNW}-(E'_{10HDNW}+W-h\nu)}{k_B T}]$,

and E'_{10HDNW} is the sub-band energy in this case which can be expressed as

$$\frac{\hbar^2(n_z\pi/d_z)^2}{2m_c}+\frac{\hbar^2(n_y\pi/d_y)^2}{2m_c}=\theta_5(E'_{10HDNW},\eta_g) \tag{21}$$

The expression the EP for QWs of III-V materials whose energy band structures are defined by the model of Palik et al. in the absence of band tailing assumes the form

$$I=\frac{\alpha_o e g_v k_B T}{\pi\hbar}\sum_{n_y=1}^{n_{y\max}}\sum_{n_z=1}^{n_{z\max}}F_0(\eta_{610}) \tag{22}$$

$$where,\ \eta_{610}\equiv[\frac{E_{F1d}-(E'_{10}+W-h\nu)}{k_BT}]$$

The EP from Nano Wires of HD II-VI Semiconductors

The 1D electron dispersion law in QW of HD II-VI semiconductors can be written as

$$\gamma_3(E,\eta_g)=a'_0[(\frac{n_x\pi}{d_x})^2+(\frac{n_y\pi}{d_y})^2]\pm\bar{\lambda}_0[(\frac{n_x\pi}{d_x})^2+(\frac{n_y\pi}{d_y})^2]^{1/2}+\frac{\hbar^2k_z^2}{2m_{\parallel}^*} \tag{23}$$

The EP in this case is given by

$$I=\frac{\alpha_o e g_v k_B T}{2\pi\hbar}\sum_{n_y=1}^{n_{y\max}}\sum_{n_z=1}^{n_{z\max}}F_0(\eta_{613HD}) \tag{24}$$

$$where,\ \eta_{613HD}\equiv[\frac{E_{F1HDNW}-(E'_{13HDNW}+W-h\nu)}{k_BT}]$$

and E'_{13HDNW} is the sub-band energy in this case which can be expressed as

$$\gamma_3(E'_{13HDNW},\eta_g)=a'_0[(\frac{n_x\pi}{d_x})^2+(\frac{n_y\pi}{d_y})^2]\pm\bar{\lambda}_0[(\frac{n_x\pi}{d_x})^2+(\frac{n_y\pi}{d_y})^2]^{1/2} \tag{25}$$

The EP photocurrent from QWs of II-VI materials is given by

$$I=\frac{\alpha_0 e g_v k_B T}{2\pi\hbar}\sum_{n_x=1}^{n_{x\max}}\sum_{n_y=1}^{n_{y\max}}\{F_0\{(k_BT)^{-1}[E_{F1d}-[G_{3,+}(n_x,n_y)+W-h\nu]]\} \\ +F_0\{(k_BT)^{-1}[E_{F1d}-[G_{3,-}(n_x,n_y)+W-h\nu]]\}\} \tag{26}$$

The EP from Nano Wires of HD IV-VI Semiconductors

(i) Dimmock Model

The 1D electron dispersion law in QW of HD IV-VI semiconductors can be expressed following (1.174) as

$$\gamma_2(E,\eta_g)+\alpha\gamma_3(E,\eta_g)(\frac{\hbar^2}{2x_4}(\frac{n_x\pi}{d_x})^2+\frac{\hbar^2}{2x_5}(\frac{n_y\pi}{d_y})^2)+\alpha\gamma_3(E,\eta_g) \\ \frac{\hbar^2}{2x_6}k_z^2-(1+\alpha\gamma_3(E,\eta_g))(\frac{\hbar^2}{2x_1}(\frac{n_x\pi}{d_x})^2+\frac{\hbar^2}{2x_2}(\frac{n_y\pi}{d_y})^2) \\ -\alpha(\frac{\hbar^2}{2x_1}(\frac{n_x\pi}{d_x})^2+\frac{\hbar^2}{2x_2}(\frac{n_y\pi}{d_y})^2)(\frac{\hbar^2}{2x_4}(\frac{n_x\pi}{d_x})^2+\frac{\hbar^2}{2x_5}(\frac{n_y\pi}{d_y})^2) \\ -\alpha(\frac{\hbar^2}{2x_1}(\frac{n_x\pi}{d_x})^2+\frac{\hbar^2}{2x_2}(\frac{n_y\pi}{d_y})^2)\frac{\hbar^2}{2x_6}k_z^2-(1+\alpha\gamma_3(E,\eta_g)) \\ \frac{\hbar^2}{2x_3}k_z^2-\alpha\frac{\hbar^2}{2x_3}k_z^2(\frac{\hbar^2}{2x_4}(\frac{n_x\pi}{d_x})^2+\frac{\hbar^2}{2x_5}(\frac{n_y\pi}{d_y})^2)-\alpha\frac{\hbar^4k_z^4}{4x_3x_6} \\ =\frac{\hbar^2}{2m_1}(\frac{n_x\pi}{d_x})^2+\frac{\hbar^2}{2m_2}(\frac{n_y\pi}{d_y})^2+\frac{\hbar^2}{2m_3}k_z^2 \tag{27}$$

Equation (27) can be written as

$$k_z=T_{36}(E,\eta_g,n_x,n_y) \tag{28}$$

$$where\ T_{36}(E,\eta_g,n_x,n_y)=[(2C_{22})^{-1}[-B_{HD}(E,\eta_g,n_x,n_y) + \sqrt{B^2{}_{HD}(E,\eta_g,n_x,n_y)+4C_{22}A_{HD}(E,\eta_g,n_x,n_y)}\,]]^{1/2}$$

$$C_{22}=(\alpha\frac{\hbar^4}{4x_3x_6}), B_{HD}(E,\eta_g,n_x,n_y)=[\alpha(\frac{\hbar^2}{2x_1}(\frac{n_x\pi}{d_x})^2+\frac{\hbar^2}{2x_2}(\frac{n_y\pi}{d_y})^2)\frac{\hbar^2}{2x_6} +(1+\alpha\gamma_3(E,\eta_g))\frac{\hbar^2}{2x_3}-\alpha\gamma_3(E,\eta_g)\frac{\hbar^2}{2x_6}$$

$$+\frac{\hbar^2}{2m_3}+\alpha\frac{\hbar^2}{2x_3}(\frac{\hbar^2}{2x_4}(\frac{n_x\pi}{d_x})^2+\frac{\hbar^2}{2x_5}(\frac{n_y\pi}{d_y})^2)]$$

and

$$A_{HD}(E,\eta_g,n_x,n_y)=[-[\frac{\hbar^2}{2m_1}(\frac{n_x\pi}{d_x})^2+\frac{\hbar^2}{2m_2}(\frac{n_y\pi}{d_y})^2]\gamma_2(E,\eta_g)+\alpha\gamma_3(E,\eta_g) (\frac{\hbar^2}{2x_4}(\frac{n_x\pi}{d_x})^2+\frac{\hbar^2}{2x_5}(\frac{n_y\pi}{d_y})^2)-\alpha(\frac{\hbar^2}{2x_1}(\frac{n_x\pi}{d_x})^2 +\frac{\hbar^2}{2x_2}(\frac{n_y\pi}{d_y})^2)(\frac{\hbar^2}{2x_4}(\frac{n_x\pi}{d_x})^2+\frac{\hbar^2}{2x_5}(\frac{n_y\pi}{d_y})^2) -(1+\alpha\gamma_3(E,\eta_g))(\frac{\hbar^2}{2x_1}(\frac{n_x\pi}{d_x})^2+\frac{\hbar^2}{2x_2}(\frac{n_y\pi}{d_y})^2)]$$

The EP in this case is given by

$$I=\frac{\alpha_o e g_v k_B T}{\pi\hbar}\sum_{n_x=1}^{n_{x_{\max}}}\sum_{n_y=1}^{n_{y_{\max}}}F_0(\eta_{614HD}) \tag{29}$$

$$where,\ \eta_{614HD}\equiv[\frac{E_{F1HDNW}-(E'_{14HDNW}+W-h\nu)}{k_BT}]$$

and E'_{14HDNW} is the sub-band energy in this case which can be expressed as

$$0=T_{36}(E'_{14HDNW},\eta_g,n_x,n_y) \tag{30}$$

The 1D electron dispersion law in QW of IV-VI semiconductors in the absence of band tails can be expressed as

$$
\begin{aligned}
&E(1+\alpha E)+\alpha E\left(\frac{\hbar^2}{2x_4}\left(\frac{n_x\pi}{d_x}\right)^2+\frac{\hbar^2}{2x_5}\left(\frac{n_y\pi}{d_y}\right)^2\right)+\alpha E\frac{\hbar^2}{2x_6}k_z^2 \\
&\quad -(1+\alpha E)\left(\frac{\hbar^2}{2x_1}\left(\frac{n_x\pi}{d_x}\right)^2+\frac{\hbar^2}{2x_2}\left(\frac{n_y\pi}{d_y}\right)^2\right) \\
&-\alpha\left(\frac{\hbar^2}{2x_1}\left(\frac{n_x\pi}{d_x}\right)^2+\frac{\hbar^2}{2x_2}\left(\frac{n_y\pi}{d_y}\right)^2\right)\left(\frac{\hbar^2}{2x_4}\left(\frac{n_x\pi}{d_x}\right)^2+\frac{\hbar^2}{2x_5}\left(\frac{n_y\pi}{d_y}\right)^2\right) \\
&-\alpha\left(\frac{\hbar^2}{2x_1}\left(\frac{n_x\pi}{d_x}\right)^2+\frac{\hbar^2}{2x_2}\left(\frac{n_y\pi}{d_y}\right)^2\right)\frac{\hbar^2}{2x_6}k_z^2 \\
&-(1+\alpha E)\frac{\hbar^2}{2x_3}k_z^2-\alpha\frac{\hbar^2}{2x_3}k_z^2\left(\frac{\hbar^2}{2x_4}\left(\frac{n_x\pi}{d_x}\right)^2+\frac{\hbar^2}{2x_5}\left(\frac{n_y\pi}{d_y}\right)^2\right) \\
&-\alpha\frac{\hbar^4k_z^4}{4x_3x_6}=\frac{\hbar^2}{2m_1}\left(\frac{n_x\pi}{d_x}\right)^2+\frac{\hbar^2}{2m_2}\left(\frac{n_y\pi}{d_y}\right)^2+\frac{\hbar^2}{2m_3}k_z^2
\end{aligned}
\tag{31}
$$

Equation (31) can be written as

$$
k_z=T_{40}(E,n_x,n_y) \tag{32}
$$

$$
\begin{aligned}
\textit{where } T_{40}(E,n_x,n_y)&=[(2C_{22})^{-1}[-B_0(E,n_x,n_y) \\
&\quad+\sqrt{B^2{}_0(E,n_x,n_y)+4C_{22}A_0(E,n_x,n_y)}\,]]^{1/2}
\end{aligned}
$$

$$
\begin{aligned}
\textit{where } B_0(E,n_x,n_y)&=[\alpha\left(\frac{\hbar^2}{2x_1}\left(\frac{n_x\pi}{d_x}\right)^2+\frac{\hbar^2}{2x_2}\left(\frac{n_y\pi}{d_y}\right)^2\right)\frac{\hbar^2}{2x_6} \\
&\quad+(1+\alpha E)\frac{\hbar^2}{2x_3}-\alpha E\frac{\hbar^2}{2x_6} \\
&\quad+\frac{\hbar^2}{2m_3}+\alpha\frac{\hbar^2}{2x_3}\left(\frac{\hbar^2}{2x_4}\left(\frac{n_x\pi}{d_x}\right)^2+\frac{\hbar^2}{2x_5}\left(\frac{n_y\pi}{d_y}\right)^2\right)]
\end{aligned}
$$

and

$$A_0(E,n_x,n_y)=[-[\frac{\hbar^2}{2m_1}(\frac{n_x\pi}{d_x})^2+\frac{\hbar^2}{2m_2}(\frac{n_y\pi}{d_y})^2]E(1+\alpha E)$$

$$+\alpha E(\frac{\hbar^2}{2x_4}(\frac{n_x\pi}{d_x})^2+\frac{\hbar^2}{2x_5}(\frac{n_y\pi}{d_y})^2)$$

$$-\alpha(\frac{\hbar^2}{2x_1}(\frac{n_x\pi}{d_x})^2+\frac{\hbar^2}{2x_2}(\frac{n_y\pi}{d_y})^2)(\frac{\hbar^2}{2x_4}(\frac{n_x\pi}{d_x})^2+\frac{\hbar^2}{2x_5}(\frac{n_y\pi}{d_y})^2)$$

$$-(1+\alpha E)(\frac{\hbar^2}{2x_1}(\frac{n_x\pi}{d_x})^2+\frac{\hbar^2}{2x_2}(\frac{n_y\pi}{d_y})^2)]$$

The EP in this case is given by

$$I=\frac{\alpha_o e g_v k_B T}{\pi\hbar}\sum_{n_x=1}^{n_{x_{\max}}}\sum_{n_y=1}^{n_{y_{\max}}}F_0(\eta_{615}) \tag{33}$$

where, $\eta_{615}\equiv[\frac{E_{F1d}-(E'_{20}+W-h\nu)}{k_BT}]$

and E'_{20} is the sub-band energy in this case which can be expressed as

$$0=T_{40}(E'_{20},n_x,n_y) \tag{34}$$

(ii) Bangert and Kastner Model

Following (1.194d), the 1D dispersion relation in QW of IV-VI semiconductors in accordance with the present model can be written as

$$F_1(E,\eta_g)[(\frac{n_x\pi}{d_x})^2+(\frac{n_y\pi}{d_y})^2]+F_2(E,\eta_g)\mathrm{k}_z^2=1 \tag{35}$$

The (35) can be written as

$$k_z = T_{60}(E, \eta_g, n_x, n_y) \tag{36}$$

where, $T_{60}(E, \eta_g, n_x, n_y) = [[1 - F_1(E, \eta_g)[(\frac{n_x \pi}{d_x})^2 + (\frac{n_y \pi}{d_y})^2]][F_2(E, \eta_g)]^{-1}]^{1/2}$

The EP in this case is given by

$$I = \frac{\alpha_o e g_v k_B T}{\pi \hbar} \sum_{n_x=1}^{n_{x_{\max}}} \sum_{n_y=1}^{n_{y_{\max}}} F_0(\eta_{615HD}), \tag{37}$$

where, $\eta_{615HD} \equiv [\frac{E_{F1HDNW} - (E'_{15HDNW} + W - h\nu)}{k_B T}]$

and E'_{15HDNW} is the sub-band energy in this case which can be expressed as

$$0 = T_{60}(E'_{15HDNW}, \eta_g, n_x, n_y) \tag{38}$$

The 1D dispersion relation in the absence of band tailing can be written in this case as

$$\omega_1(\mathrm{E})[(\frac{\pi n_x}{d_x})^2 + (\frac{\pi n_y}{d_y})^2] + \omega_2(\mathrm{E}) k_z^2 = 1 \tag{39}$$

The (39) can be written as

$$k_z = T_{61}(E, n_x, n_y) \tag{40}$$

where, $T_{61}(E, n_x, n_y) = [[1 - \omega_1(E)[(\frac{n_x \pi}{d_x})^2 + (\frac{n_y \pi}{d_y})^2]][\omega_2(E)]^{-1}]^{1/2}$

The EP in this case is given by

$$I = \frac{\alpha_o e g_v k_B T}{\pi \hbar} \sum_{n_x=1}^{n_{x_{\max}}} \sum_{n_y=1}^{n_{y_{\max}}} F_0(\eta_{616}) \tag{41}$$

$$where,\ \eta_{616} \equiv [\frac{E_{F1d} - (E'_{21} + W - h\nu)}{k_B T}]$$

and E'_{21} is the sub-band energy in this case which can be expressed as

$$0 = T_{61}(E'_{21}, n_x, n_y) \tag{42}$$

The EP from QWs of HD Stressed Kane Type Semiconductors

The 1D dispersion relation in this case can be written following as

$$P_{11}(E,\eta_g)(\frac{\pi n_x}{d_x})^2 + Q_{11}(E,\eta_g)(\frac{\pi n_y}{d_y})^2 + S_{11}(E,\eta_g)k_z^2 = 1$$

(43)

The (43) can be written as

$$k_z = T_{70}(E,\eta_g,n_x,n_y) \tag{44}$$

$$where,\ T_{70}(E,\eta_g,n_x,n_y) = [[1 - P_{11}(E,\eta_g)(\frac{\pi n_x}{d_x})^2 + Q_{11}(E,\eta_g)(\frac{\pi n_y}{d_y})^2][S_{11}(E,\eta_g)]^{-1}]^{1/2}$$

The EP in this case is given by

$$I = \frac{\alpha_o e g_v k_B T}{\pi \hbar} \sum_{n_x=1}^{n_{x\max}} \sum_{n_y=1}^{n_{y\max}} F_0(\eta_{630HD}) \tag{45}$$

$$where,\ \eta_{630HD} \equiv [\frac{E_{F1HDNW} - (E'_{30HDNW} + W - h\nu)}{k_B T}]$$

and E'_{30HDNW} is the sub-band energy in this case which can be expressed as

$$0 = T_{70}(E'_{30HDNW}, \eta_g, n_x, n_y) \tag{46}$$

In the absence of band tailing the 1D dispersion relation in this case assumes the form

$$k_z = t_{70}(E, n_x, n_y) \tag{47}$$

$$where,\ t_{70}(E, n_x, n_y) = [[\bar{c}_0(\mathrm{E})[1 - (\frac{\pi n_x}{d_x \bar{a}_0(\mathrm{E})})^2 - (\frac{\pi n_y}{d_y \bar{b}_0(\mathrm{E})})^2]]^{1/2}$$

The EP in this case is given by

$$I = \frac{\alpha_o e g_v k_B T}{\pi \hbar} \sum_{n_x=1}^{n_{x\max}} \sum_{n_y=1}^{n_{y\max}} F_0(\eta_{642}) \tag{48}$$

$$where,\ \eta_{642} \equiv [\frac{E_{F1d} - (E'_{42} + W - h\nu)}{k_B T}]$$

and E'_{42} is the sub-band energy in this case which can be expressed as

$$0 = t_{60}(E'_{42}, n_x, n_y) \tag{49}$$

The EP from Nano Wires of HD Te

The 1D dispersion relation may be written in this case as

$$k_x = t_{72}(\mathrm{E}, \mathrm{n}_y, \mathrm{n}_z, \eta_g) \tag{50}$$

$$where,\ t_{72}(E,n_y,n_z,\eta_g)=[-(\frac{n_y\pi}{d_y})^2+\psi_{5HD}(E,\eta_g)-\psi_6(\frac{\pi n_z}{d_z})^2$$
$$\pm\psi_7[\psi_{8HD}^2(E,\eta_g)-(\frac{\pi n_z}{d_z})^2]^{1/2}]^{1/2}$$

The EP in this case is given by

$$I=\frac{\alpha_o e g_v k_B T}{2\pi\hbar}\sum_{n_y=1}^{n_{y_{\max}}}\sum_{n_z=1}^{n_{z_{\max}}}F_0(\eta_{631HD}) \tag{51}$$

$$where,\ \eta_{631HD}\equiv[\frac{E_{F1HDNW}-(E'_{31HDNW}+W-h\nu)}{k_BT}]$$

and E'_{31HDNW} is the sub-band energy in this case which can be expressed as

$$0=t_{72}(E'_{31HDNW},\eta_g,n_y,n_z) \tag{52}$$

In the absence of band tailing the 1D dispersion relation in this case assumes the form

$$k_x=H_{70}(E,n_y,n_z) \tag{53}$$

$$where,\ H_{70}(E,n_y,n_z)=[-(\frac{n_y\pi}{d_y})^2+\psi_5(E)-\psi_6(\frac{\pi n_z}{d_z})^2$$
$$\pm\psi_7[\psi_8^2(E)-(\frac{\pi n_z}{d_z})^2]^{1/2}]^{1/2}$$

The EP in this case is given by

$$I=\frac{\alpha_o e g_v k_B T}{2\pi\hbar}\sum_{n_y=1}^{n_{y_{\max}}}\sum_{n_z=1}^{n_{z_{\max}}}F_0(\eta_{644}) \tag{54}$$

$$where,\ \eta_{644} \equiv [\frac{E_{F1d} - (E'_{44} + W - h\nu)}{k_B T}]$$

and E'_{44} is the sub-band energy in this case which can be expressed as

$$0 = H_{70}(E'_{44}, n_y, n_z) \tag{55}$$

The EP from Nano Wires of HD GaP

The 1D dispersion relation may be written in this case following (1.253) as

$$k_x = u_{70}(E, n_y, n_z, \eta_g) \tag{56}$$

$$where,\ u_{70}(E, n_y, n_z, \eta_g) = [-(\frac{n_y \pi}{d_y})^2 + t_{11}\gamma_3(E, \eta_g) + t_{21} - t_{31}(\frac{n_z \pi}{d_z})^2 - t_{41}[(\frac{n_z \pi}{d_z})^2 + t_5^2(E, \eta_g)]^{1/2}]^{1/2}$$

The EP in this case is given by

$$I = \frac{\alpha_o e g_v k_B T}{\pi \hbar} \sum_{n_y=1}^{n_{y\max}} \sum_{n_z=1}^{n_{z\max}} F_0(\eta_{632HD})$$

$$\eta_{632HD} \equiv [\frac{E_{F1HDNW} - (E'_{32HDNW} + W - h\nu)}{k_B T}] \tag{57}$$

and E'_{32HDNW} is the sub-band energy in this case which can be expressed as

$$0 = u_{70}(E'_{32HDNW}, \eta_g, n_y, n_z) \tag{58}$$

In the absence of band tailing the 1D dispersion relation in this case can be written using (1.260) as

$$k_x = X_{71}(\mathrm{E}, \mathrm{n}_y, \mathrm{n}_z) \tag{59}$$

$$where,\ X_{71}(E, n_y, n_z) = [-(\frac{n_y \pi}{d_y})^2 + t_{42}(E, n_z)]^{1/2}$$

The EP in this case is given by

$$I = \frac{\alpha_o e g_v k_B T}{2\pi\hbar} \sum_{n_y=1}^{n_{y\max}} \sum_{n_z=1}^{n_{z\max}} F_0(\eta_{646}) \tag{60}$$

$$where,\ \eta_{646} \equiv [\frac{E_{F1d} - (E'_{46} + W - h\nu)}{k_B T}]$$

and E'_{46} is the sub-band energy in this case which can be expressed as

$$0 = X_{71}(E'_{46}, n_y, n_z) \tag{61}$$

The EP from Nano Wires of HD $PtSb_2$

The 1D dispersion relation may be written in this case can be written as

$$k_x = V_{70}(E, n_y, n_z, \eta_g) \tag{62}$$

$$where,\ V_{70}(E, n_y, n_z, \eta_g) = [-(\frac{n_y \pi}{d_y})^2 + A_{60}(E, \eta_g, n_y)]^{1/2}$$

The EP in this case is given by

$$I = \frac{\alpha_o e g_v k_B T}{\pi \hbar} \sum_{n_y=1}^{n_{y\max}} \sum_{n_z=1}^{n_{z\max}} F_0(\eta_{634HD}) \tag{63}$$

$$where,\ \eta_{634HD} \equiv [\frac{E_{F1HDNW} - (E'_{34HDNW} + W - h\nu)}{k_B T}]$$

and E'_{34HDNW} is the sub-band energy in this case which can be expressed as

$$0 = V_{70}(E'_{34HDNW}, \eta_g, n_y, n_z) \tag{64}$$

In the absence of band tailing the 1D dispersion relation in this case can be written as

$$k_x = D_{71}(E, n_y, n_z) \tag{65}$$

$$where,\ D_{71}(E, n_y, n_z) = [-(\frac{n_y \pi}{d_y})^2 + t_{44}(E, n_z)]^{1/2}$$

The EP in this case is given by

$$I = \frac{\alpha_o e g_v k_B T}{2\pi \hbar} \sum_{n_y=1}^{n_{y\max}} \sum_{n_z=1}^{n_{z\max}} F_0(\eta_{648}) \tag{66}$$

$$where,\ \eta_{648} \equiv [\frac{E_{F1d} - (E'_{48} + W - h\nu)}{k_B T}]$$

and E'_{48} is the sub-band energy in this case which can be expressed as

$$0 = D_{71}(E'_{48}, n_y, n_z) \tag{67}$$

The EP from Nano Wires of HD Bi_2Te_3

The dispersion relation in this case can be written as

$$k_x = J_{70}(E, n_y, n_z, \eta_g) \quad (68)$$

$$where,\ J_{70}(E, n_y, n_z, \eta_g) = [[\gamma_2(E, \eta_g) - \bar{\omega}_2(\frac{n_y \pi}{d_y})^2 - \bar{\omega}_3(\frac{n_z \pi}{d_z})^2 - 2\bar{\omega}_4 \frac{n_y n_z \pi^2}{d_y d_z}](\bar{\omega}_1)^{-1}]^{1/2}$$

The EP in this case is given by

$$I = \frac{\alpha_o e g_v k_B T}{\pi \hbar} \sum_{n_y=1}^{n_{y\max}} \sum_{n_z=1}^{n_{z\max}} F_0(\eta_{650HD}) \quad (69)$$

$$where,\ \eta_{650HD} \equiv [\frac{E_{F1HDNW} - (E'_{50HDNW} + W - h\nu)}{k_B T}]$$

and E'_{50HDNW} is the sub-band energy in this case which can be expressed as

$$0 = J_{70}(E'_{50HDNW}, \eta_g, n_y, n_z) \quad (70)$$

In the absence of band tailing the 1D dispersion relation in this case can be written as

$$k_x = B_{71}(E, n_y, n_z) \quad (71)$$

$$where,\ B_{71}(E, n_y, n_z) = [[E(1 + \alpha E) - \bar{\omega}_2(\frac{n_y \pi}{d_y})^2 - \bar{\omega}_3(\frac{n_z \pi}{d_z})^2 - 2\bar{\omega}_4 \frac{n_y n_z \pi^2}{d_y d_z}](\bar{\omega}_1)^{-1}]^{1/2}$$

The EP in this case is given by

$$I = \frac{\alpha_o e g_v k_B T}{2\pi\hbar} \sum_{n_y=1}^{n_{y\max}} \sum_{n_z=1}^{n_{z\max}} F_0(\eta_{650}) \tag{72}$$

$$where,\ \eta_{650} \equiv [\frac{E_{F1d} - (E'_{50} + W - h\nu)}{k_B T}],$$

and E'_{50} is the sub-band energy in this case which can be expressed as

$$0 = B_{71}(E'_{50}, n_y, n_z) \tag{73}$$

The EP from Nano Wires of HD Ge

(i) Model of Cardona et al.

The dispersion relation in accordance with this model in the present case can be written as

$$k_x = L_{70}(E, n_y, n_z, \eta_g) \tag{74}$$

$$where\ ,\ L_{70}(E, n_y, n_z, \eta_g) = [[\gamma_2(E, \eta_g) + \alpha[\frac{\hbar^2}{2m_{\|}^*}(\frac{n_z\pi}{d_z})^2]^2 - (1 + 2\alpha\gamma_3(E, \eta_g))\frac{\hbar^2}{2m_{\|}^*}(\frac{n_z\pi}{d_z})^2](\frac{2m_{\|}^*}{\hbar^2})]^{1/2}$$

The EP in this case is given by

$$I = \frac{\alpha_o e g_v k_B T}{\pi\hbar} \sum_{n_y=1}^{n_{y\max}} \sum_{n_z=1}^{n_{z\max}} F_0(\eta_{652HD}) \tag{75}$$

$$where,\ \eta_{652HD} \equiv [\frac{E_{F1HDNW} - (E'_{52HDNW} + W - h\nu)}{k_B T}]$$

and E'_{52HDNW} is the sub-band energy in this case which can be expressed as

$$0 = L_{70}(E'_{52HDNW}, \eta_g, n_y, n_z) \quad (76)$$

In the absence of band tailing the 1D dispersion relation in this case can be written using (1.278) as

$$k_x = B_{77}(E, n_y, n_z) \quad (77)$$

$$where,\ B_{77}(E, n_y, n_z) = [[E(1+\alpha E) + \alpha[\frac{\hbar^2}{2m_{\|}^*}(\frac{n_z\pi}{d_z})^2]^2 - (1+2\alpha E)\frac{\hbar^2}{2m_{\|}^*}(\frac{n_z\pi}{d_z})^2](\frac{2m_{\|}^*}{\hbar^2})]^{1/2}$$

The EP in this case is given by

$$I = \frac{\alpha_o e g_v k_B T}{2\pi\hbar} \sum_{n_y=1}^{n_{y\max}} \sum_{n_z=1}^{n_{z\max}} F_0(\eta_{660}) \quad (78)$$

$$where,\ \eta_{660} \equiv [\frac{E_{F1d} - (E'_{60} + W - h\nu)}{k_B T}]$$

and E'_{60} is the sub-band energy in this case which can be expressed as

$$0 = B_{77}(E'_{60}, n_y, n_z) \quad (79)$$

(ii) Model of Wang et al.

The dispersion relation in accordance with this model in the present case can be written as

$$k_x = \beta_{70}(E, n_y, n_z, \eta_g) \tag{80}$$

$$where,\ \beta_{70}(E, n_y, n_z, \eta_g) = [-(\frac{n_y \pi}{d_y})^2 + \frac{2m_\perp^*}{\hbar^2}[\bar{\alpha}_8 - \bar{\alpha}_9(\frac{\pi n_z}{d_z})^2 - \bar{\alpha}_{10}[(\frac{\pi n_z}{d_z})^4 + \bar{\alpha}_{11}(\frac{\pi n_z}{d_z})^2 + \bar{\alpha}_{12}(E, \eta_g)]^{1/2}]]^{1/2}$$

The EP in this case is given by

$$I = \frac{\alpha_o e g_v k_B T}{\pi \hbar} \sum_{n_y=1}^{n_{y\max}} \sum_{n_z=1}^{n_{z\max}} F_0(\eta_{654HD}) \tag{81}$$

$$where,\ \eta_{654HD} \equiv [\frac{E_{F1HDNW} - (E'_{54HDNW} + W - h\nu)}{k_B T}]$$

and E'_{54HDNW} is the sub-band energy in this case which can be expressed as

$$0 = \beta_{70}(E'_{54HDNW}, \eta_g, n_y, n_z) \tag{82}$$

In the absence of band tailing the 1D dispersion relation in this case can be written as

$$k_x = P_{77}(E, n_y, n_z) \tag{83}$$

$$where,\ P_{77}(E, n_y, n_z) = [[I_1(E, n_z) - \frac{\hbar^2}{2m_2^*}(\frac{n_y \pi}{d_y})^2](\frac{2m_1^*}{\hbar^2})]^{1/2}$$

The EP in this case is given by

$$I = \frac{\alpha_o e g_v k_B T}{2\pi\hbar} \sum_{n_y=1}^{n_{y\max}} \sum_{n_z=1}^{n_{z\max}} F_0(\eta_{680}) \tag{84}$$

$$where,\ \eta_{680} \equiv [\frac{E_{F1d} - (E'_{80} + W - h\nu)}{k_B T}]$$

and E'_{80} is the sub-band energy in this case which can be expressed as

$$0 = P_{77}(E'_{80}, n_y, n_z) \tag{85}$$

The EP from Nano Wires of HD GaSb

The dispersion relation of the 1D electrons in this case can be written as

$$\frac{\hbar^2 (n_z\pi / d_z)^2}{2m_c} + \frac{\hbar^2 (n_y\pi / d_y)^2}{2m_c} + \frac{\hbar^2 k_x{}^2}{2m_c} = I_{36}(E, \eta_g) \tag{86}$$

The EP in this case is given by

$$I = \frac{\alpha_o e g_v k_B T}{\pi\hbar} \sum_{n_y=1}^{n_{y\max}} \sum_{n_z=1}^{n_{z\max}} F_0(\eta_{100HD}) \tag{87}$$

$$where,\ \eta_{100HD} \equiv [\frac{E_{F1HDNW} - (E'_{100HDNW} + W - h\nu)}{k_B T}]$$

and $E'_{100HDNW}$ is the sub-band energy in this case which can be expressed as

$$\frac{\hbar^2(n_z\pi/d_z)^2}{2m_c}+\frac{\hbar^2(n_y\pi/d_y)^2}{2m_c}=I_{36}(E'_{100HDNW},\eta_g) \tag{88}$$

The expression of 1D dispersion relation, for QWs of GaSb whose energy band structures in the absence of band tailing assumes the form

$$I_{36}(E)=\frac{\hbar^2k_x^2}{2m_c}+G_2(n_y,n_z) \tag{89}$$

In this case, the quantized energy E'_{101} is given by

$$I_{36}(E'_{101})=G_2(n_y,n_z) \tag{90}$$

The EP in this case is given by

$$I=\frac{\alpha_o e g_v k_B T}{\pi\hbar}\sum_{n_y=1}^{n_{y_{\max}}}\sum_{n_z=1}^{n_{z_{\max}}}F_0(\eta_{101}) \tag{91}$$

where, $\eta_{101}\equiv[\frac{E_{F1d}-(E'_{101}+W-h\nu)}{k_BT}]$

Results and Discussion

Using the appropriate equations and taking the energy band constants as given in the Table 1.1 of chapter 1, we have plotted the normalized EP from QWs of HD $CdGeAs_2$ (an example of nonlinear optical materials) as a function of d_x as shown in plot (a) of Figure 1, in which the plot (b) corresponds to $\delta=0$. The plot (c) has been drawn in accordance with the three band model of Kane and the plot (d) refers to the two band model of

Kane together with the plot (e) exhibiting the variation in accordance with the parabolic energy bands for the overall assessments of the energy band constants on the EP in this case. The Figure 2 exhibits the plots of the normalized EP from QWs of HD $CdGeAs_2$ as a function of the normalized incident photon energy for all cases Figure 1 and Figure 3 shows the dependence of the said variable on the normalized electron degeneracy for all cases of Figure 1.

The normalized EP from QWs of HD n-InAs (an example of III-V materials) in accordance with the HD three and two band models of Kane as functions of film thickness, normalized incident photon energy and the normalized electron degeneracy have, respectively, been presented in Figures 4 to 6. The Figures 7 to 9 exhibit the variations of normalized EP from QWs of HD n-InSb as functions of film thickness, normalized incident photon energy and the normalized electron degeneracy respectively. The variations of the normalized EP from QWs of HD CdS (an example of II-VI materials) as functions of thickness, normalized incident photon energy and normalized electron degeneracy have respectively been drawn in Figures 10 to 12, where the plots for $\bar{\lambda}_0 = 0$ have further been drawn for the purpose of assessing the influence of the splitting of the two-spin states by the spin orbit coupling and the crystalline field. The thickness, normalized photon energy and the normalized electron degeneracy dependences of normalized EP from QWs of HD GaP have been shown in Figures 13 to 15 respectively. The dependence of normalized EP with reference to the aforementioned variables from QWs of HD n-Ge and $PtSb_2$ has been shown in Figures 16 to 21 in accordance with the HD models of Cardona et al. Wang and Ressler and Emtage respectively. Figures 22 to 24 manifest the variations of the normalized EP from QWs of HD stressed n-InSb as functions of the film thickness, normalized incident photon energy and the normalized electron degeneracy respectively.

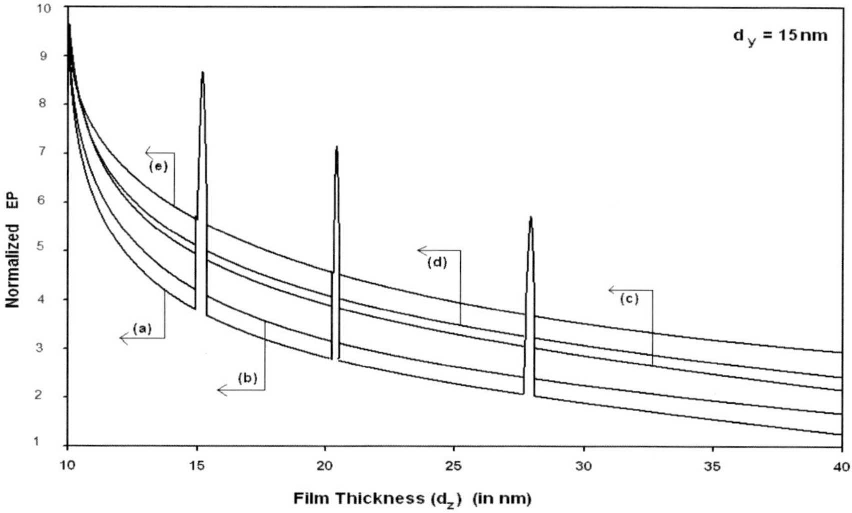

Figure 1. Plot of the normalized EP from QWs of HD $CdGeAs_2$ as a function of d_z in accordance with (a) generalized band model (b) $\delta = 0$, (c) the three-band model of Kane (d) the two band model of Kane and (e) the parabolic energy bands.

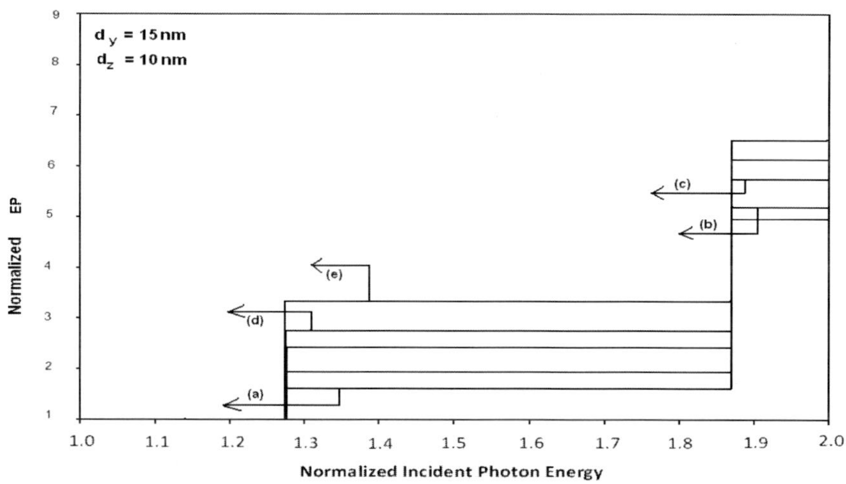

Figure 2. Plot of the normalized EP from QWs of HD $CdGeAs_2$ as a function of normalized incident photon energy for all cases of Figure 1.

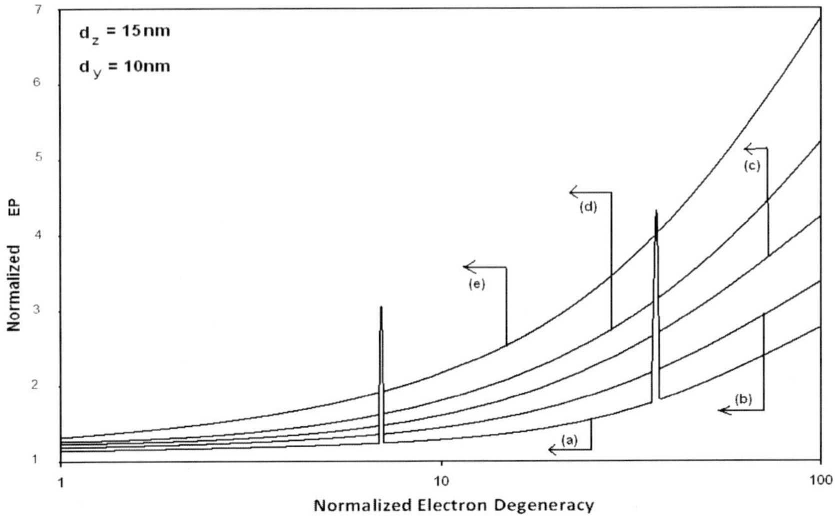

Figure 3. Plot of the normalized EP from QWs of HD $CdGeAs_2$ as a function of normalized electron degeneracy for all cases of Figure 1.

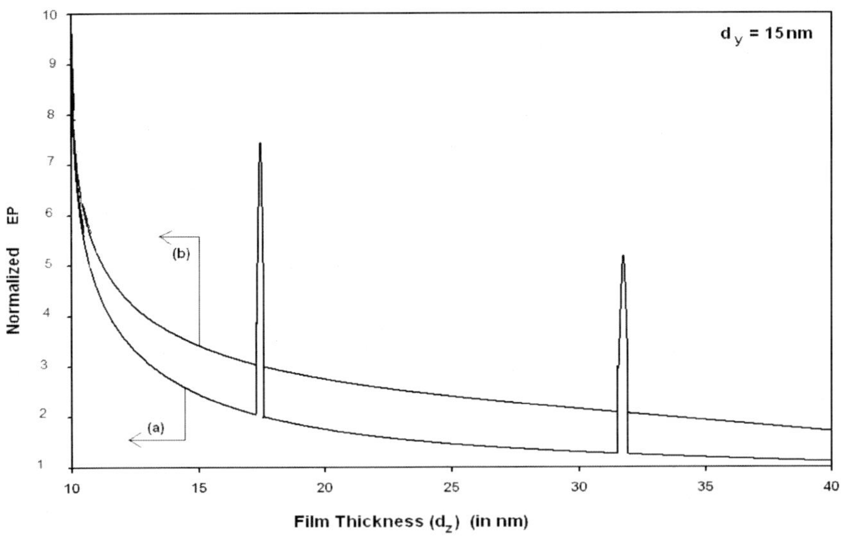

Figure 4. Plot of the normalized EP from QWs of HD n-InAs as a function of d_z in accordance with (a) the three band model of Kane and (b) the two band model of Kane.

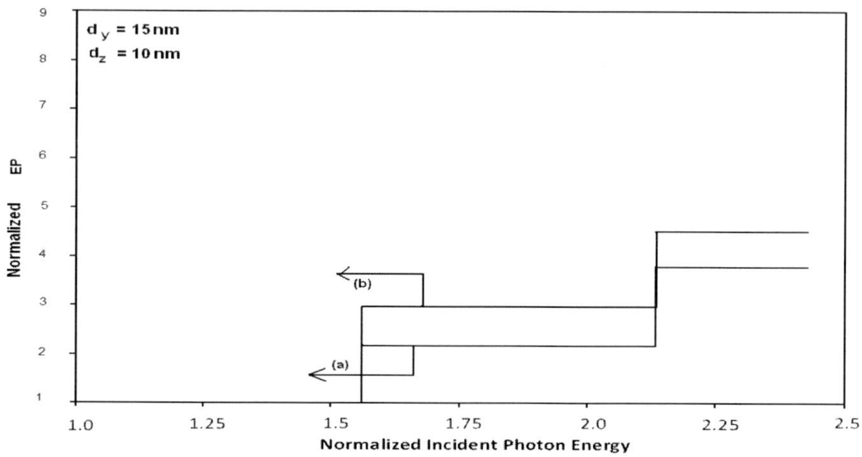

Figure 5. Plot of the normalized EP from QWs of HD n-InAs as a function of normalized incident photon energy in accordance with (a) the three band model of Kane and (b) the two band model of Kane.

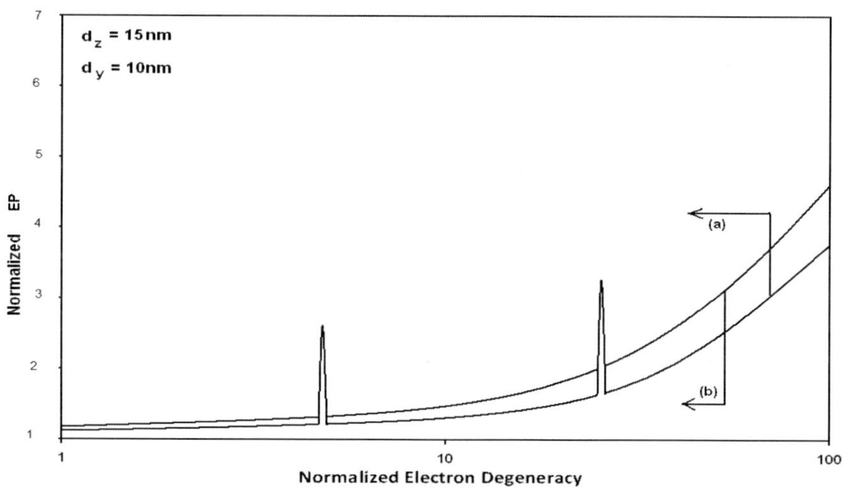

Figure 6. Plot of the normalized EP from QWs of HD n-InAs as a function of normalized electron degeneracy in accordance with (a) the three band model of Kane and (b) the two band model of Kane.

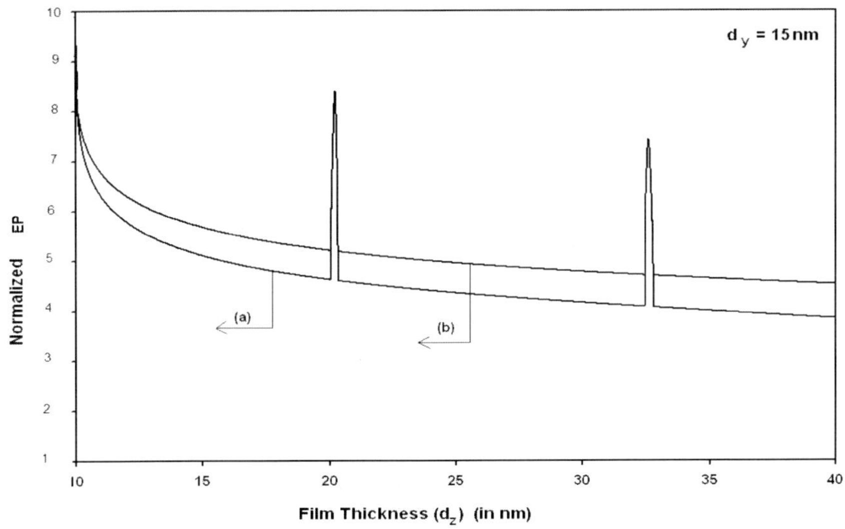

Figure 7. Plot of the normalized EP from QWs of HD n-InSb as a function of d_z in accordance with (a) the three band model of Kane and (b) the two band model of Kane.

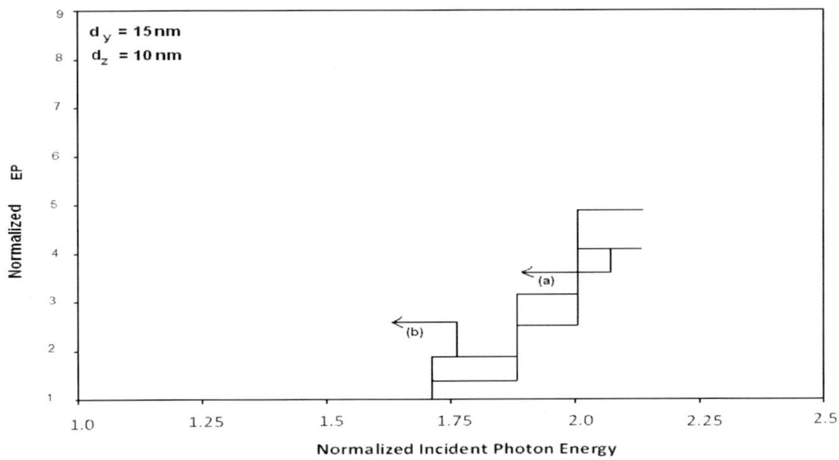

Figure 8. Plot of the normalized EP from QWs of HD n-InSb as a function of normalized incident photon energy in accordance with (a) the three band model of Kane and (b) the two band model of Kane.

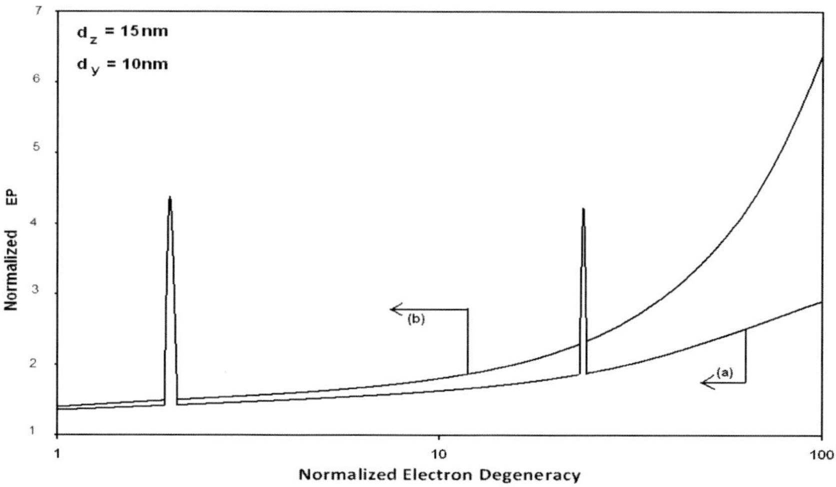

Figure 9. Plot of the normalized EP from QWs of HD n-InSb as a function of normalized electron degeneracy in accordance with (a) the three band model of Kane and (b) the two band model of Kane.

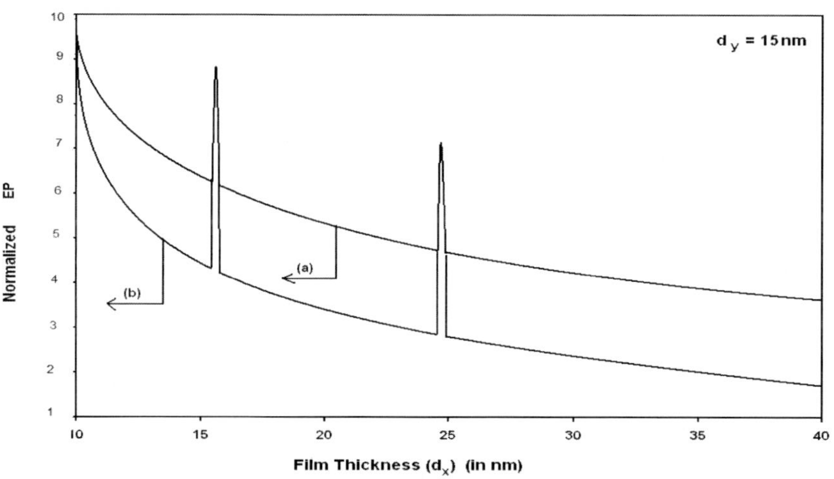

Figure 10. Plot of the normalized EP from QWs of HD CdS as a function of d_z with (a) and $\bar{\lambda}_0 \neq 0$ (b) $\bar{\lambda}_0 = 0$.

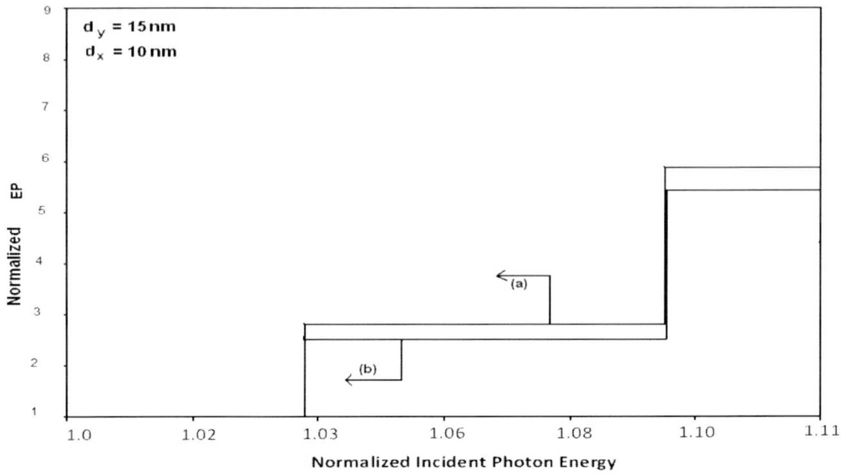

Figure 11. Plot of the normalized EP from QWs of HD CdS as a function of normalized incident photon energy with (a) $\bar{\lambda}_0 \neq 0$ and (b) $\bar{\lambda}_0 = 0$.

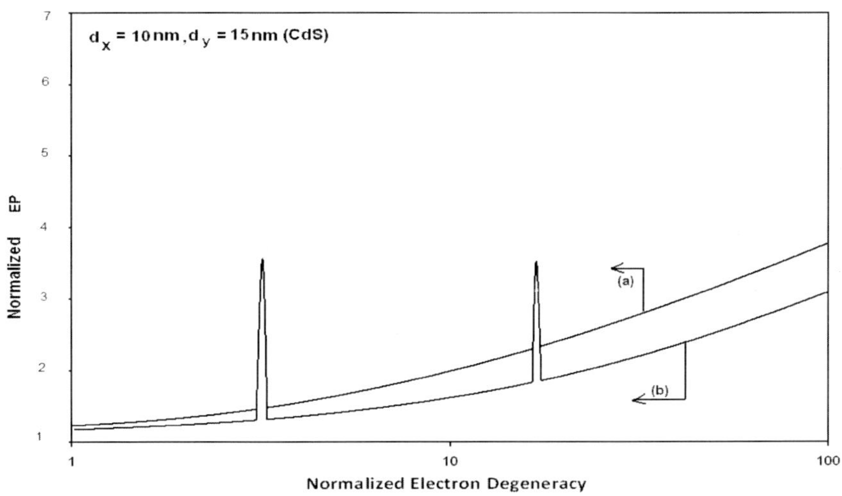

Figure 12. Plot of the normalized EP from QWs of HD CdS as a function of normalized electron degeneracy with (a) $\bar{\lambda}_0 \neq 0$ and (b) $\bar{\lambda}_0 = 0$.

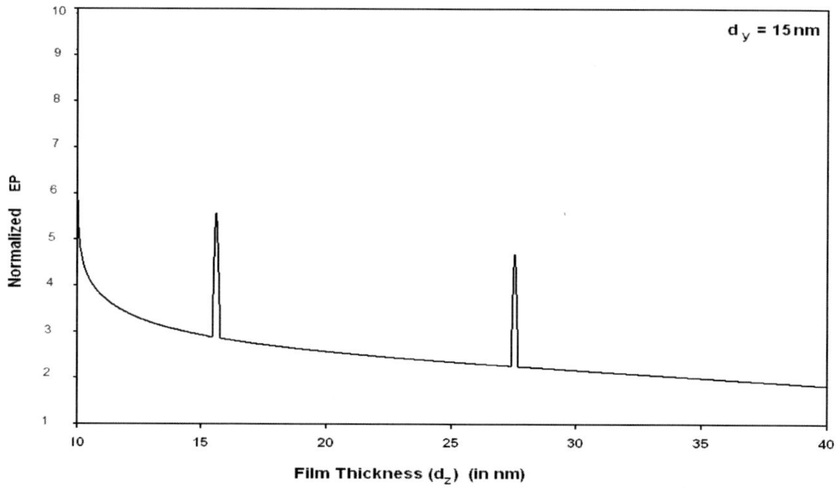

Figure 13. Plot of the normalized EP from QWs of HD n-GaP as a function of d_z.

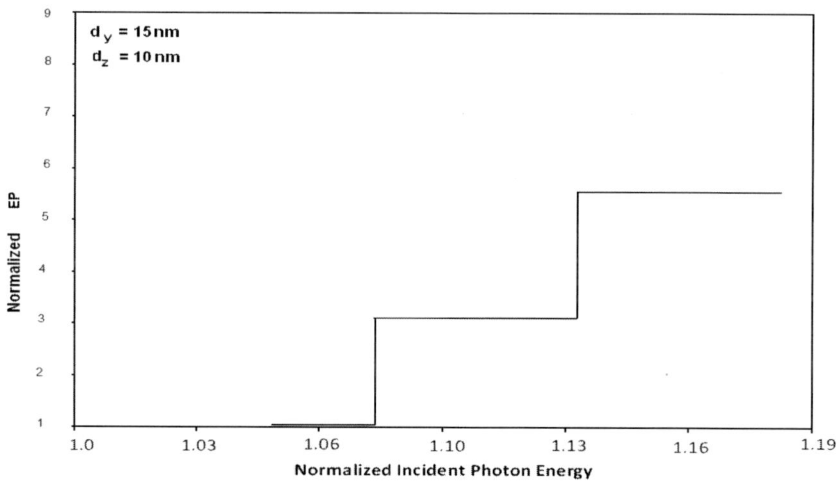

Figure 14. Plot of the normalized EP from QWs of HD n-GaP as a function of normalized incident photon energy.

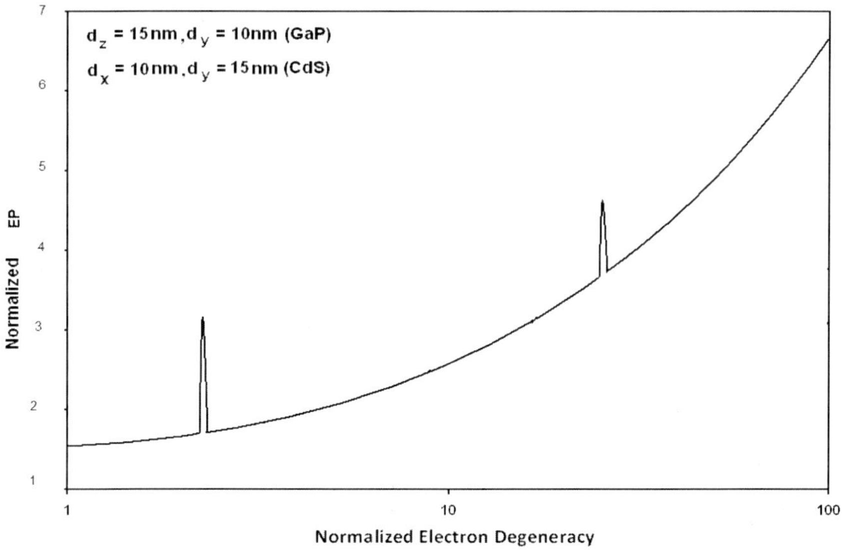

Figure 15. Plot of the normalized EP from QWs of HD n-GaP as a function of normalized electron degeneracy.

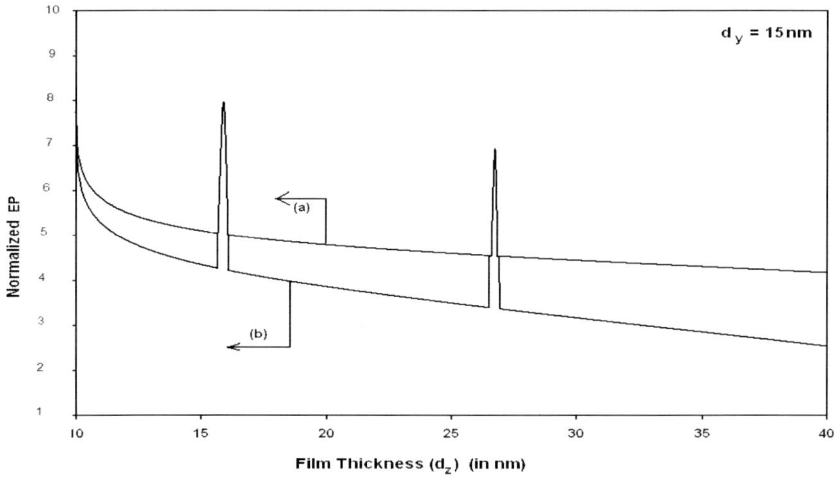

Figure 16. Plot of the normalized EP from QWs of HD n-Ge as a function of thickness in accordance with the models of (a) Cardona et al. and (b) Wang et al.

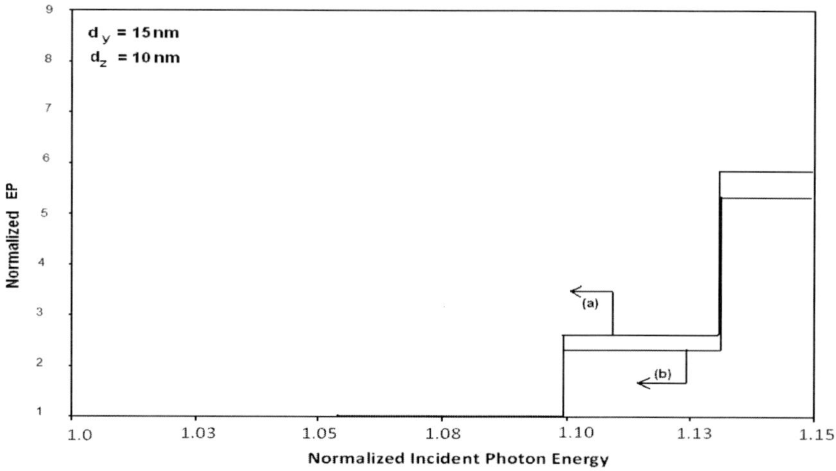

Figure 17. Plot of the normalized EP from QWs of HD n-Ge as a function of normalized incident photon energy for all the cases of Figure 43.

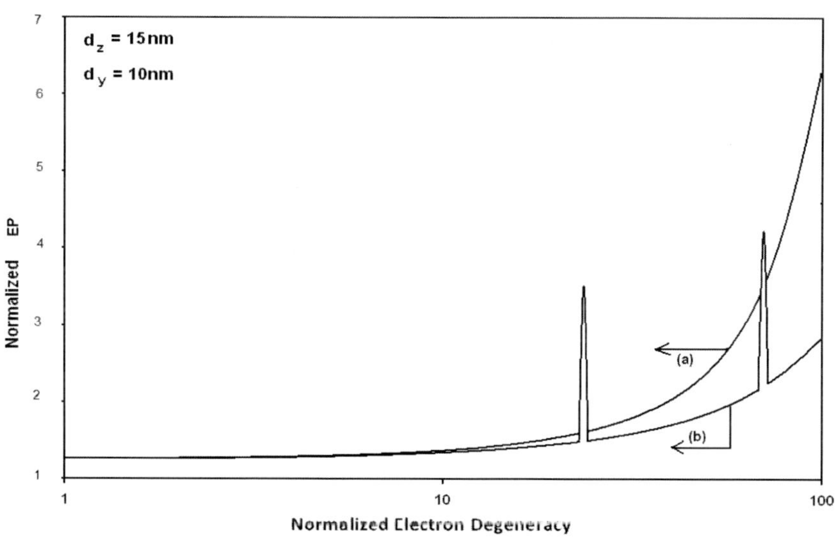

Figure 18. Plot of the normalized EP from QWs of HD n-Ge as a function of normalized electron degeneracy for all the cases of Figure 43.

The Figures 25 to 27 exhibit the normalized EP as functions of film thickness, normalized incident photon energy and normalized electron degeneracy from QWs of HD PbTe as a function of film thickness in accordance with the models of (a) the Dimmok and (b) the Bangert and Kastner respectively. The plots (c) and (d) exhibit the same for PbSe.

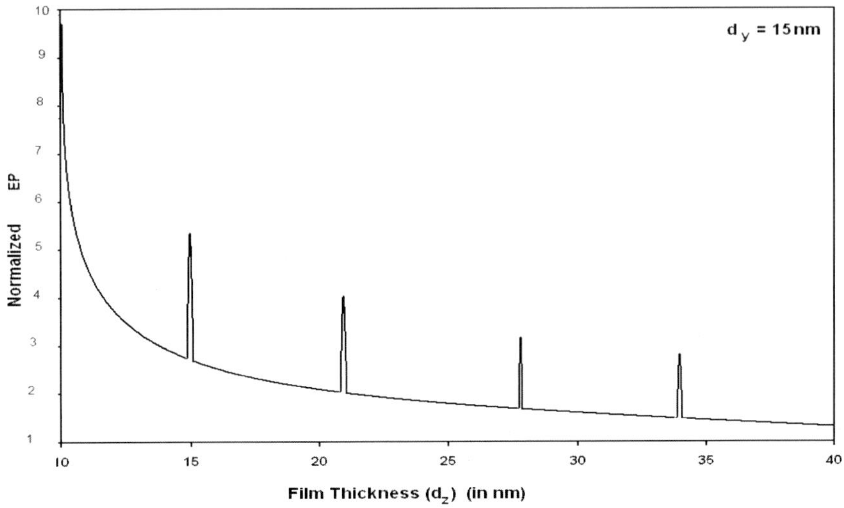

Figure 19. Plot of the normalized EP from QWs of HD n-$PtSb_2$ as a function of film thickness.

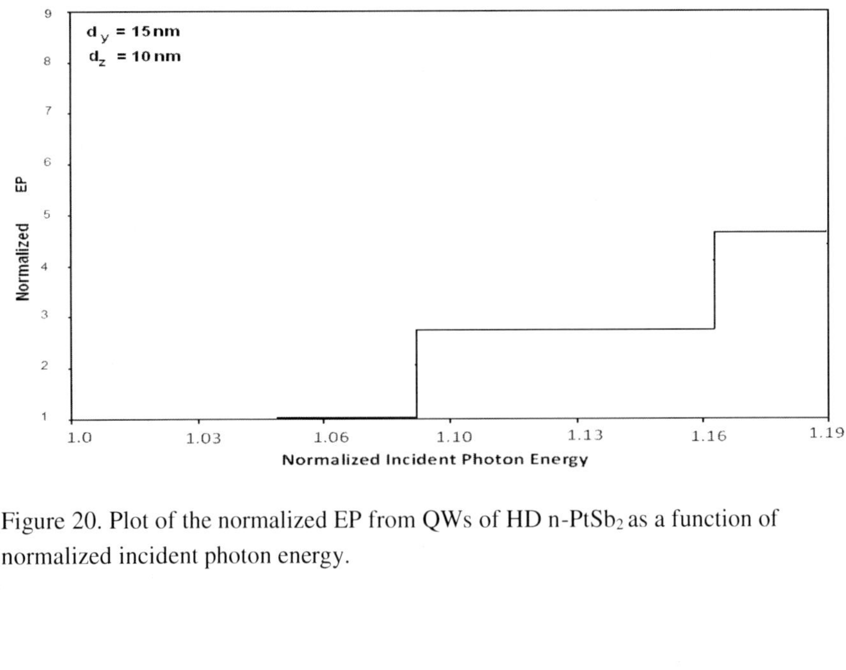

Figure 20. Plot of the normalized EP from QWs of HD n-$PtSb_2$ as a function of normalized incident photon energy.

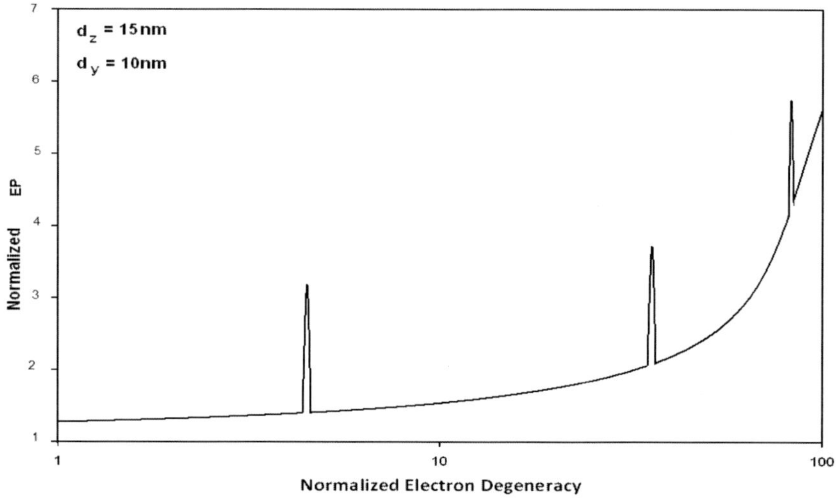

Figure 21. Plot of the normalized EP from QWs of HD n-$PtSb_2$ as a function of normalized electron degeneracy.

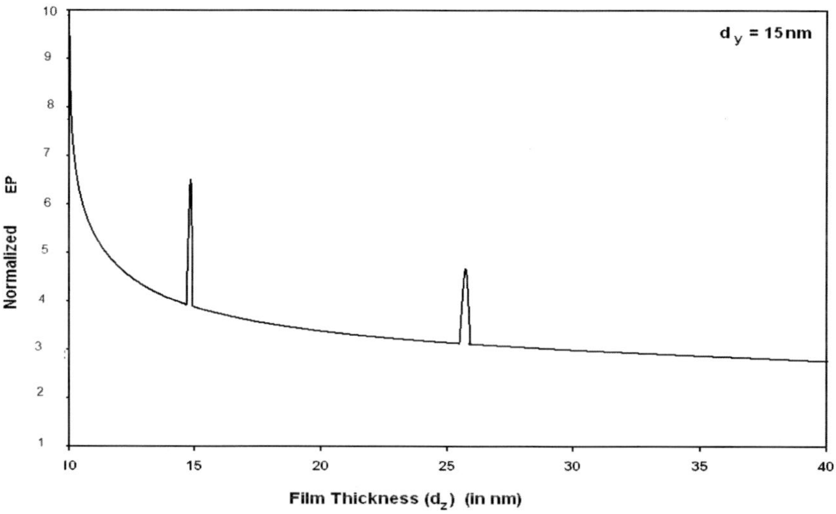

Figure 22. Plot of the normalized EP from QWs of HD stressed n-InSb as a function of film thickness.

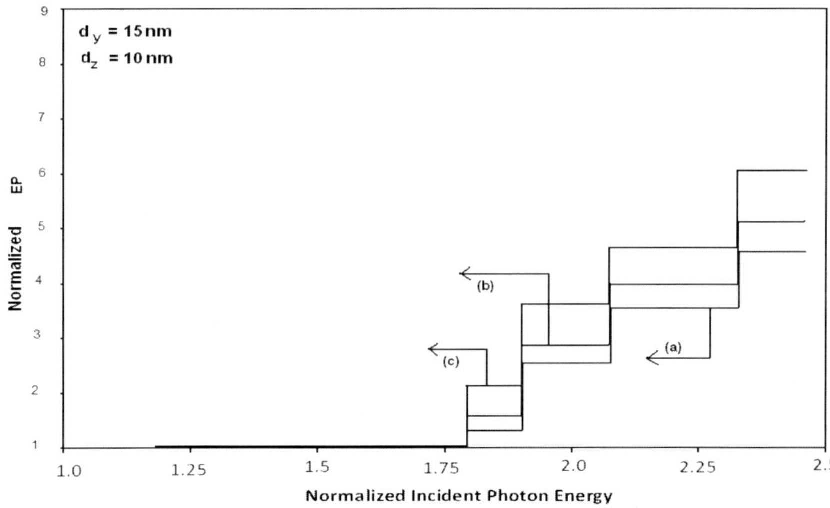

Figure 23. Plot of the normalized EP from QWs of HD stressed n-InSb as a function of normalized incident photon energy.

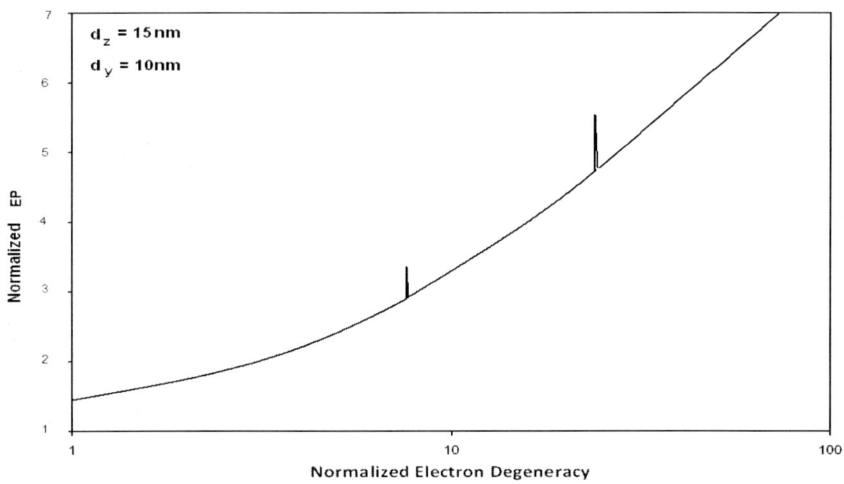

Figure 24. Plot of the normalized EP from QWs of HD stressed n-InSb as a function of normalized electron degeneracy.

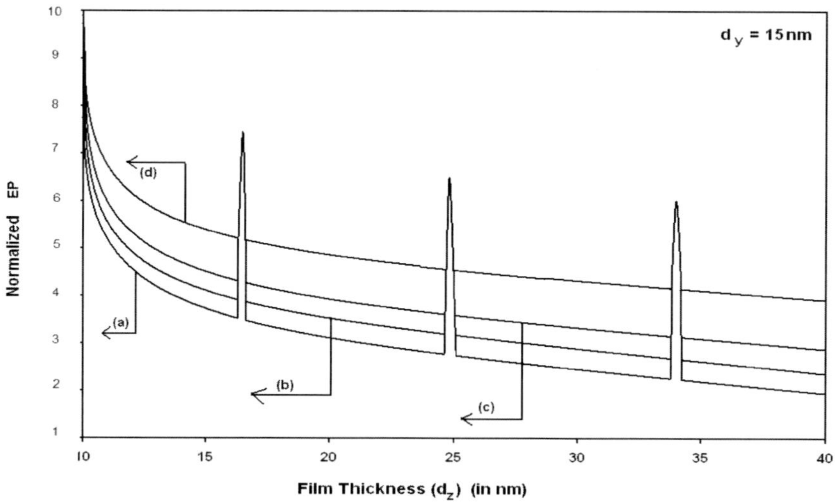

Figure 25. Plot of the normalized EP from QWs of HD PbTe as a function of film thickness in accordance with the models of (a) the Dimmok and (b) the Bangert and Kastner respectively. The plots (c) and (d) exhibit the same for PbSe.

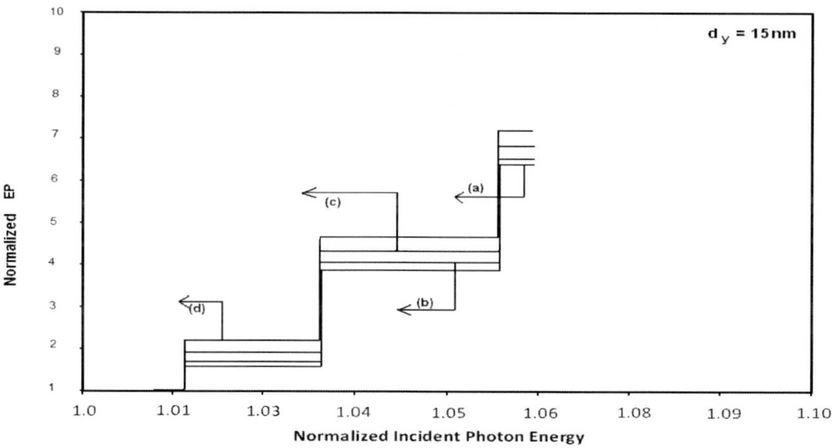

Figure 26. Plot of the normalized EP from QWs of HD PbTe as a function of normalized incident photon energy in accordance with the models of (a) the Dimmok and (b) the Bangert and Kastner respectively. The plots (c) and (d) exhibit the same for PbSe.

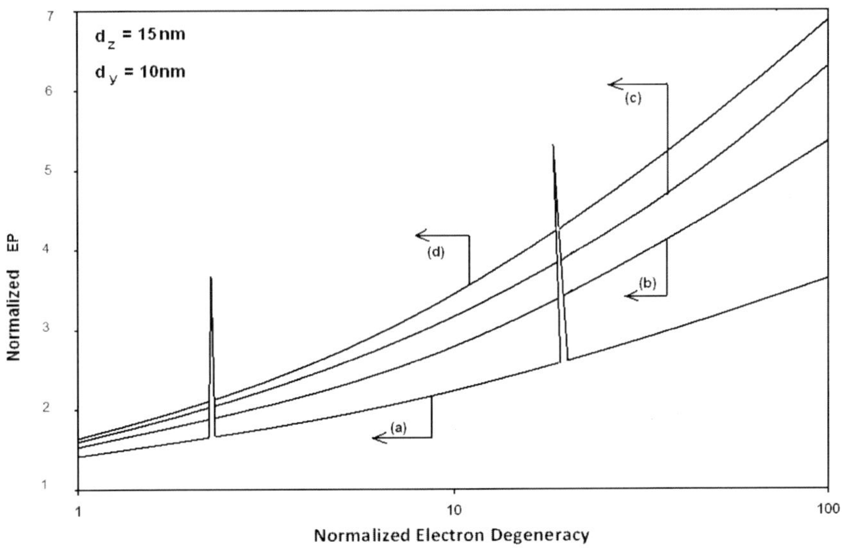

Figure 27. Plot of the normalized EP from QWs of HD PbTe as a function of electron degeneracy in accordance with the models of (a) the Dimmok and (b) the Bangert and Kastner respectively. The plots (c) and (d) exhibit the same for PbSe.

The influence of quantum confinement is immediately apparent from Figures (1), (4), (7), (10), (13), (16), (19), (22) and (25). Since the EP depends strongly on the thickness of the quantum-confined materials in contrast with the corresponding bulk specimens. The EP decreases with increasing film thickness in an oscillatory way with different numerical magnitudes for HD QWs. It appears from the aforementioned figures that the EP exhibits spikes for particular values of film thickness which, in turn, depends on the particular band structure of the specific material. Moreover, the EP from HD QWs of different compounds can become several orders of magnitude larger than of bulk specimens of the same materials, which is also a direct signature of quantum confinement. This oscillatory dependence will be less and less prominent with increasing film thickness.

REFERENCES

[1] Canham, LT. *Appl. Phys. Letts.*, 57, 1046, (1990).
[2] Lu, ZH; Lockwood, DJ; Baribeam, JM. *Nature*, 378, 258, (1995).
[3] Cullis, AG; Canham, LT; Calocott, PDO. *J. Appl. Phys.*, 82, 909, (1997).
[4] Cardona, M; Ley, L. *Photoemission in Solids 1 and 2, Topics in Applied Physics*, vols. 26, 27, (Springer-Verlag, Germany, 1978). x Hüfner, AC. *Photoelectron Spectroscopy*, (Springer, Germany, 2003). Hüfner, S. (Ed.), Very High Resolution Photoelectron Spectroscopy, *Lecture Notes in Physics*, Vol. 715, (Springer-Verlag, Germany, 2007). Lynch, DW; Olson, CG. *Photoemission Studies of High-Temperature Superconductors*, (Cambridge University Press, UK. 1999). *Photoemission and the Electronic Properties of Surfaces* ed. B. Feuerbacher, B. Fitton, R. F. Willis (Wiley, New York, 1978). Schattke, W; Hove, MAV. *Solid-State Photoemission and Related Methods: Theory and Experiment*, (Wiley, USA, 2003). Afanas'ev, VV. *Internal Photoemission Spectroscopy: Principles and Applications* (Elsevier, North Holland, 2010). Lockwood, DJ. *Light Emission in Silicon in silicon based materials and devices*, vol 2, ed H. S. Nalwa (Academic Press, San Diego, USA, 2001).
[5] Ghatak, KP; De, D; Bhattacharya, S. *Photoemission from optoelectronic materials and their nanostructures*, Springer Series in Nanostructure Science and Technology, (Springer, 2009).
[6] Mitra, B; Ghatak, KP. *Phys. Letts.*, A 141, 81 (1989). Mitra, B; Ghatak, KP. *Solid-state electronics*, 32, 810, (1989). Ghatak, KP; Mitra, B. *Phys. Letts.*, A 156, 233, (1991). Choudhury, S; Singh, LJ; Ghatak, KP. *Nanotechnology*, 15, 180, (2004), Ghatak, KP; Biswas, SN. *J. of Appl. Phys.*, 70, 299, (1991). Ghatak, KP; Chowdhury, AK; Ghosh, S; Chakravarti, AN. *Appl. Phys.*, 23, 241, (1980). Ghatak, KP; Basu, DK; Nag, B. *J. of Phys. and Chem. of Solids*, 58, 133, (1997). Chakraborty, PK; Datta, GC; Ghatak, KP. *Physica B: Condensed Matter*, 339, 198, (2003). Ghatak, KP; Ghoshal, A; Mitra,

B. *Il Nuovo Cimento*, D 14, 903, (1992). Mitra, B; Ghoshal, A; Ghatak, KP. *Phys. Stat. Sol. (b)*, 154, K147, (1989). Ghatak, KP; Mondal, M. *Jour. of Mag. and Mag Mat.*, 74, 203, (1988). Ghatak, KP; Bhattacharya, S; Pahari, S; De, D; Ghosh, S; Mitra, M. *Annalen der Physik*, 17, 195, (2008). Ghatak, KP; Mitra, B. *Physica Scripta*, 42, 103, (1990). Mitra, B; Ghatak, KP. *Solid-state electronics*, 32, 515, (1989). Ghatak, KP; Bhattacharya, S; Bhowmik, S; Benedictus, R; Choudhury, S. *J. Appl. Phys.*, 103, 094314, (2008). Nag, B; Ghatak, KP. *Phys. Scrip.*, 54, 657, (1996). Mondal, M; Bhattacharya, S; Ghatak, KP. *Appl. Phys.*, A 42, 331, (1987). Ghatak, KP; Monda, M; *l Zeitschrift für Physik B Condensed Matter*, 64, 223, (1986). Chakravarti, AN; Ghatak, KP; Dhar, A; Ghosh, KK; Ghosh, S. *Appl. Phys. A*, 26, 165, (1981).

[7] Pathria, RK. *Statistical Mechanics, 2nd ed.* (Butterworth-Heinemann, Oxford, 1996).

[8] De, D; Bhattacharya, S; Adhikari, SM; Kumar, A; Bose, PK; Ghatak, KP. *Beilstein Jour. Nanotech.*, 2, 339, (2012). De, Kumar D; Adhikari, SM; Pahari, S; Islam, N; Banerjee, P; Biswas, SK; Bhattacharya, S; Ghatak, KP. *Superlatti. and Microstruc.*, 47, 377, (2010). Ghatak, KP; Mondal, M; Biswas, SN. *J. Appl. Phys.*, 68, 3032, (1990). Mondal, M; Banik, S; Ghatak, KP. *J. Low Temp. Phys.*, 74, 423, (1989). Mitra, B; Ghoshal, A; Ghatak, KP. *Phys. Stat. Sol. (b)*, 150, K67, (1988). Ghatak, KP; Bhattacharyya, D; Nag, B; Biswas, SN. *J. Nonlin. Optics and Quant. Optics*, 13, 267, (1995). Torres, L; Lopez-Diaz, L; Iniguez, J. *Appl. Phys. Lett.*, 73, 3766, (1996). Toporov, AY; Langford, RM; Petford-Long, AK. *Appl. Phys. Lett.*, 77, 3063, (2000). Torres, L; Lopez-Diaz, L; Alejos, O; Iniguez, J. *J. Appl. Phys.*, 85, 6208, (1999). Tiwari, S; Tiwari, S. *Cryst. Res. Technol.*, 41, 78, (2006). Houdré, R; Hermann, C; Lampel, G; Frijlink, PM; Gossard, AC. *Phys. Rev. Lett.*, 55, 734, (1985).

[9] Ghatak, KP; Bhattacharya, S; Singh, KM; Choudhury, S; Pahari, S. *Physica B*, 403, 2116, (2008). Adhikari, SM; Ghatak, KP. *Jour. Adv. Phys.*, 2, 130, (2013). De, D; Bhattacharaya, S; Ghosh, S; Ghatak, KP. *Adv. Sci., Engi. and Med*, 4, 211, (2012). Kumar, A; Chowdhury,

S; Adhikari, SM; Ghosh, S; Mitra, M; De, D; Sharma, A; Bhattacharya, S; Dey, A; Ghatak, KP. *Jour. of Comput. Theo. Nanosc.*, 7, 115, (2010). Houdré, R; Hermann, C; Lampel, G; Frijlink, PM. *Surface Sci.*, 168, 538, (1986). Chiang, TC; Ludeke, R; Eastman, DE. *Phys. Rev.*, B. 25, 6518, (1982). Svensson, SP; Kanski, J; Andersson, TG; Nilsson, PO. *J. Vacuum Sci. Technol.*, B 2, 235, (1984). Alvarado, SF; Ciccacci, F; Campagna, M. *Appl. Phys. Letts.*, 39, 615, (1981). Fleming, L; Ulrich, MD; Efimenko, K; Genzer, J; Chan, ASY; Madey, TE; Oh, SJ; Zhou, O; Rowe, JE. *Vac. Sci. Technol. B*, 22, 2000, (2004). Shimoda, H; Gao, B; Tang, XP; Kleinhammes, A; Fleming, L; Wu, Y; Zhou, O. *Phys. Rev. Lett.*, 88, 015502, (2002). Maillard, M; Monchicourt, P; Pileni, MP. *Chem. Phys. Lett.*, 380, 704, (2003).

[10] Mitra, B; Ghatak, KP. *Phys. Scrip.*, 40, 776 (1989). De, D; Bhattacharya, S; Ghatak, KP. *International work shop in Physics of Semiconductor Devices*, *IEEE*, 897, (2007). Bhattacharya, S; De, D; Chowdhury, S; Karmakar, S; Basu, DK; Pahari, S; Ghatak, KP. *Jour. of Comput. Theo. Nanosc*, 3, 280, (2006).

Matsuda, I; Hasegawa, S; Konchenko, A; Nakayama, Y; Nakamura, Y; Ichikawa, M. *Phys. Rev. B*, 73, 113311, (2006). Colvin, VL; Alivisatos, AP; Tobin, JG. *Phys. Rev. Lett.*, 66, 2786, (1991). Schroeter, B; Komlev, K; Richter, W. *Mat. Sc. and Eng.*, B88, 259, (2002), Bertsch, GF; Van Giai, N; Vinh Mau, N. *Phys. Rev. A*, 61, 033202 (2000).

[11] Ghatak, KP; Biswas, SN. *Nonlinear Opt.*, 4, 39, (1993).

[12] Ghatak, KP; Biswas, SN. *SPIE, Growth and Characterization of Materials for Infrared Detectors and Nonlinear Optical Switches*, vol. 1484, (USA, 1991), p. 136.

[13] Ghatak, KP; De, B. *Polymeric materials for Integrated Optics and Information Storage*, *Materials Research Society (MRS) Symposium Proceedings*, *MRS Spring Meeting*, 1991, vol. 228, p. 237. Heyderman, LJ; Nolting, F. *Quitmann*, *Appl. Phys. Letts.*, 83, 1797, (2003). Cowburn, RP; Adeyeye, AO; Bland, JAC. *Appl. Phys. Lett.*, 70, 2309, (1997).

[14] Ghatak, KP; Nag, B; Majumdar, G. Strained Layer Expitaxy – Materials, Processing, and Device Applications, MRS Symposium Proceedings, *MRS Spring Meeting*, 1995, vol. 379, p. 85.
[15] Ghatak, KP. *SPIE, High Speed Phenomena in Photonic Materials and Optical Bistability*, USA, 1990, vol. 1280, p. 53.
[16] Ghatak, KP. Long Wave Length Semiconductor devices, Materials and Processes Symposium Proceedings, MRS Symposium Proceedings, *MRS Spring Meeting*, 1990, vol. 216, p. 469.
[17] Ghatak, KP; Ghoshal, A; Bhattacharyya, S. *SPIE, Nonlinear Optical Materials and Devices for Photonic Switching*, USA, 1990, vol. 1216, p. 282.
[18] Ghatak, KP. *SPIE, Nonlinear Optics III*, USA, vol. 1626, p. 115 (1992).
[19] Ghatak, KP; Ghoshal, A; De, B. *SPIE, Optoelectronic Devices and Applications*, USA, 1990, vol. 1338, p. 111.
[20] Ghatak, KP; Biswas, SN. in *Proceedings of the Society of Photo-optical and Instrumentation Engineers (SPIE), Nonlinear Optics II*, USA, 1991, vol. 1409, p. 28. Ghatak, KP. *SPIE, Process Module Metrology*, USA, 1992, vol. 1594, p. 110. Ghatak, KP. *SPIE, International Conference on the Application and Theory of Periodic Structures*, 1991, vol. 1545, p. 282.
[21] Ghatak, KP; Mondal, M. *Solid State Electron.*, 31, 1561, (1988).
[22] Ghatak, KP; Mondal, M. *J. Appl. Phys.*, 69, 1666, (1991).
[23] Majumdar, C; Maity, AB; Chakravarti, AN. *Phys. Stat. Sol. (b)*, 140, K7, (1987).
[24] Majumdar, C; Maity, AB; Chakravarti, AN. *Phys. Stat. Sol. (b)*, 141, K35 (1987).
[25] Das, NR; Ghosh, KK; Ghoshal, D. *Phys. Stat. Sol. (b)*, 197, 97 (1996).
[26] Majumdar, C; Maity, AB; Chakravarti, AN. *Phys. Stat. Sol. (b)*, 144, K13, (1987).
[27] Das, NR; Chakravarti, AN. *Phys. Stat. Sol. (b)*, 176, 335, (1993).
[28] Maity, AB; Majumdar, C; Chakravarti, AN. *Phys. Stat. Sol. (b)*, 144, K93, (1987).

[29] Maity, AB; Majumdar, C; Chakravarti, AN. *Phys. Stat. Sol. (b)*, 149, 565, (1988).

[30] Ziel, AVD. *Solid State Physical Electronics*, (Prentice Hall, Inc. Eaglewood Cliffs, 1957). Modinos, A. *Field, Thermionic and Secondary Electron Emission Spectroscopy*, (Plenum Press, New York, 1984).

In: Quantum Wires: An Overview
Editor: Kamakhya Prasad Ghatak
ISBN: 978-1-53617-676-6

Chapter 9

QUANTUM WIRES AND MAGNETIC SUSCEPTIBILITIES

B. Chatterjee[1,*], S. Debbarma[2] and N. Debbarma[2]

[1]Department of Computer Science and Engineering, University of Engineering and Management and Institute of Engineering and Management, Kolkata and Jaipur, Kolkata, India and Gurukul, Jaipur, Rajasthan, India

[2]Department of Computer Science and Engineering, National Institute of Technology, Agartala, Tripura, India

ABSTRACT

In this chapter we study the Landau Dia and Pauli Para magnetic susceptibilities in quantum wires of IV—VI materials under a parallel magnetic field on the basis of a newly derived one-dimensional electron dispersion law. It is found, taking quantum wires of PbS, PbSe and PbTe as examples, that both the susceptibilities increase with decreasing electron concentration and decreasing film thickness respectively. The paramagnetic-to-diamagnetic susceptibility ratio for electrons in IV-VI

[*] Corresponding Author's Email: biswajoy.chatterjee@iemcal.com.

quantum wires deviates from the well-known 1/3 rule together with the fact that there is a critical zone within which quenching of the diamagnetic susceptibility occurs.

INTRODUCTION

In recent years, with the advent of MBE, MOCVD, FLL and other experimental techniques, low-dimensional structures having quantum confinement in one, two and three dimensions, such as ultrathin films, inversion layers, quantum wires and dots, have attracted much attention, not only for their potential in uncovering new phenomena in nanostructured electronics but also for their interesting device applications [1]. Heterostructures based on various materials are being currently studied because of the enhancement of carrier mobility [2], and such quantum confined systems find extensive applications in lasers [3], heterojunction field effect transistors (FETs) [4], high speed digital networks, optical modulators, and also in other devices. In quantum wires (QWs), the electron gas is quantized in two transverse directions and the electrons can move only in the longitudinal direction. Quantum devices like quantum wire wave guides [5] and quantum-wire transistors [6] have also been fabricated. The transport properties of electrons in quantum wires may be studied by utilizing the similarity with an optical and microwave waveguide. Though con- siderable work has already been done, it appears from the literature that the magnetic susceptibility of the electrons in QWs of IV—VI semiconductors have yet to be studied. In this chapter we shall study the same for the said system having non-parabolic and non-standard energy bands in the presence of a parallel magnetic field, since in recent years there has been considerable interest in studying the parallel magnetic transport in heterostnictures [7], and also since IV—VI compounds find extensive applications in infrared detectors, thermoelectric devices, superlat- tices and quantum confined materials.

In this context, we wish to note that the susceptibility due to the free motion of electrons in the presence of a magnetic field is called the diamagnetic susceptibility and the susceptibility due to spin of the electrons is called paramagnetic susceptibility. For 2D semiconductors having parabolic energy bands, the paramagnetic-to- diamagnetic ratio is 1/3. It has been shown that both the susceptibilities depend on the density-of-states function [8, 9]. Das and Chakravarti [*1 formulated the magnetic susceptibility of electrons of semiconductor quantum wires having isotropic parabolic energy bands. Therefore in quantum wires fabricated from IV—VI semiconductors, having non-parabolic and non-standard energy bands, the susceptibilities will be rather significant since the density-of-states function in the presence of a parallel magnetic field increases much more rapidly with carrier energy than that of the monotonic variation of the same for the bulk specimens of the constituent materials.

In the following, in Section 2 we shall derive the carrier concentration in quantum wires of IV—VI materials in the presence of a parallel magnetic field on the basis of a newly formulated ID carrier dispersion law, both in the presence and absence of spin. We shall then formulate the diamagnetic and paramagnetic susceptibilities by using the appropriate carrier statistics.

We shall study the doping and thickness dependences of both the susceptibilities in quantum wires of PbS. PbTe and PbSe com- pounds respectively.

Theoretical Background

The carrier energy spectrum in bulk specimens of IV—V semiconductors can be expressed [10] as

$$E = \omega - \alpha(\omega)^2 + \alpha a_2 P_y^2 \omega - \alpha d_2 P_y^2 \omega + \alpha a_2 d_2 P_y^4 \tag{1}$$

where E is the total carrier energy as measured from the band edge; $\omega = a_1 p_x^2 + a_2 p_y^2 + a_3 p_z^2, a_i = 1/2m_i, i = 1,2 and\ 3, m_1,\ m_2\ and\ m_3$ are the effective carrier masses at the band edge along the p_x, p_y *and* p_z directions, respectively; $\alpha = 1/E_g, E_g$ is the bandgap; $a'_2 = 1/2m'_2$ and m'_2 is the longitudinal band edge effective mass of the holes for the electron dispersion law or of the electrons for the hole dispersion law.

Let us take a rectangular quantum wire, having cross- sectional dimensions d t and z. of a IV—VI semiconductor. The dimensions, comparable or less than the de Broglie wavelength of the carriers, are taken along the x and z axes respectively, of a cartesian coordinate system $(0 \le x \le d_1 \text{ and } 0 \le z \le d_2)$. The quantum wire of length L_y lies along *te* y axis and the parallel magnetic field B is also applied along the y direction. Let us choose the components of the magnetic vector potential $\vec{A}$ as $(0,\ 0, —Bx)$ so that $\vec{\nabla}\vec{A} = 0$ and $\vec{\nabla} \times \vec{A} = \vec{B}$

The total Hamiltonian can be written as

$$\hat{H} = a_1(-i\hbar(\frac{\partial}{\partial x}))^2 + a_2(-i\hbar(\frac{\partial}{\partial y}))^2 + a_3(-i\hbar(\frac{\partial}{\partial z}) - eBx)^2 - \alpha[a_1(-i\hbar(\frac{\partial}{\partial x}))^2 + a_2(-i\hbar(\frac{\partial}{\partial y}))^2 + a_3(-i\hbar(\frac{\partial}{\partial z}) - eBx)^2]^2$$
$$+\alpha a_2(-i\hbar(\frac{\partial}{\partial y}))^2[a_1(-i\hbar(\frac{\partial}{\partial x}))^2 + a_2(-i\hbar(\frac{\partial}{\partial y}))^2 + a_3(-i\hbar(\frac{\partial}{\partial z}) - eBx)^2] - \alpha a'_2(-i\hbar(\frac{\partial}{\partial y}))^2[a_1(-i\hbar(\frac{\partial}{\partial x}))^2 + a_2(-i\hbar(\frac{\partial}{\partial y}))^2$$
$$+a_3(-i\hbar(\frac{\partial}{\partial z}) - eBx)^2] + \alpha a_2 a'_2(-i\hbar(\frac{\partial}{\partial y}))^4 \tag{2a}$$

considering the band non-parabolicity and the magnetic held as perturbing parameters, the unperturbed wave function can be expressed as

$$-a_1\hbar^2(\frac{\partial^2\psi_0}{\partial x^2}) - a_2\hbar^2(\frac{\partial^2\psi_0}{\partial y^2})a_3\hbar^2(\frac{\partial^2\psi_0}{\partial z^2}) = E_0\psi_0 \tag{2b}$$

where ψ_0 and E_0, are the unperturbed wave function and unperturbed energy eigen value, respectively. From equation (2b) we get

$$E_0 = (\frac{\hbar^2}{2m_1})(\frac{L\pi}{d_1})^2 + \frac{\hbar^2 ky^2}{2m_2} + (\frac{\hbar^2}{2m_3})(\frac{n\pi}{d_2})^2 \tag{2c}$$

and

$$\psi_0 = (\frac{4}{d_1 d_2 L_y})^{1/2} \sin(\frac{L\pi x}{d_1}) + \sin(\frac{n\pi z}{d_2}) \exp(iyk_y) \tag{2d}$$

where $n(=1,2,3,.....)$and $L(=1,2,3,....)$are the size quantum numbers along the z and z axes, respectively.

The perturbed Hamiltonian assume the form

$$\hat{H} = (e^2B^2x^2 + 2i\hbar eBx(\frac{\partial^2}{\partial z})) - \alpha[-a_1\hbar^2(\frac{\partial^2}{\partial x^2}) + a_2\hbar^2(\frac{\partial^2}{\partial y^2}) + a_3(-\hbar^2(\frac{\partial^2}{\partial^2}) + e^2B^2x^2 + 2i\hbar eBx(\frac{\partial}{\partial z})]$$
$$-\alpha\hbar^2 a_2'(\frac{\partial^2}{\partial y^2})[-a_1\hbar^2(\frac{\partial^2}{\partial x^2}) - a_2\hbar^2(\frac{\partial^2}{\partial y^2}) + a_3(-\hbar^2(\frac{\partial^2}{\partial^2}) + e^2B^2x^2 2i\hbar eBx(\frac{\partial}{\partial z})] + \alpha a_2 a_2'(\hbar^4(\frac{\partial^4}{\partial y^4})) \tag{3}$$

Applying the first order perturbation theory and using together with some algebraic manipulations, the first- order perturbed energy as a consequence of first-order perturbation can be expressed as

$$< x\cot(\frac{n\pi z}{d_2}) > = 0 < x^3\cot(\frac{n\pi z}{d_2}) >$$
$$0 < x^2 > = (\frac{d_1^2}{3} - \frac{d_1^2}{2\pi^2L^2})) < x^4 >$$
$$= ((\frac{d_1^4}{5} - \frac{d_1^4}{L^2\pi^2}) + 3(\frac{d_1^4}{2\pi^4L^4}))$$
$$E_1 = -\alpha[a_1\hbar^2(\frac{L\pi}{d_1})^2 + a_2\hbar^2k_y^2 + a_3(\frac{\hbar n\pi}{d_2})^2]^2 + \alpha a_2\hbar^2k_y^2[a_1(\frac{\hbar L\pi}{d_1})^2 + a_2\hbar^2k_y^2 + a_3(\frac{\hbar n\pi}{d_2})^2]$$
$$-\alpha a_2'\hbar k_y^2[a_1(\hbar L\pi d_1)^2 + a_2\hbar^2k_y^2 + a_3(\frac{\hbar n\pi}{d_2})^2] + \alpha a_2 a_2'\hbar^4k_y^4 + a_3e^2B^2 < x^2 > + 2\alpha a_1 a_3(\hbar eB)^2$$
$$-\alpha[2a_1a_3(\frac{\hbar L\pi}{d_1})^2 + a_2a_3\hbar^2k_y^2 + 6a_3^2(\frac{\hbar n\pi}{d_2})^2] \times e^2B^2 < x^2 > - \alpha a_3^2e^4B^4 < x^4 > - \alpha a_2' a_3(\hbar eBk_y)^2 < x^2 > \tag{4}$$

where $n(=1,2,3,.....)$ and $L(=1,2,3,....)$ are the size quantum numbers along the z and x direct, respectively.

Therefore, the modified 1D carrier dispersion law for the present case up to the first order can be expressed as

$$E = \bar{E}_c + E_1 \tag{5}$$

where

$$\bar{E}_{\sigma} = (\frac{\hbar^2}{2m_1})(\frac{L\pi}{d_1})^2 + (\frac{\hbar^2 k_y^2}{2m_2}) + (\frac{\hbar^2}{2m_3})(\frac{n\pi}{d_2})^2.$$

Thus, using eqns (4) and (5), the 1D carrier energy spectrum in quantum wires of IV—VI semiconductors under a parallel magnetic field can be written as

$$E\phi(n,L) = A(n,L)k_y^2$$

where,

$$\phi(n,L) = [t_1(n,L) - \alpha t_1^2(n,L) + [a_2 e^2 B^2 < x^2 > + 2\alpha a_2 a_1 \hbar^2 e^2 B^2 - \alpha\{2a_1 a_2 (\frac{\hbar L\pi}{d_1})^2$$

$$+6a_2^2\hbar^4(\frac{n\pi}{d_2})^2\} - \alpha a_2^2 e^4 B^4 < x^4 >]],$$

$$t_1(n,L) = [a_1(\frac{L\pi\hbar}{d_1}) + a_2(\frac{\hbar n\pi}{d_2})^2]$$

and

$$A(n,L) = a_2\hbar^2 - [\alpha a_2\hbar^2 t_2(n,L) - \alpha a_2'\hbar^2 t_1(n,L) - \hbar^2\{\alpha a_2 a_2 e^2 B^2 < x^2 > + \alpha a_2' a_2 e^2 B^2 < x^2 >\}] \quad (6)$$

The density of State Function is given by

$$N(E) = (\frac{2g_v'}{\pi})\sum_{n,L} \frac{\partial ky}{\partial E} = (\frac{g_v'}{\pi}) \sum_{n,L} (\frac{1}{\sqrt{A(n,L)}})(\frac{1}{\sqrt{E-\phi(n,L)}}) \quad (7)$$

where $g_v' = (\frac{g_v}{d_2 d_1})$ and g_v is the valley degeneracy.

Combing equation (7) with the Fermi-Dirac occupation probability factor, the carrier statistics can be written as

$$n_0 = (\frac{g_v'}{\pi})\sum_{n,L} [A(n,L)]^{-1/2} \times \int_{\phi(n,L)}^{\infty} [E-\phi(n,L)]^{-1/2} f_0(E)dE$$

$$= (\sum_{n,L} F_{-1/2}(\eta)[A(n,L)]^{-1/2})C_1 \quad (8)$$

where $C_1 = (g_v'\sqrt{\pi})(\sqrt{k_B T}); k_B$ is Boltzmann's constant; T is temperature; $\eta = (\sqrt{k_B T})^{-1}[E_F - \phi(n,L)], E_F$ is the Fermi energy in this case; $f_0(E)$ is the Fermi Dirac occupation probability factor; and $f_j(\eta)$ is the one-parameter Fermi-Dirac integral of order j, which is defined as [11]

$$F_j(\eta) = (\frac{1}{\sqrt{(j+1)}})\int_0^{\infty} (\frac{x^j dx}{(1+\exp(x-\eta))} \; for \; j > -1 \tag{9}$$

in which x is a dummy variable, $\sqrt{j+1} = j\sqrt{(j)}$ *and* $\sqrt{(j)}$ is the well-known Gamma function. For all j, equation (9) can be analytically continued as a complex contour integral around the negative x axis as

$$F_j(\eta) = (\frac{\sqrt{(-j)}}{2\pi\sqrt{-1}}) \times \int_{(-\infty)}^{(0+)} (\frac{x^j dx}{(1+\exp(-x-\eta))} \tag{10}$$

It is well-known that the diamagnetic susceptibility of the electrons is given by

$$X_d = -\mu_0(\frac{\partial^2 f(B)}{\partial B^2}) \tag{11}$$

where μ_0 is the absolute permittivity of the free space and $f(B)$ is the free energy which can, in turn, be written as [8]

$$f(B) = n_0 E_F - \int_{\phi(n,L)}^{\infty} \phi_0 f_0(E) dE \tag{12}$$

in which $\phi_0 = \int N(E)dE$. From Equation (12) we get

$$f(B) = n_0 E_F - C_2 \sum_{n,L} [A(n,L)]^{-1/2} F_{1/2}(\eta) \tag{13}$$

where

$$C_2 = g'_v\left(\frac{k_B T}{\pi}\right)^{1/2}$$

Using equations (11) and (13), we can write

$$X_d = -\mu_0\left[n_0\left(\frac{\partial^2 E_F}{\partial B^2}\right) - \left(\frac{C_2}{(k_B T)^2}\right)^2 \times \sum_{n,L} [A(n,L)]^{-1/2} F_{-3/2}(\eta) \times \left[\left(\frac{\partial E_F}{\partial B}\right) - C_3 B + C_4 B^3\right]^2\right.$$
$$-\left(\frac{C_2}{(k_B T)}\right)\sum_{n,L} F_{-1/2}(\eta)(A(n,L))^{-1/2} \times \left[\left(\frac{\partial^2 E_F}{\partial B^2}\right) - C_3 + 3C_4 B^3\right]^2 - C_3 B + C_4 B^3) - \left(\frac{C_2}{2}\right)\sum_{n,L} C_5 F_{1/2}(\eta)(A(n,L))^{-3/2}$$
$$\left. -\left(\frac{3C_2}{4}\right)\sum_{n,L} F_{1/2}(\eta)(A(n,L))^{-5/2} C_5^2 B^2\right] \tag{14}$$

where

$$C_3 = 2a_3 e^2 < x^2 > +2\alpha a_1 a_3 \hbar^2 e^2,\ C_4 = 4\alpha a_3^2 e^4 < x^4 > and\ C_5 = \hbar^2[\alpha a_2 a_3 e^2 < x^2 > +\alpha a'_2 a_3 e^2 < x^2 > 1$$

In equation (14), $\frac{\partial E_F}{\partial B}$ *and* $\frac{\partial^2 E_F}{\partial B^2}$ can respectively, be expressed through the equation

$$\frac{\partial E_F}{\partial B} = \left[\sum_{n,L}\left(\frac{1}{\sqrt{A(n,L)}}\right)F_{-3/2}(\eta)\right]^{-1} \times \left[\sum_{n,L} F_{-3/2}(\eta)\left(\frac{1}{\sqrt{A(n,L)}}\right)\right][C_3 - C_4 B^3]^{-(1/2)}$$
$$\left[\sum_{n,L} A(n,L)^{-3/2} C_5 B F_{-1/2}(\eta)\right]^{-(1/2)} \tag{15}$$

and

$$\frac{\partial^2 E_F}{\partial B^2} = -\left[\left(\frac{1}{k_B^T}\right)\sum_{n,L} F_{-3/2}(\eta)\left(\frac{1}{\sqrt{A(n,L)}}\right)\right]^1 \times \left[\left(\frac{1}{k_B^T}\right)\sum_{n,L} F_{-5/2}(\eta) \times \left[\frac{\partial E_F}{\partial B} - C_3 B + C_4 B^3\right]^2\right.$$
$$\left. + \frac{1}{2}\sum_{n,L} (A(n,L)^{-3/2} C_5 F_{-1/2}(\eta)\right] + \frac{3}{4}\sum_{n,L} [A(n,L)^{-5/2} C_5^2 B^2 F_{-1/2}(\eta)] \tag{16}$$

In the presence of electron spin, the free energy can be expressed as

$$\psi(B) = n_0 E_f - \left(\frac{C_2}{2}\right)\sum_{n,L} [A(n,L)]^{-1/2} F_{1/2}(\eta_\pm) \tag{17}$$

where E_f is the Fermi energy when spin splitting is considered and can be obtained through the equation

$$n_0 = \left(\frac{C_1}{2}\right)\sum_{n,L} \left(\frac{F_{-1/2}(\eta_\pm)}{\sqrt{A(n,L)}}\right) \tag{18}$$

where $(\eta_\pm) = (k_B T)^{-1}[E_f - \phi(n,L) \pm (\frac{1}{2}) g\mu B], g$ is the magnitude of the Landsberg factor at the band edge and μ is the Bohr magnetron.

The expression of the paramagnetic susceptibility can be written as

$$\begin{aligned} X_p = \mu_0[n_0(\frac{\partial^2 E_f}{\partial B^2}) - (\frac{C_2}{2(k_B T)})^2 \sum_{n,L} (A(n,L))^{-1/2} \times F_{-3/2}(\eta_\pm)[(\frac{\partial E_f}{\partial B}) - C_3 B + C_4 B^3]^2 - (\frac{C_2}{2k_B T})\sum_{n,L} F_{-1/2}(\eta_\pm) \\ \times (A(n,L))^{-1/2}(\frac{\partial^2 E_f}{\partial B^2}) - C_3 B + 3C_4 B^2] - (\frac{C_2}{2k_B T})\sum_{n,L} C_5 B (A(n,L))^{-3/2} F_{-1/2}(\eta_\pm) \times [(\frac{\partial E_f}{\partial B}) - C_3 B + C_4 B^3] \\ -(\frac{C_2}{4})\sum_{n,L} C_5 F_{1/2}(\eta_\pm)(A(n,L))^{-3/2} - (\frac{3C_2}{8})\sum_{n,L} F_{1/2}(\eta_\pm)(A(n,L))^{-5/2} C_5^2 B^2] \end{aligned} \tag{19}$$

In equation (19), $(\frac{\partial E_f}{\partial B})$ and $(\frac{\partial^2 E_f}{\partial B^2})$ can respectively, be written as

$$\frac{\partial E_f}{\partial B} = [\sum_{n,L} (\frac{1}{\sqrt{A(n,L)}}) F_{-3/2}(\eta_\pm) \times \sum_{n,L} F_{-3/2}(\eta_\pm)(\frac{1}{\sqrt{A(n,L)}})(C_3 B - C_4 B^3) - (\frac{1}{2})\sum_{n,L} [A(n,L)]^{3/2} C_5 B F_{-\frac{1}{2}}(\eta_\pm)] \tag{20}$$

$$\begin{aligned} \frac{\partial^2 E_f}{\partial B^2} = [\sum_{n,L} (\frac{1}{k_B T}) F_{-3/2}(\eta_\pm)(\frac{1}{\sqrt{A(n,L)}}) \times (\frac{1}{k_B T})\sum_{n,L} F_{-\frac{5}{2}}(\eta_\pm)] \times [(\frac{\partial E_f}{\partial B}) - C_3 B - C_4 B^3 (\frac{1}{\sqrt{A(n,L)}}) \\ +(\frac{1}{k_B T})\sum_{n,L} (A(n,L))^{-3/2} C_5 B F_{-\frac{3}{2}}(\eta_\pm) \times [(\frac{\partial E_f}{\partial B}) - C_3 B - C_4 B^3] + \frac{1}{2}\sum_{n,L} (A(n,L))^{-3/2} C_5 F_{-\frac{1}{2}}(\eta_\pm) \\ + \frac{3}{4}\sum_{n,L} (A(n,L))^{-5/2} C_5^2 B^2 F_{-\frac{1}{2}}(\eta_\pm) \end{aligned} \tag{21}$$

RESULTS AND DISCUSSION

Using the appropriate equation together with the parameters [12]

$T = 4.2\ K,\ m_1 = m_3 = 0.080 m_0,$

$m_2' = 0.105 m_0,\ E_g = 0.29 eV,\ (dE_g / dT) = 4.7 \times 10^{-4} eVK^{-1}$

for $n - PbSe,\ m_1 = m_3 = 0.040 m_0,$

$m_2 = 0.07 m_0, m_2' = 0.068 m_0,\ E_g = 0.15 eV, (dE_g / dT) = 4.7 \times 10^{-4} eVK^{-1}$

for $n - PbSe,$

$m_2' = 0.31 m_0,\ E_g = 0.19 eV, (dE_g / dT) = 4.7 \times 10^{-4} eVK^{-1},\ m_1 = m_3 = 0.024 m_0$

for n-PbTe, we have plotted the X_d and X_p under strong degeneracy and in the quantum limit $(n = L = 1)$ versus n_0 for these materials as shown in figure 1, in which the same dependence has also been plotted in the absence of non-parabolicity for the purpose of relative comparison. In Figure 9.2. we have plotted the X_p and X_d versus d_1 for all the cases of Figure 9.1. for$n_0 = 5.9 \times 10^{24} m^{-3}$, and have further plotted the same dependence for$n_0 = 6 \times 10^{23} m^{-3}$ in Figure 9.3.

It appears from Figure 9.1. that in general $X_d$and X_p both decrease with increasing concentrations. The X_d which normally takes only negative values, becomes positive below a certain critical value of n_0 i.e., the diamagnetism disappears in this range. When an external magnetic field is applied to a semiconductor, the electrons can move only in the quantum mechanically allow orbits. These paths are in a plane perpendicular to the direction of the magnetic field which is opposite to the direction of this field.

Thus, the net magnetic field inside the system is less than that applied externally. This is the well-known phenomena of diamagnetism. In a quantum wire, the values of the magnetic field and the dimensions of the system are such that the electrons cannot move on complete orbits. They hit the boundary walls somewhere and get reflected to move in a curved path. The electron trying to move along a circular are (clockwise according to Lenz's Law) hits the system somewhere at the boundary, gets reftec1ed and repeats its course, crossing the initial position. This is equivalent to an anticlockwise rotation of the electrons in the presence of a magnetic field. This sense of rotation is opposite to that which gives rise to diamagnetism.

We wish to state that the individual are contributes to diamagnetism to some extent. When the contribution to magnetic moment, arising from the equivalent anticlockwise rotation, exceeds the total contribution to the individual arcs, diamagnetism cannot occur. The predominance of one contribution over the other is controlled by the individual are lengths and the electron density of the allowed levels. For this reason, the crawling motion of the electrons in the less populated higher energy states, around the boundary of a 3D or2D system, cannot influence the diamagnetism which is due to the cyclic motion of the electrons in the densely populated lower energy states. Thus the diamagnetism is not quenched for the aforementioned systems.

It appears from the figures that the susceptibility ratio deviates from 1/3. Unlike the case wilh parabolic semiconductors. In Figure 9.2 the variation of the susceptibilities with the dimensional d1 is shown, keeping the other dimension fixed. In Figures 9.2 and 9.3. two different electron concentrations have been assumed. With the increase of d" the susceptibilities decrease. In Figure 9.2, it may be seen that there is a critical value of d1, below which the diamagnetism disappears for the whole range of thickness as considered here. Thus, there is a critical relation between the electron concentration and the dimensions of the. Quantum wire to determine whether the diamagnetic susceptibility will appear or not.

The numerical values of Xd and Xp in n-PbS is greater than that in n·PbTe. Since the values of the band parameters of PbS enhance the numerical values of the Fermi energy in the presence of a parallel magnetic field in n-·PbS significantly more in comparison with the EF of PbTc, although both the materials are described by the same dispersion law. Besides the aforementioned fact is also true in the absence of band non-parabolicity. Therefore, from all t!he figures we can compare not only the influence of band non-parabolicity but also the impact of the material constants on the numerical magnitudes of susceptiblities for the nanostructured materials in the present case.

The influence of quantum confinement is immediately apparent from Figures 9.2 and 9. 3, since and depend strongly on the nanothickness of the quantum wires of IV-VI materials in constant with the bulk specimens of the said compounds, and exhibit strong dependences with the nano-thickness and in the presence of a parallel magnetic field can become several orders of magnitude larger than in bulk specimens of the same material. Which is also a direct consequence of nano-structures. The band non-parabolicity significantly alters the numerical values of and which is evident from all the figures. In the presence of many body effect and also under arbitrary orientation of the magnetic field, we shall obtain different numerical values for and of quantum wires, although the nature of the variations will be unaltered. Thus the trends of the curves of and with and film thickness as shown here would be similar for all other types of non-parabolic materials.

It may be noted that the variations of and are band structure dependent. We have not plotted and with respect to other physical variables or considered other materials for the purpose of our condensed presentation. It may be remarked that the results for and for bulk specimens of wide gap materials are well-known in the literature. The expression for the above quantities for quantum wires of VI-IV materials in the presence of a parallel magnetic field have yet to be derived. In this chapter we have first formulated the expressions for and for the present system on the basis of

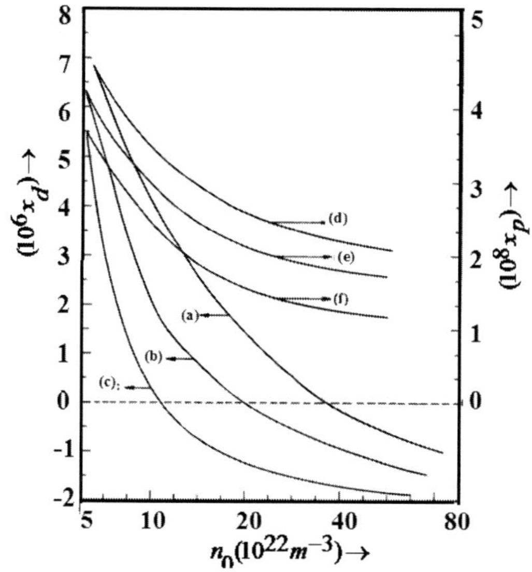

Figure 9.1. Plots of X_d and X_p versus n_0 in quantum wire IV-VI materials (a) PbS, (b) PbSe (c) PbTe (d), € and (f) are valid for $E_g = \infty$.

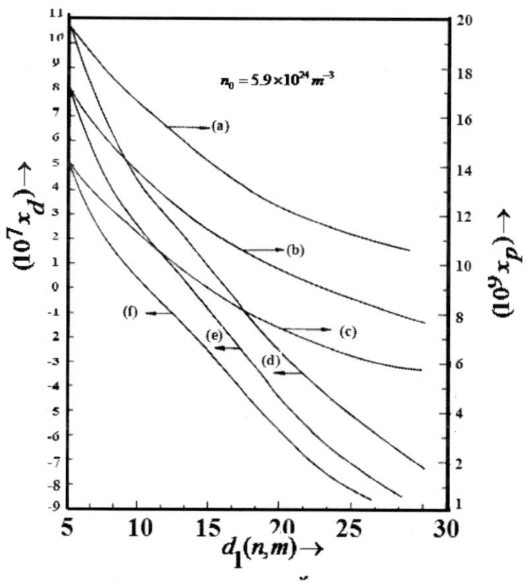

Figure 9.2. Plots of X_d and X_p versus d_1 for all cases of figure 1. $d_2 = 10nm, n_0 = 5.9 \times 10^{24} m^{-3}$.

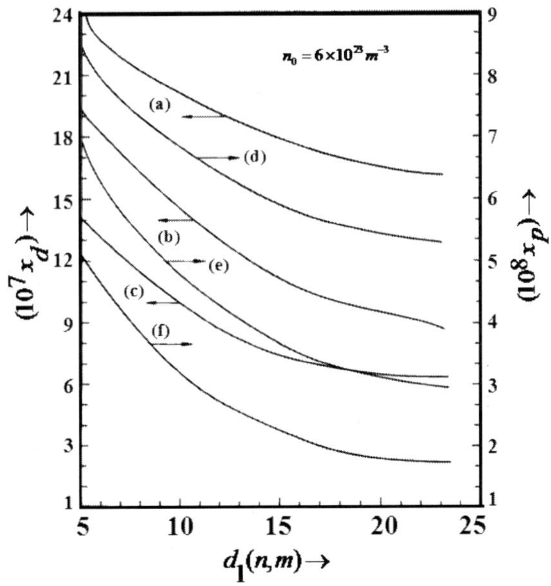

Figure 9.3. Plots of X_d and X_p versus d_1 for all cases of figure 1. $d_2 = 10nm, n_0 = 5.9 \times 10^{24} m^{-3}$.

newly formulated dispersion law. The well-known expressions of and can be obtained as a special case of our generalized results. Finally, it may be noted that the basic aim of this chapter is not only to formulate the expressions for magnetic susceptibilities and but also to develop the expressions for, since the carrier transport and the formulations of different physical properties of nanodevices are directly based on the temperature dependent electron statistics in such heterostructures.

REFERENCES

[1] Ghatak, K. P. and Mondal, M. *Journal of Applied Physics*, 69, 1666 (1991).

[2] Linch, N. T. *Journal of Festkorperprobleme*, 23, 227 (1985).

[3] Scifres, D. R., Lindstrom, C., Burnham, R. D., Strefier, W. and Paoli, T. L. *Journal of Electron Letter*, 19, 170 (1983).

[4] Schlesimger, J. E. and Kuech, T. *Journal of Applied Physics Letter,* 49, 519 (1980).

[5] Beamount, S. P. and Torres, C. M. S. *Science and engineering of one and zero dimensional semiconduvtors,* Plenum press, New York (1990).

[6] Tusbaki, K., Fukui, T., Tokura, T., Saito, H. and Susa, N. *Journal of Electron Letter,* 24,1267 (1988).

[7] Das, M. R. and chakrabarti, A. N. *Journal of Physica status solidi (b),* 172, 603 (1990).

[8] Ghatak, K. P. and Mondal, M. *Journal of Physica status solidi (b),* 160, 673 (1990).

[9] Lovett, D. R. *Semimetals and Narrow Band-gap Semiconductors,* Pion Limited, London, (1977).

[10] Ghatak, K. P. and Bhattacharya, D. *Journal of Physica Scripita,* 52, 343 (1995).

[11] Nag, B. R. *Electron Transport compound semiconductors*, Springer-Verlag, Germany, (1980).

[12] K. P. Ghatak, S. Bhattacharya and D. De, *Einstein Relation in Compound Semiconductors and their Nanostructures*, Springer Series in Materials Science, Vol. 116, Springer Verlag, Germany, (2009).

In: Quantum Wires: An Overview
Editor: Kamakhya Prasad Ghatak
ISBN: 978-1-53617-676-6

Chapter 10

THE HEISENBERG'S UNCERTAINTY PRINCIPLE AND THE CARRIER CONTRIBUTION TO THE ELASTIC CONSTANTS IN HEAVILY DOPED OPTO ELECTRONIC QUANTUM WIRES

S. K. Biswas**, A. K. Pradhan, S. Maiti, K. Bhattacharyya, M. Maiti and S. Gangopadhyay
Department of Physics, University of Calcutta, Kolkata, India

ABSTRACT

In this chapter we study the second and third order carrier contribution to the elastic constants (ΔC_{44} *and* ΔC_{456}) in heavily doped quantum wires of optoelectronic materials in the presence of light waves taking $n-InAs$, $InSb, n-Hg_{1-x}Cd_xTe$, $In_{1-x}Ga_xAs_yP_{1-y}$ and lattice matched to InP as examples. We have also suggested an experimental method of

* Correseponding Author's Email: skbiswas23@gmail.com.

determining ΔC_{44} and ΔC_{456} for materials having arbitrary dispersion laws. The influence of light is immediately apparent from the plots since ΔC_{44} and ΔC_{456} depend strongly on the light intensity (I) and wave length (λ) which is in direct contrast as compared with I = 0. The variations of ΔC_{44} and ΔC_{456} reflect the direct signature of the light wave on the electronic, optic and the other band structure-dependent properties of HD semiconducting materials in the presence of light waves and the photon assisted transport for the corresponding semiconductor devices. Although both ΔC_{44} and ΔC_{456} tend to increase with intensity and the wave length, the rate of increase is totally band structure dependent. We wish to note that in view of large changes of the elastic constants with n_0 detailed experimental work on second and third-order elastic constants as functions of n_0 would be interesting for the present case. It may be suggested that the experiments on the velocity of sound involving the shear mode as function of n_0 may exhibit the carrier contribution to the elastic constants for materials having arbitrary carrier energy spectra. This fact again suggests another experimental determination of ΔC44 and ΔC_{456} beside the suggested methods of determining them as said above.

INTRODUCTION

In the last chapter we shall study the carrier contribution to the elastic constants (CEC) of the nano materials which has been studied in the last thirty years mainly by the group of Ghatak et all. [1-28] and few others [29-30]. In this regard, we wish to note that the theory for determining the carrier contributions to the elastic constants in p-type Si already exists [29]. It has been shown that the carrier contribution of the second and third order elastic constants depends on the density- of states function (DOS) [29]. Sreedhar and Gupta [30] formulated the same for small gap materials whose energy band structures are defined by the two-band model of Kane. It has therefore different values in various materials and varies with the electron concentration, with doping, with the thickness of ultra-thin films and with temperature for semiconductors and their hetero-structures having various carrier energy spectra. The nature of these variations has been investigated in the literature. [1-30] some of the significant features, which have emerged from these studies, are:

a) The CEC changes monotonically with electron concentration in bulk materials.
b) The nature of the variations is significantly affected by the band non - parabolicity.

The said contribution has significantly different values in ultra-thin films.

In this chapter an attempt is made to investigate the CEC in heavily doped quantum wires of opto-electronic materials in the presence of light waves and for the purpose of relative comparison we have also studied the same in bulk specimens in the section of theoretical background. We have plotted the 2nd and third order elastic constants in heavily doped quantum wires of n-InAs, n-InSb, $n-Hg_{1-x}Cd_xTe, In_{1-x}Ga_xAs_yP_{1-y}$ and lattice matched to InP in section 10.3 with different physical variables which also contains results and discussion pertinent to this chapter.

Theoretical Background

The CEC in the Presence of Light Waves in HD III-V, Ternary and Quaternary Semiconductors

Thus, the simplified electron energy spectrum in III-V, ternary and quaternary materials up to the second order in the presence of light waves whose unperturbed electron dispersion relations obey three and two band models of Kane together with parabolic energy bands can, respectively approximately be written as

$$\frac{\hbar^2 k^2}{2m^*} = \beta_0(E,\lambda) \tag{1}$$

$$\frac{\hbar^2 k^2}{2m^*} = \tau_0(E,\lambda) \quad (2)$$

$$\frac{\hbar^2 k^2}{2m^*} = \rho_0(E,\lambda) \quad (3)$$

where the notations are defined in [31]

$$\frac{\hbar^2 k^2}{2m_c} = U_\lambda I_{11}(E) - P_\lambda \quad (4)$$

$$\frac{\hbar^2 k^2}{2m_c} = t_{1\lambda} E + t_{2\lambda} E^2 - \delta_\lambda \quad (5)$$

and

$$\frac{\hbar^2 k^2}{2m_c} = t_{1\lambda} E - \delta_\lambda \quad (6)$$

where,

$$U_\lambda = (1+\theta_\lambda), \theta_\lambda = \frac{C_0}{A}(t_\lambda + \frac{BJ_\lambda}{A}), C_0 = [\frac{|e|^2}{96\pi c^3 m_r} \frac{I\lambda^2 E_{g_0}(E_{g_0}+\Delta)}{\sqrt{\varepsilon_{sc}\varepsilon_0}(E_{g_0}+\frac{2}{3}\Delta)} \frac{\beta^2}{4}(1+\frac{\rho}{\sqrt{2}})^2],$$

$$A = E_{g_0}, B = [1+\frac{m^*}{m_V}], G_\lambda = [\frac{2B}{(A+\delta')^3} - \frac{BC_\lambda}{(A+\delta')}], C_\lambda = [(E_{g_0}+\delta')^{-1} + (E_{g_0}+\delta')(E_{g_0}-\delta')^{-2}](A+\delta')^{-1},$$

$$P_\lambda = \frac{C_0}{A} J_\lambda, J_\lambda = (D_\lambda + 2(E_{g_0}-\delta')\sqrt{f_\lambda}), D_\lambda = (1+\frac{2(E_{g_0}-\delta')}{(A+\delta')}), f_\lambda = [\frac{1}{(A+\delta')^2} + \frac{1}{(E_{g_0}-\delta')^2} - C_\lambda],$$

$$t_{1\lambda} = (1+\frac{3m_c}{m_r}\alpha\delta_\lambda), \alpha = \frac{1}{E_{g_0}}, \delta_\lambda = \frac{|e|^2 I\lambda^2}{96 m_r \pi c^3 \sqrt{\varepsilon_{sc}\varepsilon_0}} \text{ and } t_{2\lambda} = \alpha t_{1\lambda}$$

It is well known that the band tails are being formed in the forbidden zone of HDS and can be explained by the overlapping of the impurity band with the conduction and valence bands [32]. Kane [33] and Bonch Bruevich [34] have independently derived the theory of band tailing for semiconductors having unperturbed parabolic energy bands. Kane's model [33] was used to explain the experimental results on tunneling [35] and the optical absorption edges [36, 37] in this context. Halperin and Lax [38] developed a model for band tailing applicable only to the deep tailing states. Although Kane's concept is often used in the literature for the investigation of band tailing [39, 40], it may be noted that this model [32, 41] suffers from serious assumptions in the sense that the local impurity potential is assumed to be small and slowly varying in space coordinates [38]. In this respect, the local impurity potential may be assumed to be a constant. In order to avoid these approximations, we have developed in this book, the electron energy spectra for HDS for studying the EP based on the concept of the variation of the kinetic energy [38] of the electron with the local point in space coordinates. This kinetic energy is then averaged over the entire region of variation using a Gaussian type potential energy. It may be noted that, a more general treatment of many-body theory for the DOS of HDS merges with one-electron theory under macroscopic conditions [31]. Also, the experimental results for the Fermi energy and others are the average effect of this macroscopic case. So, the present treatment of the one-electron system is more applicable to the experimental point of view and it is also easy to understand the overall effect in such a case [42]. In a HDS, each impurity atom is surrounded by the electrons, assuming a regular distribution of atoms, and it is screened independently [39, 41, 43]. The interaction energy between electrons and impurities is known as the impurity screening potential. This energy is determined by the inter-impurity distance and the screening radius, which is known as the screening length. The screening radius changes with the electron concentration and the effective mass. Furthermore, these entities are important for HDS in characterizing the semiconductor properties [44, 45] and the modern electronic devices [39, 46]. The works on Fermi energy and the screening length in an n-type GaAs have already been initiated in the literature [47,48], based on Kane's model. Incidentally, the limitations of Kane's model [32, 40], as mentioned above, are also present in their studies.

The Gaussian distribution *F(V)* of the impurity potential is given by [32-33].

$$F(V) = (\pi\eta_g^2)^{-1/2} \exp(-V^2/\eta_g^2) \tag{7}$$

where η_g is the impurity scattering potential. It appears from equation (7) that the variance parameter η_g is not equal to zero, but the mean value is zero. Further, the impurities are assumed to be uncorrelated and the band mixing effect has been neglected in this simplified theoretical formalism.

Under the condition of heavy doping, using the method of averaging the kinetic energy of the electron, the HD dispersion relations in this case in the presence of light waves can be written as

$$\frac{\hbar^2 k^2}{2m_c} = T_1(E,\eta_g,\lambda) \tag{8}$$

$$\frac{\hbar^2 k^2}{2m_c} = T_2(E,\eta_g,\lambda) \tag{9}$$

$$\frac{\hbar^2 k^2}{2m_c} = T_3(E,\eta_g,\lambda) \tag{10}$$

$$where,\ T_1(E,\eta_g,\lambda) = [U_\lambda[T_{31}(E,\eta_g) + iT_{32}(E,\eta_g)] - P_\lambda],$$

$$T_{31}(E,\eta_g) \equiv \left(\frac{2}{1+Erf(E/\eta_g)}\right)\left[\frac{\alpha b}{c}\theta_0(E,\eta_g) + \left[\frac{\alpha c + bc - \alpha b}{c^2}\right]\gamma_0(E,\eta_g) + \right.$$

$$+\frac{1}{c}\left(1-\frac{\alpha}{c}\right)\left(1-\frac{b}{c}\right)\frac{1}{2}\left[1+Erf\left(\frac{E}{\eta_g}\right)\right]$$

$$-\frac{1}{c}(1-\frac{\alpha}{c})(1-\frac{b}{c})\frac{2}{c\eta_g\sqrt{\pi}}\exp(-u_2^2)[\sum_{p=1}^{\infty}\frac{\exp(-p^2/4)}{p}sinh(pu_2)],$$

$$b\equiv(\frac{1}{E_g+\Delta}),c\equiv(\frac{1}{E_g+\frac{2}{3}\Delta}),u_2\equiv\frac{1+cE}{c\eta_g},$$

$$T_{32}(E,\eta_g)\equiv(\frac{2}{1+Erf(E/\eta_g)})\frac{1}{c}(1-\frac{\alpha}{c})(1-\frac{b}{c})\frac{\sqrt{\pi}}{c\eta_g}\exp(-u_2^2).$$

$$\theta_0(E,\eta_g)=\frac{\eta_g E}{2\sqrt{\pi}}\exp(\frac{-E^2}{\eta_g^2})+\frac{1}{4}(\eta_g^2+2E^2)[1+Erf(\frac{E}{\eta_g})],$$

$$\gamma_0(E,\eta_g)=\eta_g\exp(-E^2/\eta_g^2)(2\sqrt{\pi})^{-1}+\frac{E}{2}(1+Erf(E/\eta_g)),$$

$$T_2(E,\eta_g,\lambda)=[U_{1\lambda}\gamma_3(E,\eta_g)+(t_{2\lambda})2\theta_0(E,\eta_g)[1+\mathrm{Erf}(E/\eta_g)]^{-1}-\delta_\lambda]$$

$$\gamma_3(E,\eta_g)\equiv[\frac{2}{(1+Erf(E/\eta_g))}]\gamma_0(E,\eta_g)$$

$$and\ T_3(E,\eta_g,\lambda)=[\mathrm{t}_{1\lambda}\,\gamma_3(E,\eta_g)-\delta_\lambda]$$

The HUP can be written as

$$\Delta p_i\Delta i\sim A\hbar \tag{11}$$

where $i = x, y\ and\ z$, p is the momentum, $\Delta's$ are the errors in measuring $p_i\ and\ i, \hbar$ is the Dirac's constant and and A is the dimension less constant.

Since $p_i = \hbar k_i\ where\ k_i$ is the electron wave vector, from (12) we can write

$$\Delta k_x\Delta k_y\Delta k_z=\frac{A^3}{\Delta V} \tag{12}$$

where $\Delta V = \Delta x \Delta y \Delta z$

The use HUP tells us each electron occupies at least a volume ΔV and this electron must exists in either of the two possible spin orientations due to Pauli's exclusion principle. If n_{03D} is the Electron Statistics (ES) in unit volume then $\Delta V = \frac{2}{n_{03D}}$ and the combination of which with (12) leads to the result

$$n_{03D} = \frac{2}{C_{3D}}(\Delta k_x \Delta k_y \Delta k_z) \text{ where } C_{3D} = A^3 \quad (13)$$

If n_{03D} is the valley degeneracy, we can write

$$n_{03D} = \frac{2g_v}{C_{3D}}(\Delta k_x \Delta k_y \Delta k_z) \quad (14)$$

For two and one dimensions we get

$$n_{02D} = \frac{2g_v}{C_{2D}}(\Delta k_x \Delta k_y) \quad (15)$$

and

$$n_{01D} = \frac{2g_v}{C_{1D}}(\Delta k_x) \quad (16)$$

where n_{02D} and n_{01D} are ES per unit area per sub band and per unit length per sub band respectively and C_{2D} and C_{1D} are two dimension less constants is the respective cases.

In accordance with HUP

$$V(E_F) = (\Delta k_x \Delta k_y \Delta k_z) \tag{17}$$

Using (8) we get

$$V(E_{FHDL}) = \frac{4\pi}{3}[\frac{2m_c}{\hbar^2}]^{3/2}[T_1(E_{FHDL}, \eta_g \lambda)]^{\frac{3}{2}} \tag{18}$$

where E_{FHDL} *is the Fermi energy in this case.*

Using (13) and (18), the ES can be written as

$$n_0 = \frac{8\pi g_v}{3C_{3D}}[\frac{2m_c}{\hbar^2}]^{3/2} \text{Re}al\ Part\ of[T_1(E_{FHDL}, \eta_g \lambda)]^{3/2} \tag{19}$$

Under the substitution $C_{3D} = 1$, (19) assumes the form

$$n_0 = \frac{8\pi g_v}{3}[\frac{2m_c}{\hbar^2}]^{3/2} \text{Re}al\ Part\ of[T_1(E_{FHDL}, \eta_g \lambda)]^{3/2} \tag{20}$$

Equation (20) is the expression of electron statistics in HD III-V, ternary and quaternary semiconductors in accordance with the three band model of Kane in the presence of intense light waves under the condition of extreme degeneracy with the formation of band tails without using the DOS function approach and by directly applying the HUP.

Similarly for perturbed two band model of Kane and that for parabolic energy bands we can write

$$n_0 = \frac{8\pi g_v}{3}[\frac{2m_c}{\hbar^2}]^{3/2}[T_2(E_{FHDL}, \eta_g \lambda)]^{3/2} \tag{21}$$

$$n_0 = \frac{8\pi g_v}{3}[\frac{2m_c}{\hbar^2}]^{3/2}[T_3(E_{FHDL}, \eta_g \lambda)]^{3/2} \tag{22}$$

In the absence of Heavy Doping, the ES can be written as

$$n_0 = \frac{g_v}{3\pi^2}(\frac{2m_c}{h^2})^{3/2} [\beta_0(E_{FL}, \lambda)]^{3/2} \tag{23}$$

$$n_0 = \frac{g_v}{3\pi^2}(\frac{2m_c}{h^2})^{3/2} [\tau_0(E_{FL}, \lambda)]^{3/2} \tag{24}$$

$$n_0 = \frac{g_v}{3\pi^2}(\frac{2m_c}{h^2})^{3/2} [\rho_0(E_{FL}, \lambda)]^{3/2} \tag{25}$$

In the absence of Heavy Doping and photons, the ES in this case can be written as

$$n_0 = \frac{g_v}{3\pi^2}(\frac{2m_c}{h^2})^{3/2} [I_{11}(E_F)]^{3/2} \tag{26}$$

$$n_0 = \frac{g_v}{3\pi^2}(\frac{2m_c}{h^2})^{3/2} [E_F(1+\alpha E_F)]^{3/2} \tag{27}$$

$$n_0 = \frac{g_v}{3\pi^2}(\frac{2m_c}{h^2})^{3/2} [E_F]^{3/2} \tag{28}$$

The equations (1.53b) and (1.53c) are well known in the literature [31].

The electronic contribution to the second- and third- order elastic constants for HD materials can be written as [51-63]

$$\Delta C_{44} = \frac{-G_0^2 Z_x}{9}[\frac{\partial n_0}{\partial(E_{FHD} - E_{0HD})}] \tag{29}$$

and

$$\Delta C_{456} = \frac{G_0^3 Z_x}{27}[\frac{\partial^2 n_0}{\partial (E_{FHD} - E_{0HD})^2}] \tag{30}$$

where G_0 is the deformation potential constant, $Z_x = (\text{Re}\,al\,Part)^x$, E_{FHD} is the Fermi energy in HD materials in the presence of band-tails, E_{0HD} is obtained from the corresponding HD dispersion relation under the conditions $E = E_{0HD}$ when $k = 0$.

Using (29), (30) and (31) and $x = 1$ successively we can study $\Delta C_{44}\, and\, \Delta C_{456}$ for HD materials in the presence of light waves whose energy band structures in the absence of any field are given by three band model of Kane. Besides again using (29), (30) and (31) and (22) together with, $x = 0$ we can study the same for two band Kane models together with parabolic energy bands respectively in this case.

In the absence of band-tails we can write

$$\Delta C_{44} = \frac{-G_0^2}{9}[\frac{\partial n_0}{\partial E_{FL}}] \tag{31}$$

and

$$\Delta C_{456} = \frac{G_0^3}{27}[\frac{\partial^2 n_0}{\partial E_{FL}{}^2}] \tag{32}$$

where E_{FL} the corresponding Fermi energy in this case.

Using (31), (32) and (23), (24) and (25) successively, the expressions for $\Delta C_{44}\, and\, \Delta C_{456}$ for materials in the absence of band tails and in the presence of light waves whose energy band structures in the absence of any field are given by three and two band Kane models together with parabolic energy bands can, respectively, be written as

$$\Delta C_{44} = \frac{-G_0^2 g_v}{27\pi^2}(\frac{2m_c}{h^2})^{3/2} [[[\beta_0(E_{FL},\lambda)]^{3/2}]']] \quad (33)$$

$$\Delta C_{44} = \frac{-G_0^2 g_v}{27\pi^2}(\frac{2m_c}{h^2})^{3/2} [[[\tau_0(E_{FL},\lambda)]^{3/2}]']] \quad (34)$$

$$\Delta C_{44} = \frac{-G_0^2 g_v}{27\pi^2}(\frac{2m_c}{h^2})^{3/2} [[[\rho_0(E_{FL},\lambda)]^{3/2}]']] \quad (35)$$

and

$$\Delta C_{456} = \frac{G_0^3 g_v}{81\pi^2}(\frac{2m_c}{h^2})^{3/2} [[[\beta_0(E_{FL},\lambda)]^{3/2}]'']] \quad (36)$$

$$\Delta C_{456} = \frac{G_0^3 g_v}{81\pi^2}(\frac{2m_c}{h^2})^{3/2} [[[\tau_0(E_{FL},\lambda)]^{3/2}]'']] \quad (37)$$

$$\Delta C_{456} = \frac{G_0^3 g_v}{81\pi^2}(\frac{2m_c}{h^2})^{3/2} [[[\rho_0(E_{FL},\lambda)]^{3/2}]'']] \quad (38)$$

where the primes denote the differention with respect to Fermi energy.

The ΔC_{44} *and* ΔC_{456} for bulk materials in the absence of any field can be expressed as

$$\Delta C_{44} = \frac{-G_0^2}{9}[\frac{\partial n_0}{\partial E_F}] \quad (39)$$

and

$$\Delta C_{456} = \frac{G_0^3}{27}[\frac{\partial^2 n_0}{\partial E_F{}^2}] \quad (40)$$

where E_F is the corresponding Fermi energy.

Using (39), (40) and (26) and (27) successively, the expressions for ΔC_{44} *and* ΔC_{456} for bulk materials in the absence of any field whose energy band structures in the absence of any field are given by three and two band Kane models can, respectively, be written as

$$\Delta C_{44} = \frac{-G_0^2 g_v}{27\pi^2}\left(\frac{2m_c}{\hbar^2}\right)^{3/2} [[[I_{11}(E_F)]^{3/2}]'] \tag{41}$$

$$\Delta C_{44} = \frac{-G_0^2 g_v}{27\pi^2}\left(\frac{2m_c}{\hbar^2}\right)^{3/2} [[[E_F(1+\alpha E_F)]^{3/2}]'] \tag{42}$$

and

$$\Delta C_{456} = \frac{G_0^3 g_v}{81\pi^2}\left(\frac{2m_c}{\hbar^2}\right)^{3/2} [[[I_{11}(E_F)]^{3/2}]''] \tag{43}$$

$$\Delta C_{456} = \frac{G_0^3 g_v}{81\pi^2}\left(\frac{2m_c}{\hbar^2}\right)^{3/2} [[[E_F(1+\alpha E_F)]^{3/2}]''] \tag{44}$$

Under the conditions $\Delta >> E_g$ *or* $\Delta << E_g$ together with the inequality $\alpha E_F << 1$, the electron concentration in bulk specimens of III-V, ternary and quaternary semiconductors whose energy band structures are defined by two band model of Kane can be expressed in the presences of finite temperature as

$$n_0 = N_c[F_{1/2}(\eta) + (\frac{15ak_BT}{4})F_{3/2}(\eta)] \tag{45}$$

where,

$N_C \equiv g_{v.} 2(\frac{2\pi m_c k_B T}{h^2})^{3/2}$, $\eta \equiv \frac{E_F}{k_B T}$ and $F_j(\eta)$ is the one parameter Fermi-Dirac integral of order j which can be written as

$$F_j(\eta) = (\frac{1}{\Gamma(j+1)})\int_0^{\infty} y^j (1+\exp(y-\eta)^{-1} dy, j > -1 \tag{46}$$

where, $\Gamma(j+1)$ is the complete Gamma function or for all j, analytically continued as a complex integral around the negative axis

$$F_j(\eta) = A_j \int_0^{(0+)} y^j (1+\exp(-y-\eta)^{-1} dy, \tag{47}$$

in which $A_j \equiv \frac{\Gamma(-j)}{2\pi\sqrt{-1}}$

Using (39), (40) and (45) the ΔC_{44} *and* ΔC_{456} under the condition $\alpha E \ll 1$ can be written for two band model of Kane at a finite temperature as

$$\Delta C_{44} = \frac{-G_0^2}{9k_B T} N_c [F_{-1/2}(\eta) + (\frac{15ak_B T}{4}) F_{1/2}(\eta)] \tag{48}$$

$$\Delta C_{456} = \frac{G_0^3}{27(k_B T)^2} N_c [F_{-3/2}(\eta) + (\frac{15ak_B T}{4}) F_{-1/2}(\eta)] \tag{49}$$

Under the condition of non-degeneracy (48) and (49) assume the forms

$$\Delta C_{44} = \frac{-G_0^2}{9k_B T} n_0 \tag{50}$$

$$\Delta C_{456} = \frac{G_0^3}{27(k_B T)^2} n_0 \tag{51}$$

The (50) and (51) are well-known in the literature.

(a) The 1D DR in NWs of HD III-V, ternary and quaternary materials, whose unperturbed band structure is defined by the three band model of Kane in the absence of any field, in the presence of light waves can be expressed as

$$\frac{\hbar^2 (n_z \pi / d_z)^2}{2m_c} + \frac{\hbar^2 (n_y \pi / d_y)^2}{2m_c} + \frac{\hbar^2 k_x^2}{2m_c} = T_1(E, \eta_g, \lambda) \tag{52}$$

Using (16) and summing over both n_y and n_z together with the substitution of $C_{1D} = \pi$, the use of HUP leads to the expression of ES as

$$n_0 = \frac{2g_v}{\pi} \mathrm{Re}\,al\ part\ of \sum_{n_y=1}^{n_{y_{\max}}} \sum_{n_z=1}^{n_{z_{\max}}} [T_{3L1}(E_{F1HDNWL1}, n_y, n_z, \eta_g, \lambda)] \tag{53}$$

where,

$$T_{3L1}(E, n_y, n_z, \eta_g, \lambda) = [[T_1(E, \eta_g, \lambda) - [\frac{\hbar^2 (n_z \pi / d_z)^2}{2m_c} + \frac{\hbar^2 (n_y \pi / d_y)^2}{2m_c}]] \frac{2m_c}{\hbar^2}]^{\frac{1}{2}}$$

and $E_{F1HDNWL1}$ is the Fermi energy in this case.

The electronic contribution to the second- and third- order elastic constants for HD materials in this case can be written as

$$\Delta C_{44} = \frac{-G_0^2}{9 d_y d_z} \mathrm{Re}\,al\ part\ of [\frac{\partial n_0}{\partial (E_{F1HDNWL1} - E'_{3HDNWL1})}] \tag{54}$$

and

$$\Delta C_{456} = \frac{G_0^3}{27 d_y d_z} \mathrm{Re}\, al\ part\ of [\frac{\partial^2 n_0}{\partial (E_{F1HDNWL1} - E_{3HDNWL1}^{'})^2}] \tag{55}$$

Thus Using (54), (55) and (53) we can find $\Delta C_{44}\ and\ \Delta C_{456}$ in this case.

The 1D DR for NWs of III-V materials whose energy band structures are defined by the three-band model of Kane in the absence of band tailing assumes the form

$$\frac{\hbar^2 (n_z \pi / d_z)^2}{2m_c} + \frac{\hbar^2 (n_y \pi / d_y)^2}{2m_c} + \frac{\hbar^2 k_x^2}{2m_c} = \beta_0(E, \lambda) \tag{56}$$

$$f_{12L1}(E, n_y, n_z, \lambda) = [[\beta_0(E, \lambda) - [\frac{\hbar^2 (n_z \pi / d_z)^2}{2m_c} + \frac{\hbar^2 (n_y \pi / d_y)^2}{2m_c}]] \frac{2m_c}{\hbar^2}]^{\frac{1}{2}}$$

The electron statistics per unit length can be written as

$$n_0 = \frac{2g_v}{\pi} \sum_{n_y=1}^{n_{y\max}} \sum_{n_z=1}^{n_{z\max}} [f_{12L1}(E_{F1NWL2}, \mathrm{n}_y, \mathrm{n}_z, \lambda)] \tag{57}$$

where,

$$f_{12L1}(E_{F1NWL2}, n_y, n_z, \lambda) = [[\beta_0(E_{F1NWL2}, \lambda) - [\frac{\hbar^2 (n_z \pi / d_z)^2}{2m_c} + \frac{\hbar^2 (n_y \pi / d_y)^2}{2m_c}]] \frac{2m_c}{\hbar^2}]^{\frac{1}{2}}$$

and E_{F1NWL2} is the Fermi energy in this case.

The electronic contribution to the second- and third- order elastic constants for HD materials in this case can be written as

$$\Delta C_{44} = \frac{-G_0^2}{9d_y d_z}[\frac{\partial n_0}{\partial E_{F1NWL2}}] \tag{58}$$

and

$$\Delta C_{456} = \frac{G_0^3}{27d_y d_z}[\frac{\partial^2 n_0}{\partial E_{F1NWL2}^2}] \tag{59}$$

Thus Using (58), (59) and (57) we can find ΔC_{44} *and* ΔC_{456} in this case.

(b) The 1D DR in NWs of HD III-V, ternary and quaternary materials, whose unperturbed band structure is defined by the two band model of Kane in the absence of any field, in the presence of light waves can be expressed as

$$\frac{\hbar^2(n_z\pi / d_z)^2}{2m_c} + \frac{\hbar^2(n_y\pi / d_y)^2}{2m_c} + \frac{\hbar^2 k_x^2}{2m_c} = T_2(E, \eta_g, \lambda) \tag{60}$$

The electron statistics per unit length can be written as

$$n_0 = \frac{2g_v}{\pi} \sum_{n_y=1}^{n_{y\max}} \sum_{n_z=1}^{n_{z\max}} [T_{3L2}(E_{F1HDNWL2}, n_y, n_z, \eta_g, \lambda)] \tag{61}$$

The electronic contribution to the second- and third- order elastic constants for HD materials in this case can be written as

$$\Delta C_{44} = \frac{-G_0^2}{9d_y d_z}[\frac{\partial n_0}{\partial(E_{F1HDNWL2} - E'_{3HDNWL2})}] \tag{62}$$

and

$$\Delta C_{456} = \frac{G_0^3}{27d_y d_z}[\frac{\partial^2 n_0}{\partial(E_{F1HDNWL2} - E'_{3HDNWL2})^2}] \quad (63)$$

Thus using (62), (63) and (61) we can find ΔC_{44} *and* ΔC_{456} in this case.

The 1D DR, for NWs of III-V materials whose energy band structures are defined by the two-band model of Kane in the absence of band tailing assumes the form

$$\frac{\hbar^2(n_z\pi / d_z)^2}{2m_c} + \frac{\hbar^2(n_y\pi / d_y)^2}{2m_c} + \frac{\hbar^2 k_x^2}{2m_c} = \tau_0(E,\lambda) \quad (64)$$

The electron statistics per unit length can be written as

$$n_0 = \frac{2g_v}{\pi}\sum_{n_y=1}^{n_{y_{max}}}\sum_{n_z=1}^{n_{z_{max}}}[f_{12L2}(E_{F1NWL21}, n_y, n_z, \lambda)] \quad (65)$$

where

$$f_{12L2}(E_{F1NWL21}, n_y, n_z, \lambda) = [[\tau_0(E_{F1NWL21}, \lambda) - [\frac{\hbar^2(n_z\pi / d_z)^2}{2m_c} + \frac{\hbar^2(n_y\pi / d_y)^2}{2m_c}]]\frac{2m_c}{\hbar^2}]^{\frac{1}{2}}$$

and $E_{F1NWL21}$ *is the Fermi energy in this case.*

The electronic contribution to the second- and third- order elastic constants for HD materials in this case can be written as

$$\Delta C_{44} = \frac{-G_0^2}{9d_y d_z}[\frac{\partial n_0}{\partial E_{F1NWL21}}] \quad (66)$$

and

$$\Delta C_{456} = \frac{G_0^3}{27 d_y d_z}[\frac{\partial^2 n_0}{\partial E_{F1NWL21}^2}] \quad (67)$$

Thus Using (66), (67) and (65) we can find ΔC_{44} *and* ΔC_{456} in this case.

(c) The 1D DR in NWs of HD III-V, ternary and quaternary materials, whose unperturbed band structure is defined by the parabolic energy bands in the absence of any field, in the presence of light waves can be expressed as

$$\frac{\hbar^2 (n_z \pi / d_z)^2}{2m_c} + \frac{\hbar^2 (n_y \pi / d_y)^2}{2m_c} + \frac{\hbar^2 k_x^2}{2m_c} = T_3(E, \eta_g, \lambda) \quad (68)$$

The electron statistics per unit length can be written as

$$n_0 = \frac{2g_v}{\pi} \sum_{n_y=1}^{n_{y\max}} \sum_{n_z=1}^{n_{z\max}} [T_{3L3}(E_{F1HDNWL3}, n_y, n_z, \eta_g, \lambda)] \quad (69)$$

where

$$T_{3L3}(E_{F1HDNWL3}, n_y, n_z, \eta_g, \lambda) = [[T_3(E_{F1HDNWL3}, \eta_g, \lambda) -$$

$$-[\frac{\hbar^2 (n_z \pi / d_z)^2}{2m_c} + \frac{\hbar^2 (n_y \pi / d_y)^2}{2m_c}]] \frac{2m_c}{\hbar^2}]^{\frac{1}{2}}$$

and $E_{F1HDNWL3}$ *is the Fermi energy in this case*

The electronic contribution to the second- and third- order elastic constants for HD materials in this case can be written as

$$\Delta C_{44} = \frac{-G_0^2}{9 d_y d_z}[\frac{\partial n_0}{\partial (E_{F1HDNWL3} - E'_{3HDNWL3})}] \quad (70)$$

and

$$\Delta C_{456} = \frac{G_0^3}{27d_y d_z}[\frac{\partial^2 n_0}{\partial(E_{F1HDNWL3} - E'_{3HDNWL3})^2}] \tag{71}$$

Thus Using (70), (71) and (69) we can find ΔC_{44} *and* ΔC_{456} in this case.

The 1D DR for NWs of III-V materials whose energy band structures are defined by the parabolic energy bands in the absence of band tailing assumes the form

$$\frac{\hbar^2(n_z\pi / d_z)^2}{2m_c} + \frac{\hbar^2(n_y\pi / d_y)^2}{2m_c} + \frac{\hbar^2 k_x^2}{2m_c} = \rho_0(E,\lambda) \tag{72}$$

The electron statistics per unit length can be written as

$$n_0 = \frac{2g_v}{\pi}\sum_{n_y=1}^{n_{y_{\max}}}\sum_{n_z=1}^{n_{z_{\max}}}[f_{12L3}(E_{F1NWL3}, n_y, n_z, \lambda)] \tag{73}$$

The electronic contribution to the second- and third- order elastic constants for HD materials in this case can be written as

$$\Delta C_{44} = \frac{-G_0^2}{9d_y d_z}[\frac{\partial n_0}{\partial E_{F1NWL3}}] \tag{74}$$

and

$$\Delta C_{456} = \frac{G_0^3}{27d_y d_z}[\frac{\partial^2 n_0}{\partial E_{F1NWL3}^2}] \tag{75}$$

Thus Using (74), (75) and (73) we can find ΔC_{44} *and* ΔC_{456} in this case.

Suggestion for the Experimental Determination of CECs

The magnitude of the thermoelectric power G can be written as [46]

$$G = \frac{1}{|e| T n_0} \int_{-\infty}^{\infty} (E - E_F) R(E) [-\frac{\partial f_0}{\partial E}] dE \tag{76}$$

where, $R(E)$ is the total number of states.

The (76) can be written under the condition of carrier degeneracy [45] as

$$G = (\frac{\pi k_B^2 T}{3|e| n_0})(\frac{\partial n_0}{\partial E_F}) \tag{77}$$

For inversion layers and NIPI structures, under the condition of electric quantum limit, (77) assumes the form

$$G = \frac{\pi^2 k_B^2 T}{3 e n_{02D}} [\frac{\partial n_{02D}}{\partial (E_{F2D} - E_{02D})}] \tag{78}$$

where n_{02D}, E_{F2D} and E_{02D} are the surface electron concentration, the Fermi energy and the sub band energy for the said 2D systems at the electric quantum limit

For heavily doped semiconductors (76) assumes the form

$$G = \frac{\pi^2 k_B^2 T}{3 e n_{0HD}} [\frac{\partial n_{0HD}}{\partial (E_{FHD} - E_{0HD})}] \tag{79}$$

where n_{0HD}, E_{FHD} and E_{0HD} are the electron concentration, the Fermi energy and the band tail energy for the heavily doped semiconductors where E_{0HD} should be obtained from the heavily doped dispersion relation of the semiconductor under the conditions $E = E_{0HD}$ and $k = 0$.

The knowledge of the carrier contribution to the elastic constants is important in studying the mechanical properties of the materials and has been investigated in the literature [60-63]. The electronic contribution to the second- and third- order elastic constants for HD materials can be written as [60-63]

$$\Delta C_{44} = \frac{-G_0^2}{9}[\frac{\partial n_{0HD}}{\partial(E_{FHD} - E_{0HD})}] \tag{80}$$

$$\Delta C_{456} = \frac{G_0^3}{27}[\frac{\partial^2 n_{0HD}}{\partial(E_{FHD} - E_{0HD})^2}] \tag{81}$$

where G_0 is the deformation potential constant. Thus, using (79), (80) and (81), we can write

$$\Delta C_{44} = [\frac{-n_0(\bar{G}_0)^2 |e| G}{(3\pi^2 k_B^2 T)}] \tag{82}$$

and

$$\Delta C_{456} = (\frac{n_0 |e| (\bar{G}_0)^3 G^2}{(3\pi^4 k_B^3 T)})(1 + \frac{n_0}{G}\frac{\partial G}{\partial n_0}) \tag{83}$$

Therefore, by using (82) and (83) we can investigate ΔC_{44} and ΔC_{456} for all the cases of this monograph. Besides, the experimental graph of G

versus n_0 allows us to determine the electronic contribution to the elastic constants for materials having arbitrary spectra.

RESULTS AND DISCUSSION

Using the appropriate equations we have plotted the normalized ΔC_{44} and ΔC456 as functions of electron concentration n0 (for a given wave to 1.4 and 1.5 to 1.8 by taking HD specimens of n-InAs ($E_{g_o} = 0.36eV, \Delta = 0.43eV, m^* = 0.026m_0, g_v = 1, \varepsilon_{sc} = 12.25\varepsilon_0[13]), n-$ $- InSb(E_{g_0} = 0.2352eV,\ \Delta = 0.81eV, m^* = 0.01359m_0, g_v = 1,$ length where we have considered red light for which $\lambda = 660nm and\ I =$ $= 1nWm^{-}2)\ at\ T = 4.2K$ in Figures 10.1 $\varepsilon_{sc} = 15.56\varepsilon_0[13],[n - Hg_{1-x}Cd_xTe$ $(E_{go} = -0.302 + 1.93x + 5.35\times10^{-4}(1-2x)T - 0.810x^2 + 0.832x^3)eV, \Delta =$ $= (0.63 + 0.24x - 0.27x^2)eV,\ m^* = 0.1m_0E_{go}(eV)^{-1},\ g_v = 1, \varepsilon_x = [20.262 - 14.812x +$ $+ 5.22795x^2]\varepsilon_0[13])$ $and\ n - In_{1-x}Ga_xAs_yP_{1-y}$ $n - In_1\_xGa_xAs_yP_{1-y}$ *lattice matched to* $InP(E_{g_0} = 1.337$ $(-0.73y + 0.13y^2)eV$, $\Delta = (0.114 + 0.26y - 0.22y^2)eV$, $m^* = (0.08 - 0.039y)m_0$ $y = (0.1896 - 0.4052x)/(0.1896 - 0.0123x)$, $g_v = 1$, $\varepsilon_{sc} = [10.65 + 0.1320y]\varepsilon_0$) [13] in accordance with the perturbed three and two band models of Kane and that of perturbed parabolic energy bands respectively. In Figures 10.9-10.12 and 10.13-10.16 we have plotted the normalized ΔC44 and ΔC456 as functions of I, taking $\lambda = 660nm$ and $n_0 = 1\times10^{25}m^{-3}$ for the purpose of numerical computations. In Figures 10.17-10.20 and 10.21-10.24 we have plotted the normalized ΔC44 and ΔC456 as functions of the wave length. In Figures 10.25 and 10.26, we have plotted the normalized ΔC44 and ΔC456 as functions of alloy composition for HD n-Hg1-x CdxTe while the Figures 1.27 and 1.28 exhibit the same as functions of y for HD n-In1-x Gax Asy P1-y lattice matched to InP respectively.

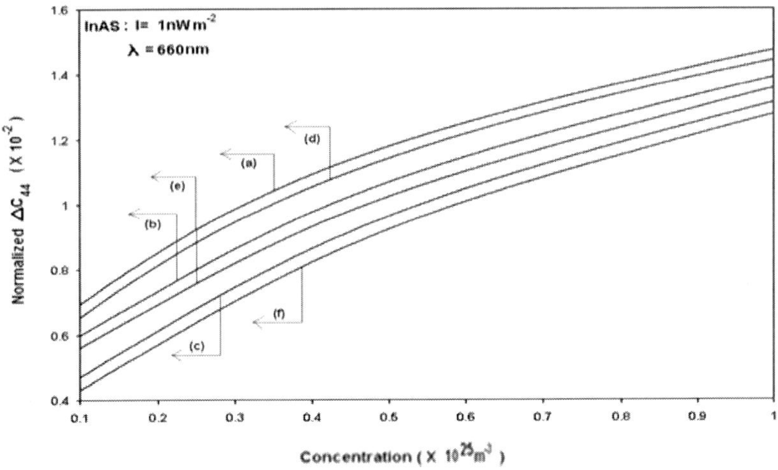

Figure 10.1. Plot of the normalized ΔC_{44} as a function of electron concentration for HD n-InAs in the presence of light waves in which the curves (a), (b) and (C) represent the perturbed three and two band models of Kane and that of the perturbed parabolic energy bands respectively. The curves (d), (e) and (f) represent the same in the absence of external photo-excitation.

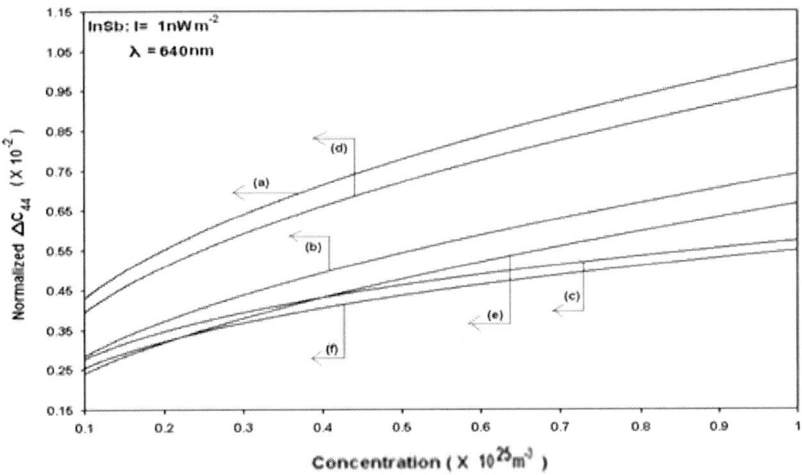

Figure 10.2. Plot of the normalized ΔC_{44} as a function of electron concentration for HD n-InSb in the presence of light waves in which the curves (a) (b) and (C) represent the perturbed three and two band models of Kane and that of the perturbed parabolic energy bands respectively. The curves (d), (e) and (f) represent the same in the absence of external photo-excitation.

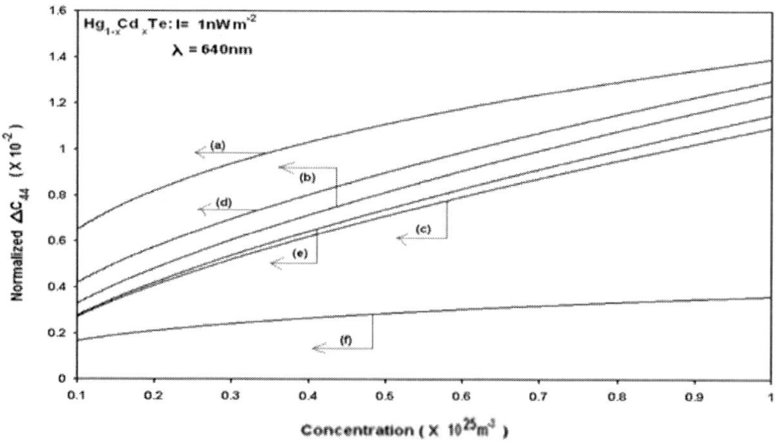

Figure 10.3. Plot of the normalized ΔC_{44} as a function of electron concentration for HD n-$Hg_{1-x}Cd_xTe$ in the presence of light waves in which the curves (a) (b) and (C) represent the perturbed three and two band models of Kane and that of the perturbed parabolic energy bands respectively. The curves (d), (e) and (f) represent the same in the absence of external photo-excitation.

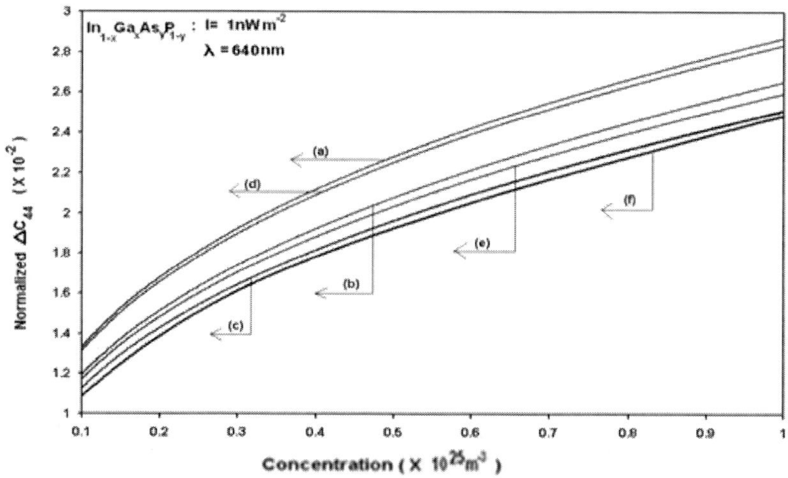

Figure 10.4. Plot of the normalized ΔC_{44} as a function of electron concentration for HD n-$In_{1-x}\,Ga_x\,As_y\,P_{1-y}$ in the presence of light waves in which the curves (a) (b) and (C) represent the perturbed three and two band models of Kane and that of the perturbed parabolic energy bands respectively. The curves (d), (e) and (f) represent the same in the absence of external photo-excitation.

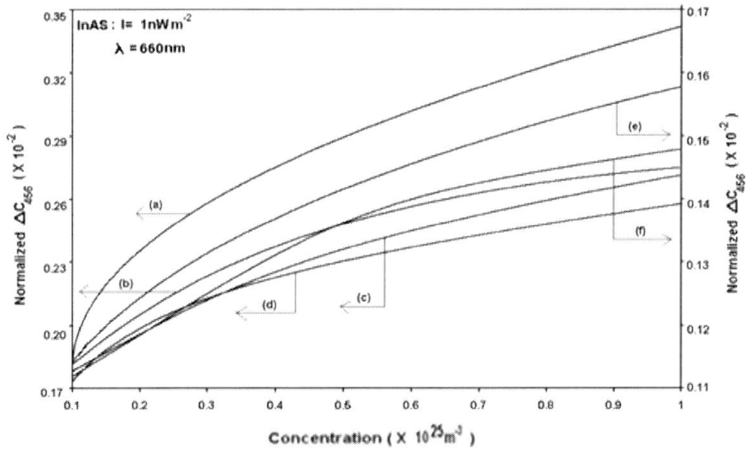

Figure 10.5. Plot of the normalized ΔC_{456} as a function of electron concentration for HD n-InAs in the presence of light waves in which the curves (a) (b) and (C) represent the perturbed three and two band models of Kane and that of the perturbed parabolic energy bands respectively. The curves (d), (e) and (f) represent the same in the absence of external photo-excitation.

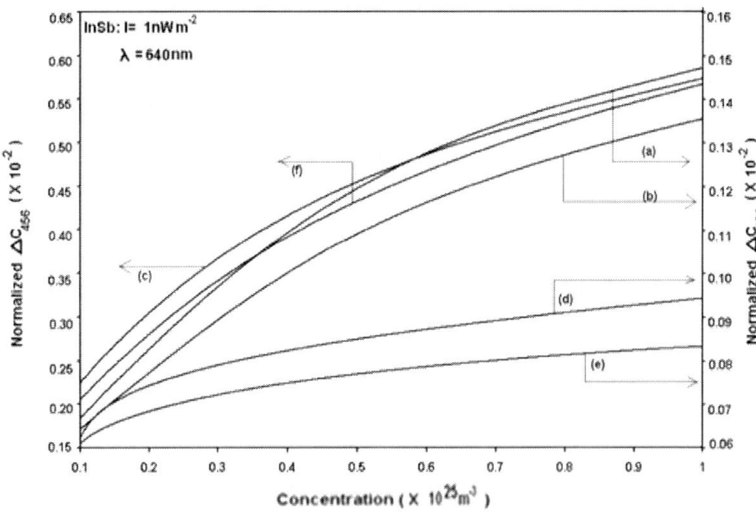

Figure 10.6. Plot of the normalized ΔC_{456} as a function of electron concentration for HD n-InSb in the presence of light waves in which the curves (a) (b) and (C) represent the perturbed three and two band models of Kane and that of the perturbed parabolic energy bands respectively. The curves (d), (e) and (f) represent the same in the absence of external photo-excitation.

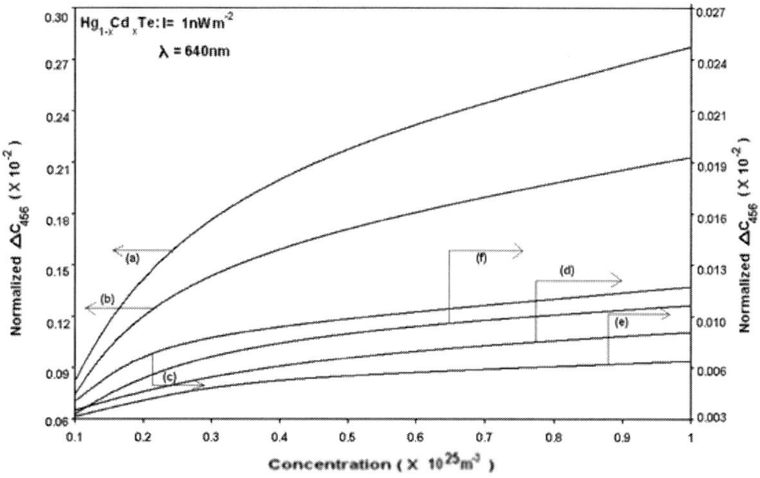

Figure 10.7. Plot of the normalized ΔC_{456} as a function of electron concentration for HD n-Hg_{1-x} Cd_xTe in the presence of light waves in which the curves (a) (b) and (C) represent the perturbed three and two band models of Kane and that of the perturbed parabolic energy bands respectively. The curves (d), (e) and (f) represent the same in the absence of external photo-excitation.

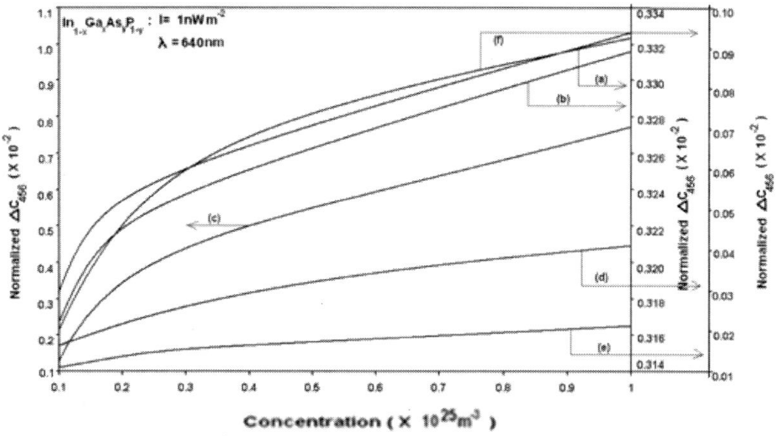

Figure 10.8. Plot of the normalized ΔC_{456} as a function of electron concentration for HD n-In_{1-x} Ga_x As_y P_{1-y} in the presence of light waves in which the curves (a) (b) and (C) represent the perturbed three and two band models of Kane and that of the perturbed parabolic energy bands respectively. The curves (d), (e) and (f) represent the same in the absence of external photo-excitation.

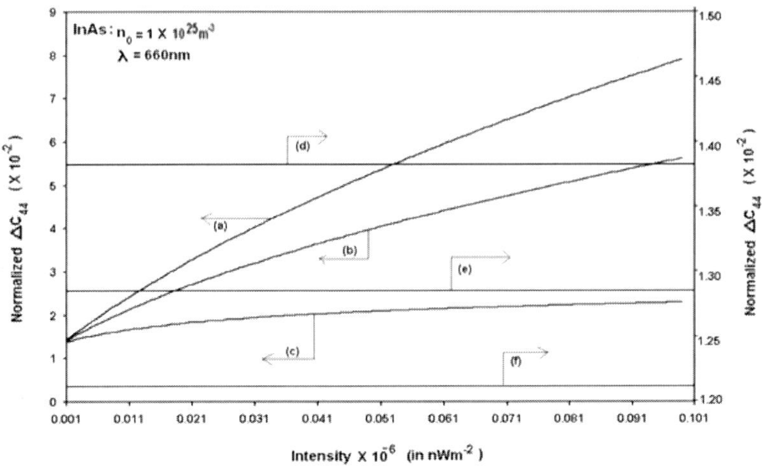

Figure 10.9. Plot of the normalized ΔC_{44} as a function of light intensity for HD n-InAs in which the curves (a) (b) and (C) represent the perturbed three and two band models of Kane and that of the perturbed parabolic energy bands respectively. The curves (d), (e) and (f) represent the same in the absence of external photo-excitation.

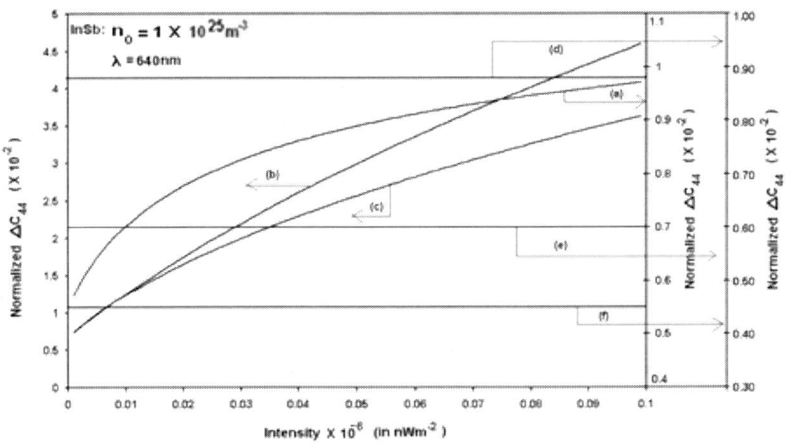

Figure 10.10. Plot of the normalized ΔC_{44} as a function of light intensity for HD n-InSb in which the curves (a) (b) and (C) represent the perturbed three and two band models of Kane and that of the perturbed parabolic energy bands respectively. The curves (d), (e) and (f) represent the same in the absence of external photo-excitation.

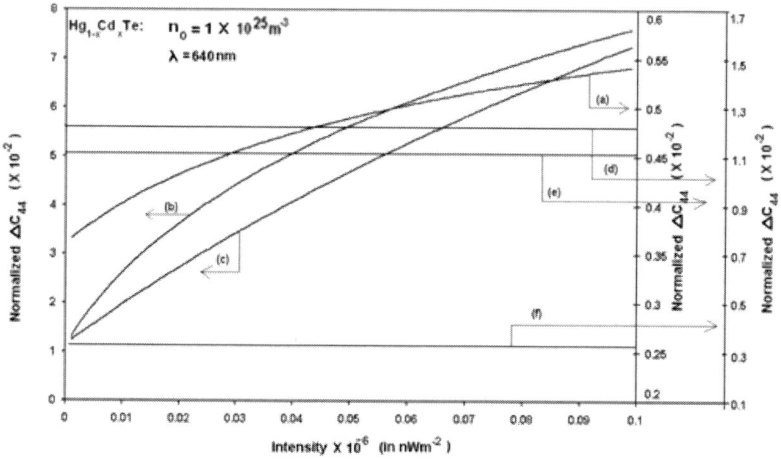

Figure 10.11. Plot of the normalized ΔC_{44} as a function of light intensity for HD n-$Hg_{1-x}Cd_xTe$ in which the curves (a) (b) and (C) represent the perturbed three and two band models of Kane and that of the perturbed parabolic energy bands respectively. The curves (d), (e) and (f) represent the same in the absence of external photo-excitation.

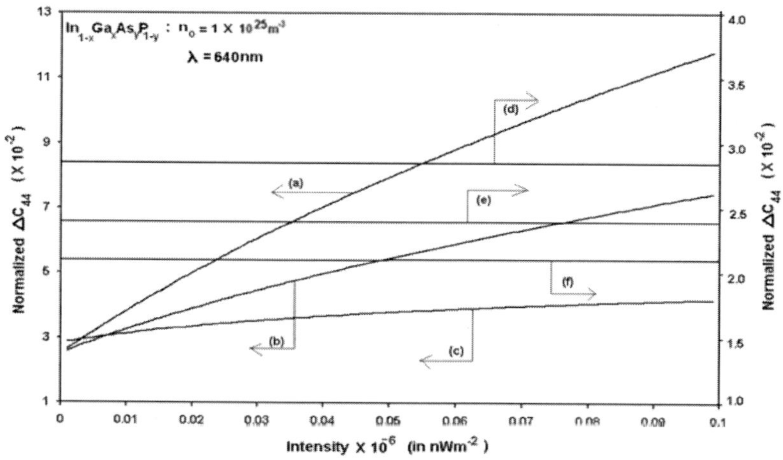

Figure 10.12. Plot of the normalized ΔC_{44} as a function of light intensity for HD n-$In_{1-x}Ga_x\,As_yP_{1-y}$ in which the curves (a) (b) and (C) represent the perturbed three and two band models of Kane and that of the perturbed parabolic energy bands respectively. The curves (d), (e) and (f) represent the same in the absence of external photo-excitation.

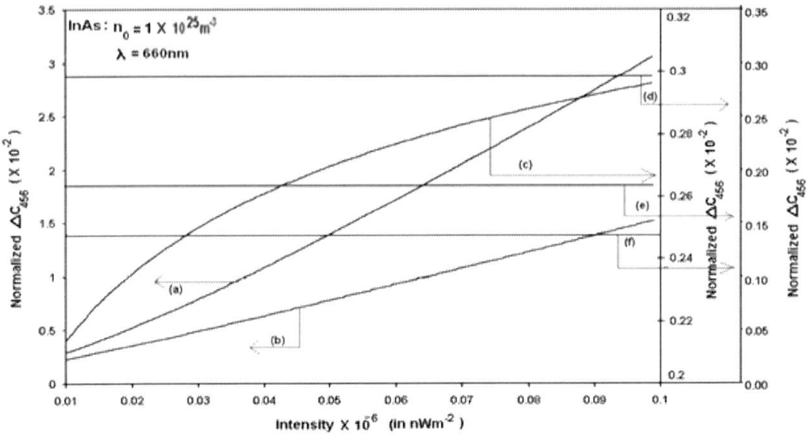

Figure 10.13. Plot of the normalized ΔC_{456} as a function of light intensity for HD n-InAs in which the curves (a) (b) and (C) represent the perturbed three and two band models of Kane and that of the perturbed parabolic energy bands respectively. The curves (d), (e) and (f) represent the same in the absence of external photo-excitation.

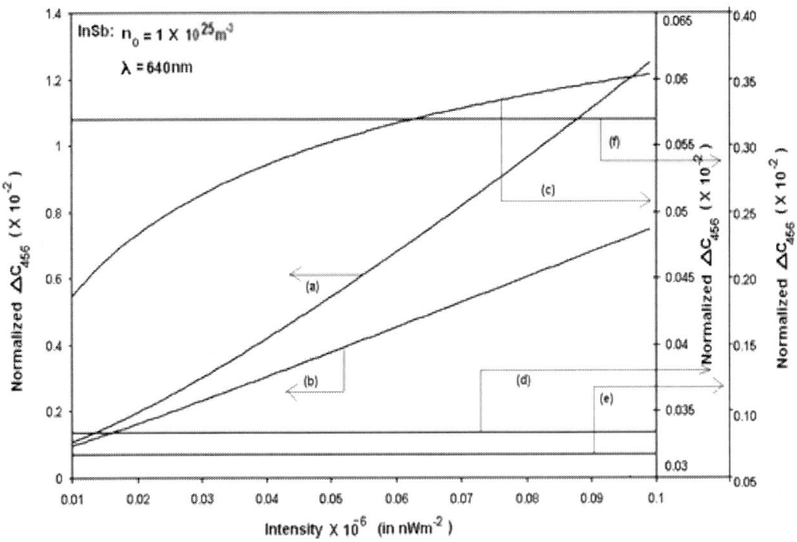

Figure 10.14. Plot of the normalized ΔC_{456} as a function of light intensity for HD n-InSb in which the curves (a) (b) and (C) represent the perturbed three and two band models of Kane and that of the perturbed parabolic energy bands respectively. The curves (d), (e) and (f) represent the same in the absence of external photo-excitation.

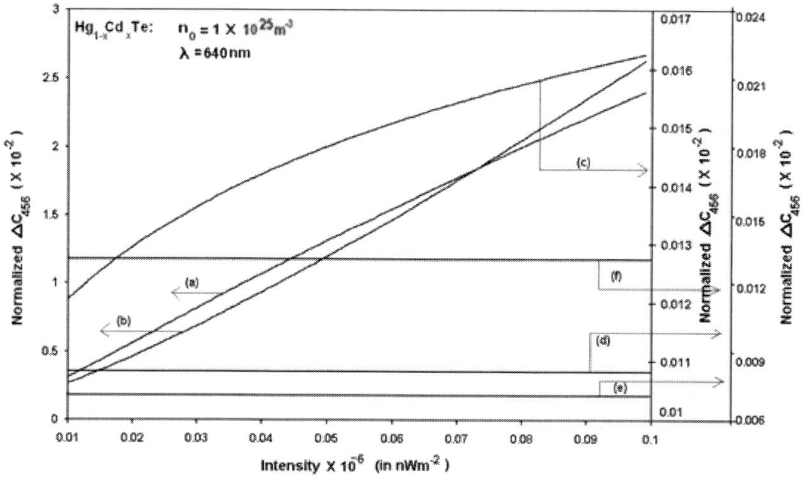

Figure 10.15. Plot of the normalized ΔC_{456} as a function of light intensity for HD n-$Hg_{1-x}Cd_xTe$ in which the curves (a) (b) and (C) represent the perturbed three and two band models of Kane and that of the perturbed parabolic energy bands respectively. The curves (d), (e) and (f) represent the same in the absence of external photo-excitation.

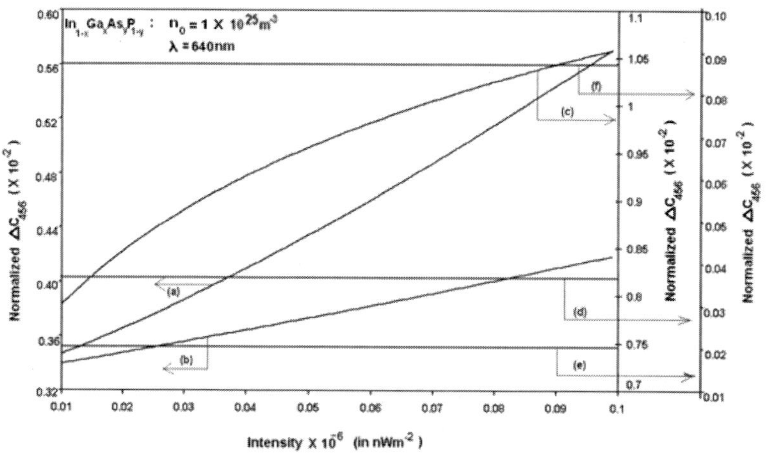

Figure 10.16. Plot of the normalized ΔC_{456} as a function of light intensity for HD n-$In_{1-x}Ga_x As_y P_{1-y}$ in which the curves (a) (b) and (C) represent the perturbed three and two band models of Kane and that of the perturbed parabolic energy bands respectively. The curves (d), (e) and (f) represent the same in the absence of external photo-excitation.

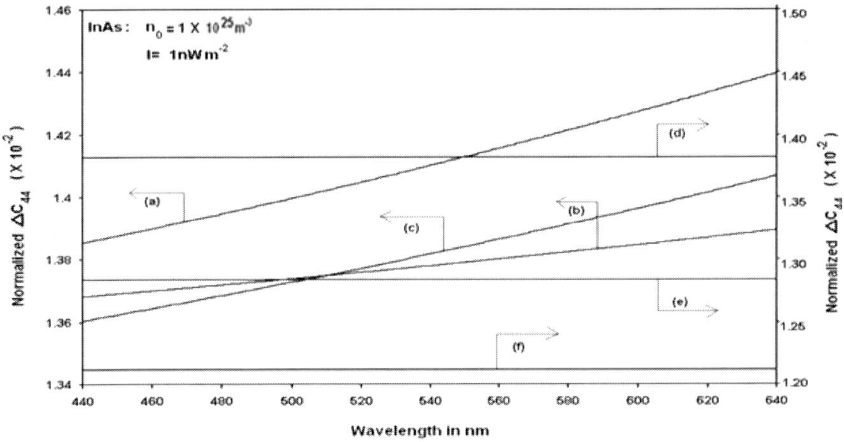

Figure 10.17. Plot of the normalized ΔC_{44} as a function of Wave length for HD n-InAs in which the curves (a) (b) and (C) represent the perturbed three and two band models of Kane and that of the perturbed parabolic energy bands respectively. The curves (d), (e) and (f) represent the same in the absence of external photo-excitation.

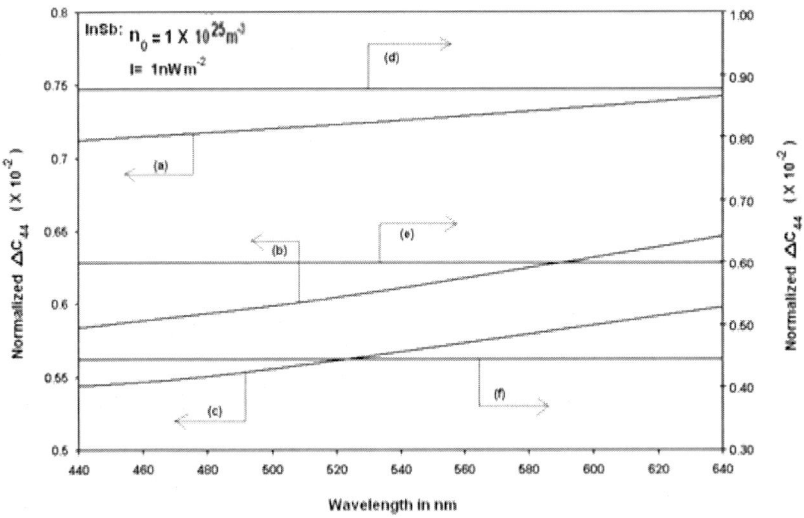

Figure 10.18. Plot of the normalized ΔC_{44} as a function of Wave length for HD n-InSb in which the curves (a) (b) and (C) represent the perturbed three and two band models of Kane and that of the perturbed parabolic energy bands respectively. The curves (d), (e) and (f) represent the same in the absence of external photo-excitation.

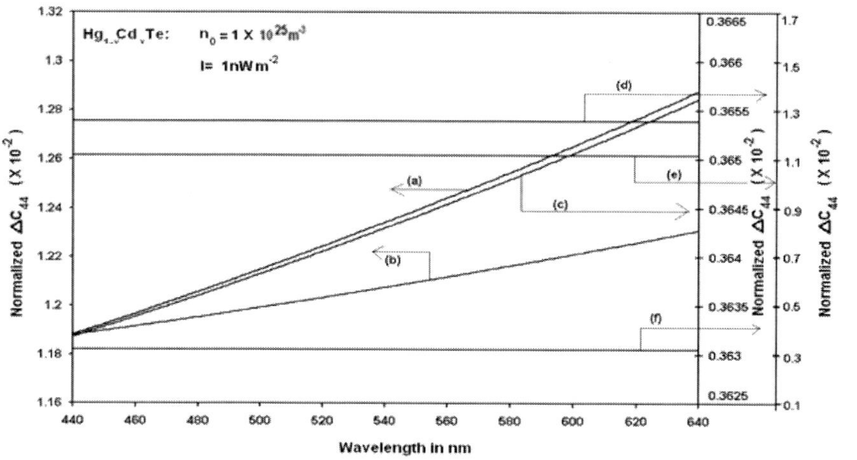

Figure 10.19. Plot of the normalized ΔC_{44} as a function of Wave length for HD n-$Hg_{1-x}Cd_xTe$ in which the curves (a) (b) and (C) represent the perturbed three and two band models of Kane and that of the perturbed parabolic energy bands respectively. The curves (d), (e) and (f) represent the same in the absence of external photo-excitation.

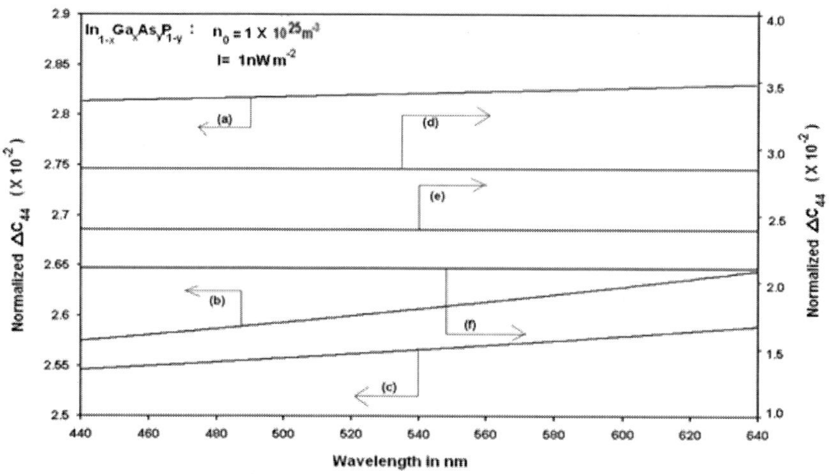

Figure 10.20. Plot of the normalized ΔC_{44} as a function of Wave length for HD n-$In_{1-x}Ga_x As_y P_{1-y}$ in which the curves (a) (b) and (C) represent the perturbed three and two band models of Kane and that of the perturbed parabolic energy bands respectively. The curves (d), (e) and (f) represent the same in the absence of external photo-excitation.

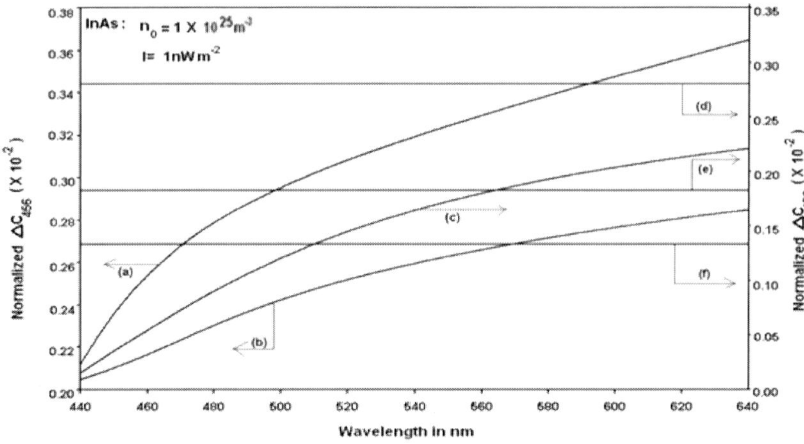

Figure 10.21. Plot of the normalized ΔC_{456} as a function of Wave length for HD n-InAs in the presence of light waves in which the curves (a) (b) and (C) represent the perturbed three and two band models of Kane and that of the perturbed parabolic energy bands respectively. The curves (d), (e) and (f) represent the same in the absence of external photo-excitation.

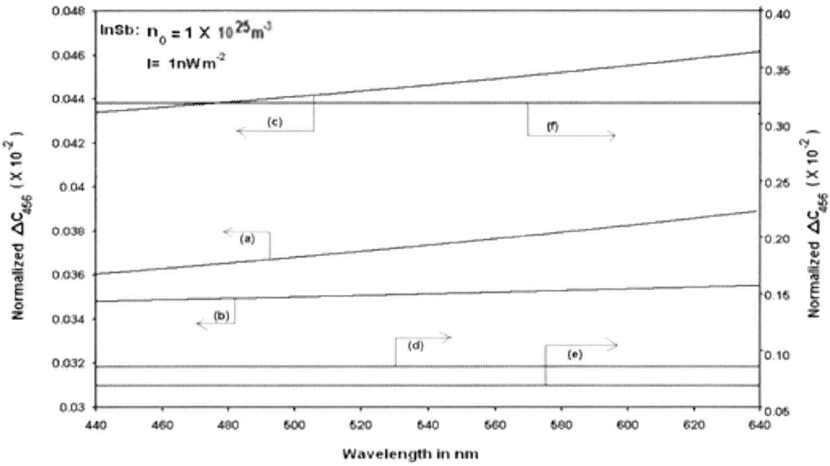

Figure 10.22. Plot of the normalized ΔC_{456} as a function of Wave length for HD n-InSb in the presence of light waves in which the curves (a) (b) and (C) represent the perturbed three and two band models of Kane and that of the perturbed parabolic energy bands respectively. The curves (d), (e) and (f) represent the same in the absence of external photo-excitation.

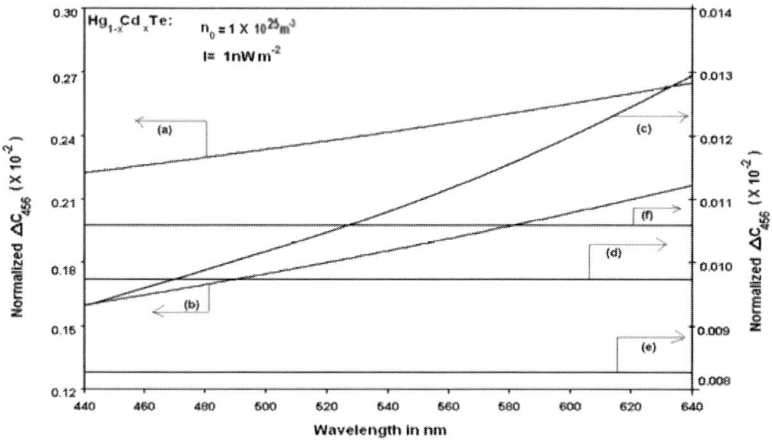

Figure 10.23. Plot of the normalized ΔC_{456} as a function of Wave length for HD n-$Hg_{1-x}Cd_xTe$ in the presence of light waves in which the curves (a) (b) and (C) represent the perturbed three and two band model of Kane and that of the perturbed parabolic energy bands respectively. The curves (d), (e) and (f) represent the same in the absence of external photo-excitation.

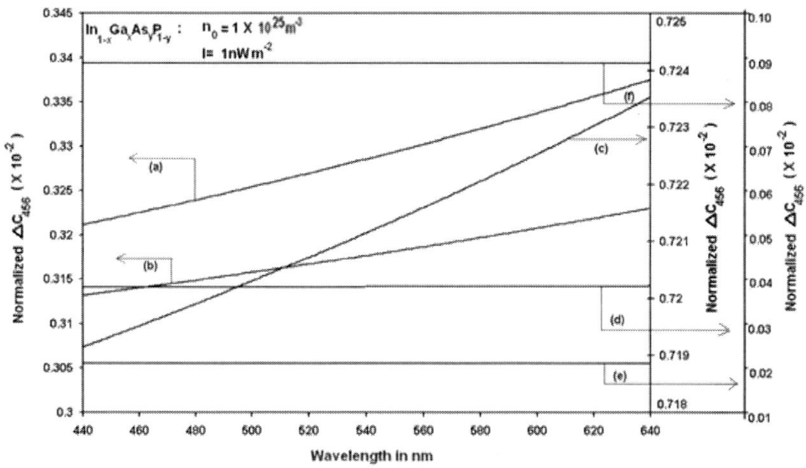

Figure 10.24. Plot of the normalized ΔC_{456} as a function of Wave length for HD n-$In_{1-x}Ga_x As_y P_{1-y}$ in the presence of light waves in which the curves (a) (b) and (C) represent the perturbed three and two band models of Kane and that of the perturbed parabolic energy bands respectively. The curves (d), (e) and (f) represent the same in the absence of external photo-excitation.

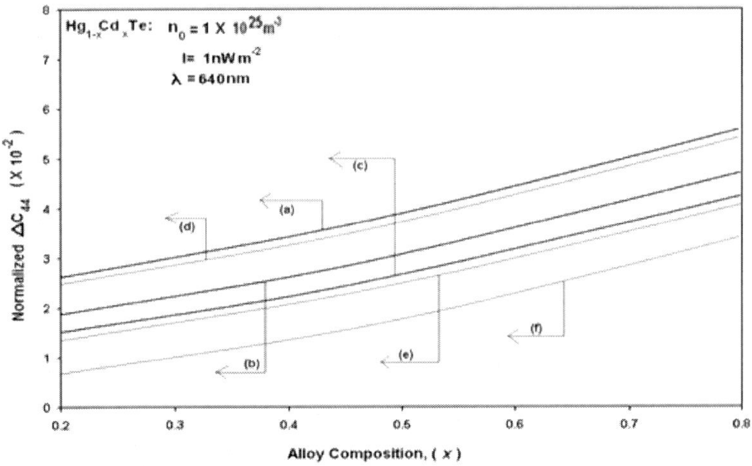

Figure 10.25. Plot of the normalized ΔC_{44} as a function of alloy composition for HD n-$Hg_{1-x}Cd_xTe$ in which the curves (a) (b) and (C) represent the perturbed three and two band models of Kane and that of the perturbed parabolic energy bands respectively. The curves (d), (e) and (f) represent the same in the absence of external photo-excitation.

From Figures 10.1-10.4 and 10.1-10.4, we observe that both ΔC_{44} and ΔC_{456} increase with increasing electron concentration and the numerical values of them in the presence of light waves for all the materials are large as compared with I = 0. The combined influence of the energy band constants on ΔC_{44} and ΔC_{456} for HD n-InAs and HD n- InSb can easily be assessed from Figures 10.1 and 10.2 and 10.1 and 10.2. For the purpose of relative assessment, all the plots in the absence of light waves have further been drawn. From Figures 1.10-10.12 and 10.13-10.16 we can write that both ΔC_{44} and ΔC_{456} increase with increasing light intensity for all the materials. It appears from the Figures 10.17-10.20 and 10.21-10.24 that ΔC_{44} and ΔC_{456} increases as the wave length shifts from red to violet color. The plots of the Figures 10.25 and 10.26 are valid for x>0.17, since for x<0.17 the band gap becomes negative in HD n-$Hg_{1-x}Cd\ _xTe$ leading to semi-metallic state. The plots of the Figure 10.27 and 10.28exhibits the variation of the normalized ΔC_{44} and ΔC_{456} with the alloy composition y, for HD n-$In\ _{1-}xGa_xAs_yP_{1-y}$ lattice matched to InP.

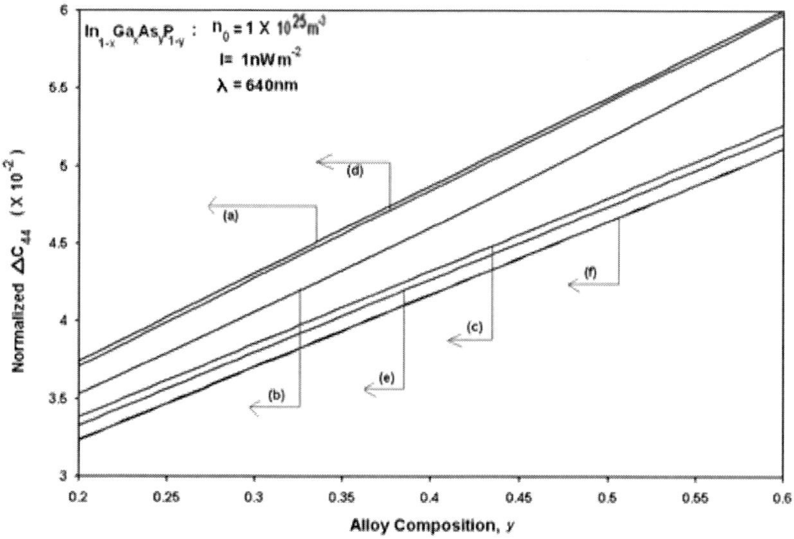

Figure 10.26. Plot of the normalized ΔC_{44} as a function of Wave length for HD n-$In_{1-x}Ga_x As_yP_{1-y}$ in which the curves (a) (b) and (C) represent the perturbed three and two band models of Kane and that of the perturbed parabolic energy bands respectively. The curves (d), (e) and (f) represent the same in the absence of external photo-excitation.

The influence of light is immediately apparent from the plots in the Figures 10.9-10.24, since ΔC_{44} and ΔC_{456} depend strongly on I and λ which is in direct contrast as compared with I = 0. The variations of ΔC_{44} and ΔC_{456} in the Figures 10.9-10.24 reflects the direct signature of the light wave on the electronic, optic and the other band structure-dependent properties of HD semiconducting materials in the presence of light waves and the photon assisted transport for the corresponding semiconductor devices. Although both ΔC_{44} and ΔC_{456} tend to increase with intensity and the wave length, the rate of increase is totally band structure dependent. We wish to note that in view of large changes of the elastic constants with n_0 detailed experimental work on second and third-order elastic constants as functions of n_0 would be interesting for the present case. It may be suggested that the experiments on the velocity of sound involving the shear mode as function of n_0 may exhibit the carrier contribution to the elastic constants for materials having arbitrary carrier energy spectra. This fact again suggests another experimental determination of ΔC44 and ΔC_{456} beside the suggested methods of determining them as given by Eqs, (82) and (83) respectively.

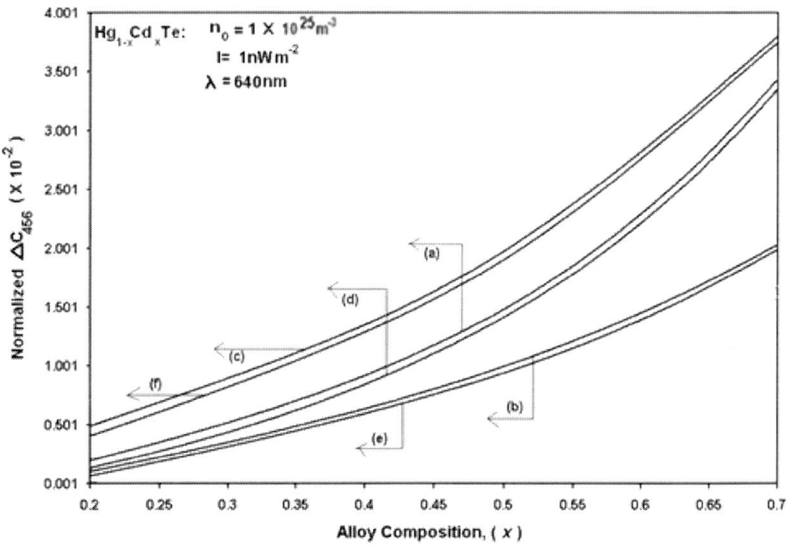

Figure 10.27. Plot of the normalized ΔC_{456} as a function of alloy composition for HD n-$Hg_{1-x}Cd_xTe$ in the presence of light waves in which the curves (a) (b) and (C) represent the perturbed three and two band model of Kane and that of the perturbed parabolic energy bands respectively. The curves (d), (e) and (f) represent the same in the absence of external photo-excitation.

From all the Figures of (10.1) to (10.28), we observe that both the elastic constants increases from the light off case to the light on case, since the value of the Fermi energy in the presence of light waves becomes larger due to the increase in the carrier concentration as compared with the same as in the absence of photo-excitation. Therefore the numerical magnitude of ΔC_{44} and ΔC_{456} in the presence of light is larger as compared with the same in the light off case in the whole range of the concentration considered, although the same increases with increase in carrier degeneracy.

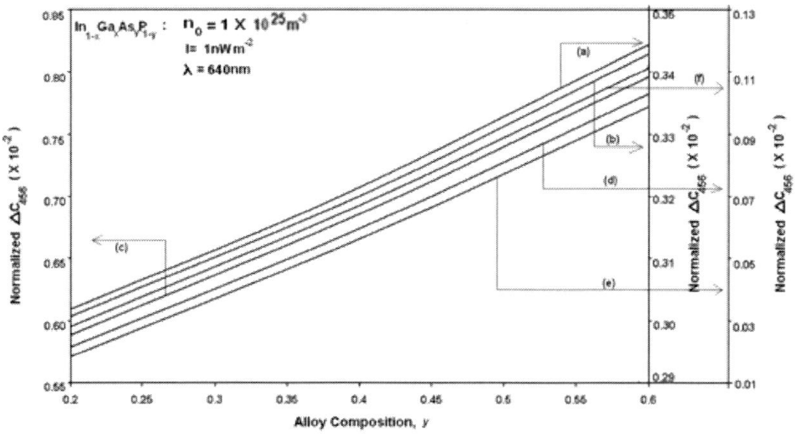

Figure 10.28. Plot of the normalized ΔC_{456} as a function of alloy composition for HD n-$In_{1-x}Ga_xAs_yP_{1-y}$ in the presence of light waves in which the curves (a) (b) and (C) represent the perturbed three and two band model of Kane and that of the perturbed parabolic energy bands respectively. The curves (d), (e) and (f) represent the same in the absence of external photo-excitation.

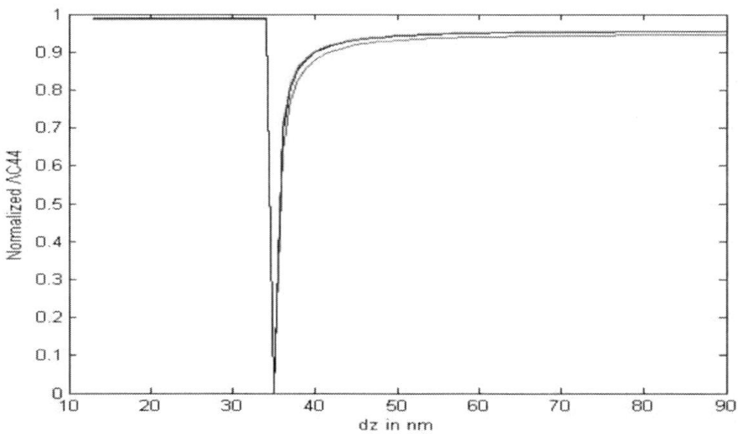

Figure 10.29. Plot of the normalized ΔC_{44}as a function of d_z in the presence of light waves in Nano Wires of HD n-InAs (red), HD $Hg_{1-x}Cd_xTe$ (green)and HD $In_{1-x}Ga_xAs_yP_{1-y}$ lattice matched to InP (black) where the unperturbed electrons obey the three band model of Kane. ($\lambda = 660$ nm, $I = 1nWm^{-2}$, $n_o = 10^{11}$ m^{-1} and $d_y = 10$ nm)

It is worth remarking that our basic equations.(10.8) to (10.10) cover various materials having different energy band structures. In this chapter, the concentration, the light intensity, the wave length and the alloy composition dependences of the CECs' for bulk specimens of HD n-InAs, HD n- InSb, HD n-$Hg_{1-x}Cd_xTe$ and HD n-$In_{1-x}Ga_xAs_yP_{1-y}$ lattice matched to InP have been studied in Figures (10.1) to (10.28). Thus we have covered a wide class of compounds whose energy band structures are defined by the three and two band model of Kane in the absence of photon field. Under certain limiting conditions all the results of ΔC_{44} and ΔC_{456} lead to the well-known expressions of ΔC_{44} and ΔC_{456} for non-degenerate compounds having parabolic energy bands. This indirect test not only exhibits the mathematical compatibility of our formulation but also shows the fact that our simple analysis is a more generalized one, since one can obtain the corresponding results for the relatively wide gap materials having parabolic energy bands under certain limiting conditions from our present derivation.

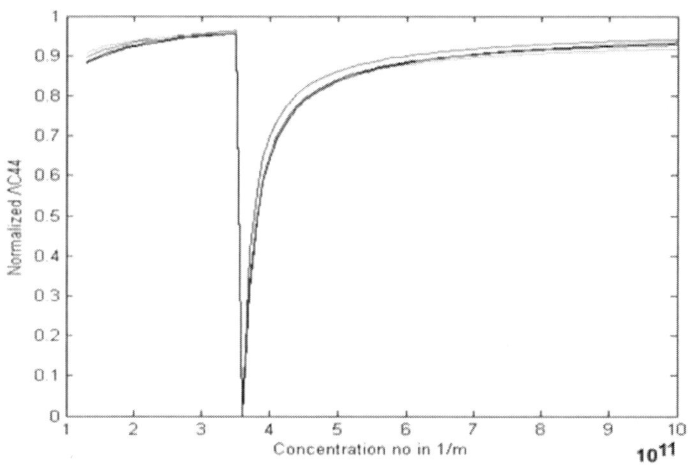

Figure 10.30. Plot of the normalized ΔC_{44} as a function of electron concentration in the presence of light waves in Nano Wires of HD n-InAs (red), HD $Hg_{1-x}Cd_xTe$ (green)and HD $In_{1-x}Ga_xAs_yP_{1-y}$ lattice matched to InP (black) where the unperturbed electrons obey the three band model of Kane. ($\lambda = 660$ nm, $I = 1nWm^{-2}$, $d_z = 10$ nm, $d_y = 10$ nm)

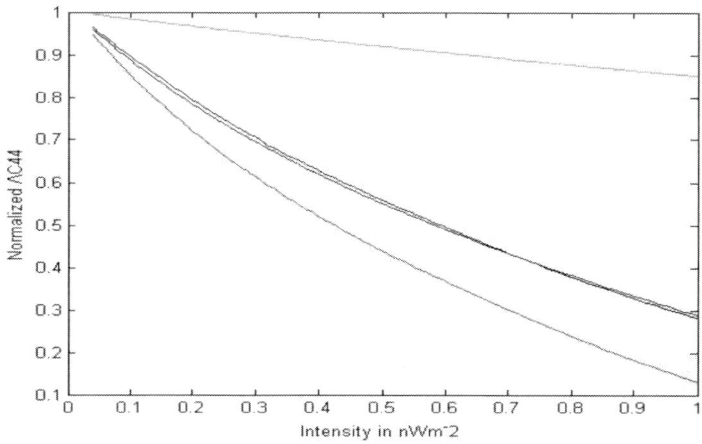

Figure 10.31. Plot of the normalized ΔC_{44} as a function of light intensity in Nano Wires of HD n-InSb (blue), HD n-InAs (red), HD $Hg_{1-x}Cd_xTe$ (green) and HD $In_{1-x}Ga_xAs_yP_{1-y}$ lattice matched to InP (black) where the unperturbed electrons obey the three band model of Kane. ($\lambda = 660$ nm, $d_z = 10$ nm, $d_y = 10$ nm and $n_o = 10^{11} m^{-1}$)

The plot 10.29 exhibits the normalized ΔC_{44}as a function of $d_z$in the presence of light waves in Nano Wires of HD n-InSb (blue), HD n-InAs (red), HD $Hg_{1-x}Cd_xTe$ (green) and HD $In_{1-x}Ga_xAs_yP_{1-y}$ lattice matched to InP (black) where the unperturbed electrons obey the three band model of Kane ($\lambda = 660$ nm, $I = 1nWm^{-2}$, $n_o = 10^{11}$ m^{-1} and $d_y = 10$ nm). It appears that in Nano Wires the ΔC_{44} becomes invariant of the film thickness for the small value of film thickness, exhibits sharp fall at a particular value of thickness manifesting the quantum size effects in ΔC_{44} in Nano Wires and again increases rapidly with increasing thickness. The normalized ΔC_{44}as a function of electron concentration per unit length has been plotted in Figure 10.30 the presence of light waves in Nano Wires of HD n-InSb (blue), HD n-InAs (red), HD $Hg_{1-x}Cd_xTe$ (green) and HD $In_{1-x}Ga_xAs_yP_{1-y}$ lattice matched to InP (black) where the unperturbed electrons obey the three band model of Kane ($\lambda = 660$ nm, $I = 1nWm^{-2}$, $d_z = 10$ nm, $d_y = 10$ nm). It is interesting to note that ΔC_{44} exhibits more or less the same type of variation with electron statistics as observed earlier in Figure 10.29 in the case of the quantized variation of ΔC_{44} with d_z. The Figure 10.35 shows the plot of the normalizedΔC_{44}as a function of light intensity in Nano

Wires of HD n-InSb (blue), HD n-InAs (red), HD $Hg_{1-x}Cd_xTe$ (green) and HD $In_{1-x}Ga_xAs_yP_{1-y}$ lattice matched to InP (black) where the unperturbed electrons obey the three band model of Kane ($\lambda = 660$ nm, $d_z = 10$ nm, $d_y = 10$ nm and $n_o = 10^{11}\ m^{-1}$). The light intensity attenuates the ΔC_{44} which decreases as intensity increases although the amount of attenuation is different for different materials. The Figure 10.32 shows the normalized ΔC_{44}as a function of wavelength in Nano Wires of HD n-InSb (blue), HD n-InAs (red), HD $Hg_{1-x}Cd_xTe$ (green) and HD $In_{1-x}Ga_xAs_yP_{1-y}$ lattice matched to InP (black) where the unperturbed electrons obey the three band model of Kane ($I = 1nWm^{-2}$, $d_z = 10$ nm, $d_y = 10$ nm and $n_o = 10^{11}\ m^{-1}$). The influence of wave length on the ΔC_{44} decreases as with the increment of wave length of the external photo excitation for the whole range of wave lengths. The Figure 10.33 shows the plot of the normalized ΔC_{44}as a function of alloy composition in Nano Wires of HD $Hg_{1-x}Cd_xTe$ (green) and HD $In_{1-x}Ga_xAs_yP_{1-y}$ lattice matched to InP (black) where the unperturbed electrons obey the three band model of Kane ($n_o = 10^{11}\ m^{-1}$, $d_y = 20$ nm and $d_z = 15$ nm). As usual, the ΔC_{44} decreases with increasing alloy composition for all the quantized materials.

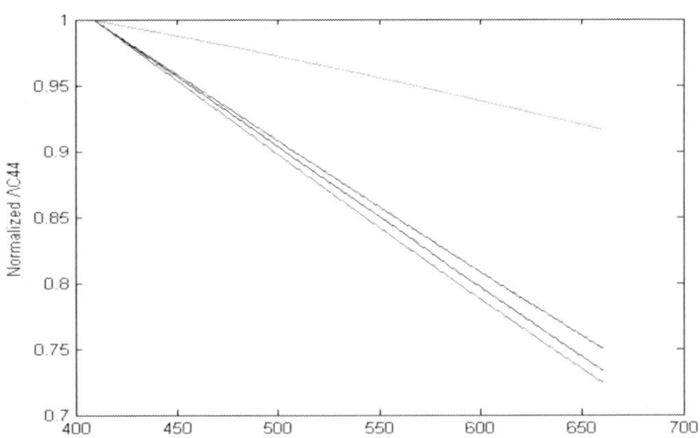

Figure 10.32. Plot of the normalized ΔC_{44} as a function of wavelength in Nano Wires of HD n-InSb (blue), HD n-InAs (red), HD $Hg_{1-x}Cd_xTe$ (green)and HD $In_{1-x}Ga_xAs_yP_{1-y}$ lattice matched to InP (black) where the unperturbed electrons obey the three band model of Kane. ($I = 1nWm^{-2}$, $d_z = 10$ nm, $d_y = 10$ nm and $n_o = 10^{11}\ m^{-1}$)

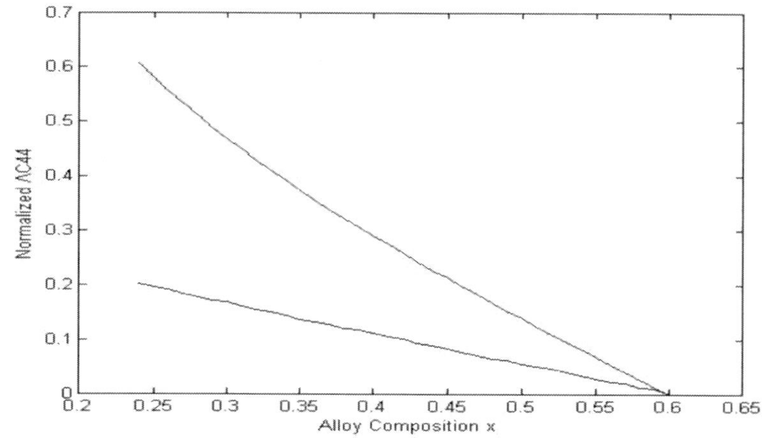

Figure 10.33. Plot of the normalized ΔC_{44} as a function of alloy composition in Nano Wires of HD $Hg_{1-x}Cd_xTe$ (green)and HD $In_{1-x}Ga_xAs_yP_{1-y}$ lattice matched to InP (black) where the unperturbed electrons obey the three band model of Kane. ($n_o = 10^{11}m^{-1}$, $d_y = 20$ nm and $d_z = 15$ nm)

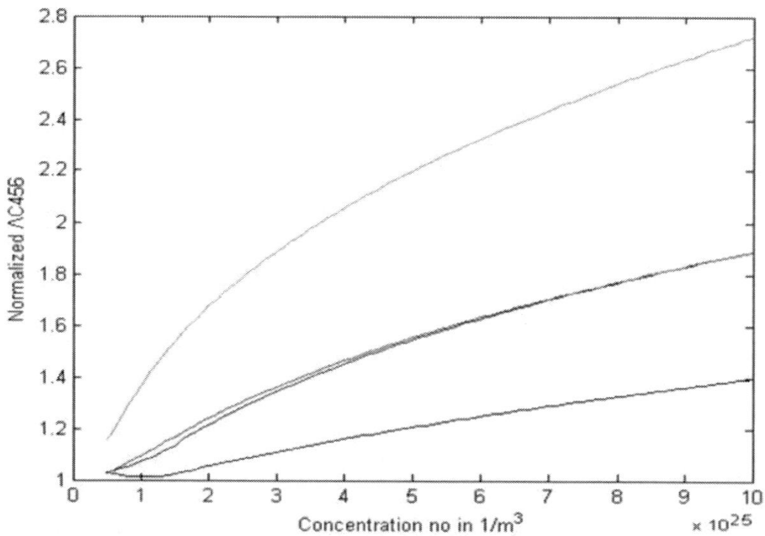

Figure 10.34. Plot of the normalized ΔC_{456} as a function of electron concentration per unit volume in the presence of light waves for bulk specimens of HD n-InSb (blue), HD n-InAs (red), HD $Hg_{1-x}Cd_xTe$ (green) and HD $In_{1-x}Ga_xAs_yP_{1-y}$ lattice matched to InP (black) respectively where the unperturbed electrons obey the three band model of Kane. ($I = 10^{-4}nWm^{-2}$ and $\lambda = 660$ nm)

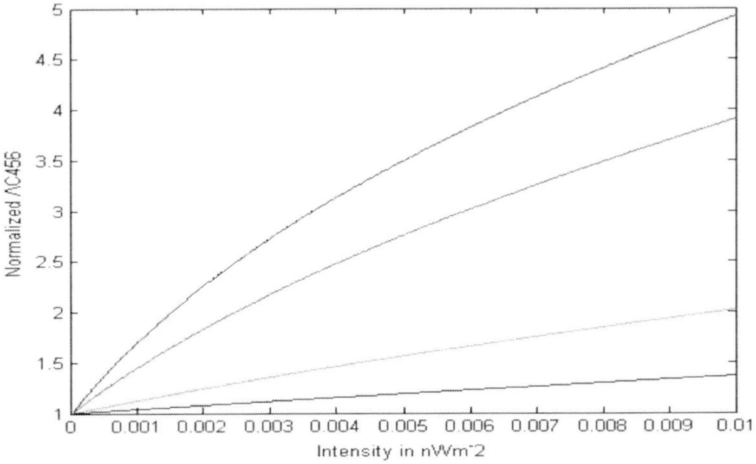

Figure 10.35. Plot of the normalized ΔC_{456} as a function of light intensity for bulk specimens of HD n-InSb (blue),HD n-InAs (red), HD $Hg_{1-x}Cd_{x}Te$ (green) and HD $In_{1-x}Ga_xAs_yP_{1-y}$ lattice matched to InP (black) respectively where the unperturbed electrons obey the three band model of Kane. ($n_o = 10^{25} m^{-3}$ and $\lambda = 660$ nm)

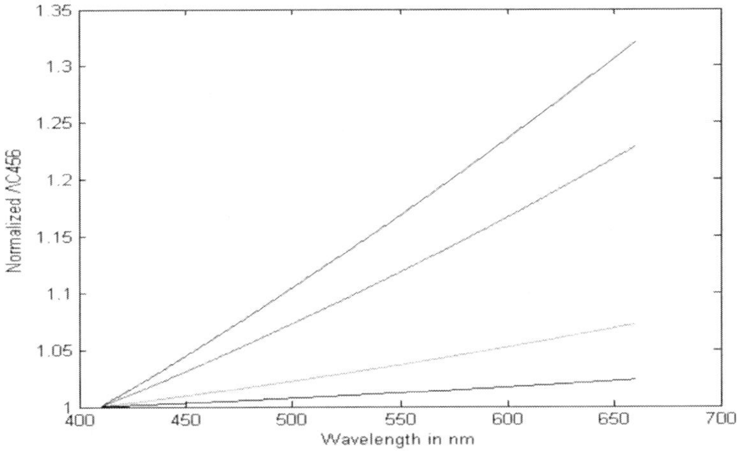

Figure 10.36. Plot of the normalized ΔC_{456} as a function of wave length for bulk specimens of HD n-InSb (blue), HD n-InAs (red), HD $Hg_{1-x}Cd_{x}Te$ (green) and HD. $In_{1-x}Ga_xAs_yP_{1-y}$ lattice matched to InP (black) respectively where the unperturbed electrons obey the three band model of Kane. ($n_o = 10^{25}$ m^{-3} and $I = 10^{-4}$ nWm^{-2})

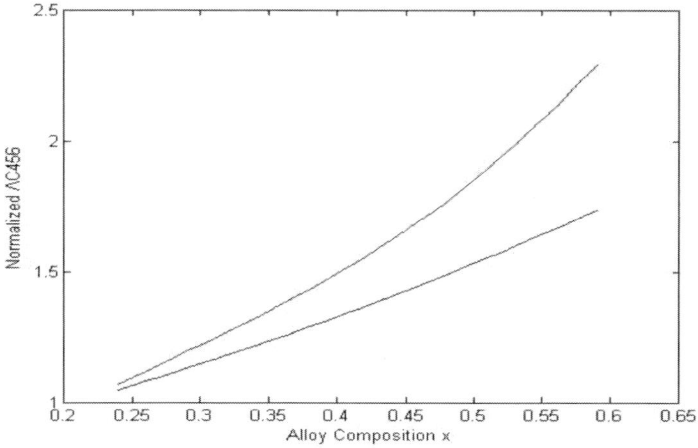

Figure 10.37. Plot of the normalized ΔC_{456} as a function of alloy composition x in the presence of light wave for bulk specimens of HD $Hg_{1-x}Cd_xTe$ (blue) and HD $In_{1-x}Ga_xAs_yP_{1-y}$ lattice matched to InP (black) where the unperturbed electrons obey the three band model of Kane. ($n_o = 10^{25}m^{-3}$, $I = 10^{-4}$ nWm^{-2} and $\lambda = 660$ nm)

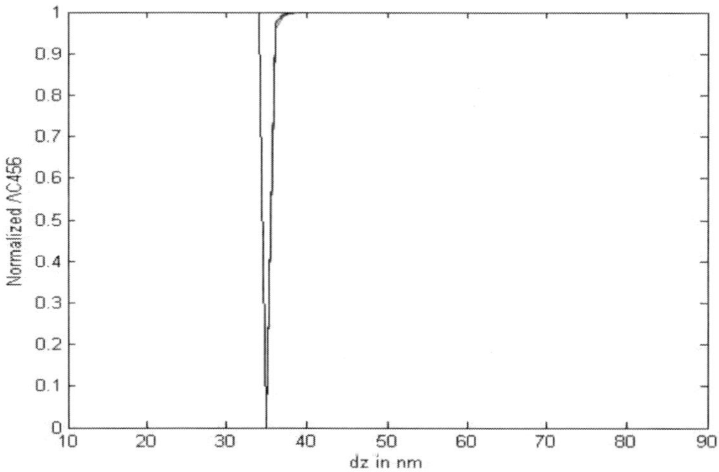

Figure 10.38. Plot of the normalized ΔC_{456} as a function of $d_z$in the presence of light waves in Nano Wires of HD n-InSb (blue), HD n-InAs (red), HD $Hg_{1-x}Cd_xTe$ (green) and HD $In_{1-x}Ga_xAs_yP_{1-y}$ lattice matched to InP (black) where the unperturbed electrons obey the three band model of Kane. ($\lambda = 660$ nm, $I = 1nWm^{-2}$, $n_o = 10^{11}$ m^{-1} and $d_y = 10$ nm)

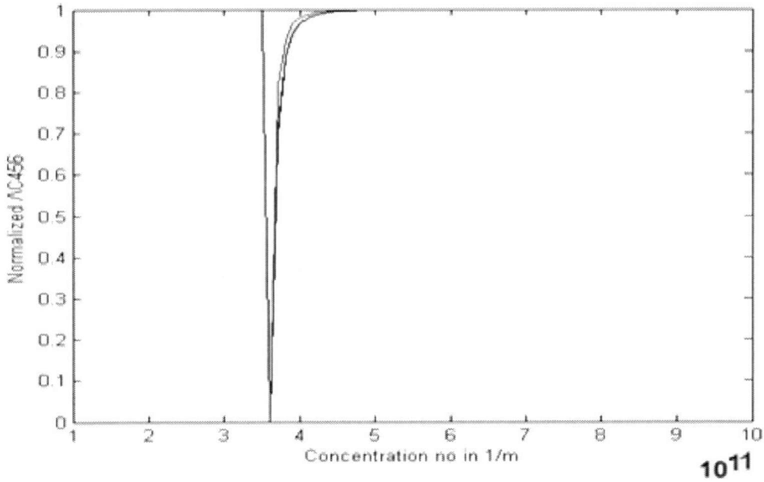

Figure 10.39. Plot of the normalized ΔC_{456}as a function of electron concentration in the presence of light waves in Nano Wires of HD n-InSb (blue), HD n-InAs (red), HD $Hg_{1-x}Cd_xTe$ (green) and HD $In_{1-x}Ga_xAs_yP_{1-y}$ lattice matched to InP (black) where the unperturbed electrons obey the three band model of Kane. ($\lambda = 660$ nm, I = 1nWm^{-2}, $d_z = 10$ nm, $d_y = 10$ nm)

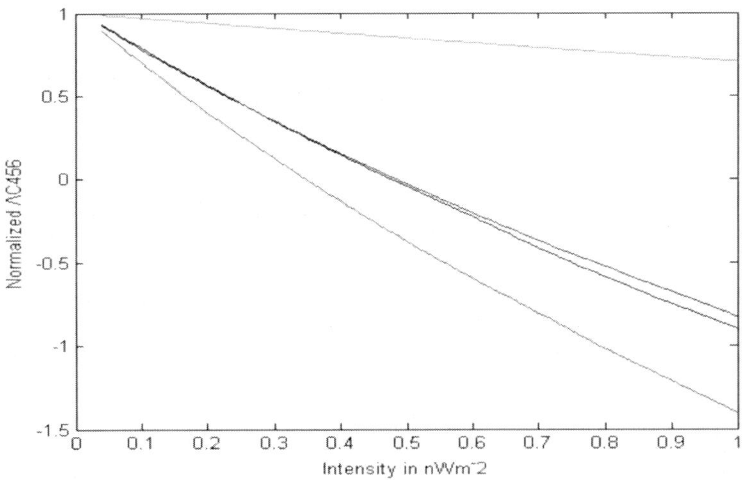

Figure 10.40. Plot of the normalizedΔC_{456}as a function of light intensity in Nano Wires of HD n-InSb (blue), HD n-InAs (red), HD $Hg_{1-x}Cd_xTe$ (green) and HD $In_{1-x}Ga_xAs_yP_{1-y}$ lattice matched to InP (black) where the unperturbed electrons obey the three band model of Kane. ($\lambda = 660$ nm, $d_z = 10$ nm, $d_y = 10$ nm and $n_o = 10^{11}$ m^{-1})

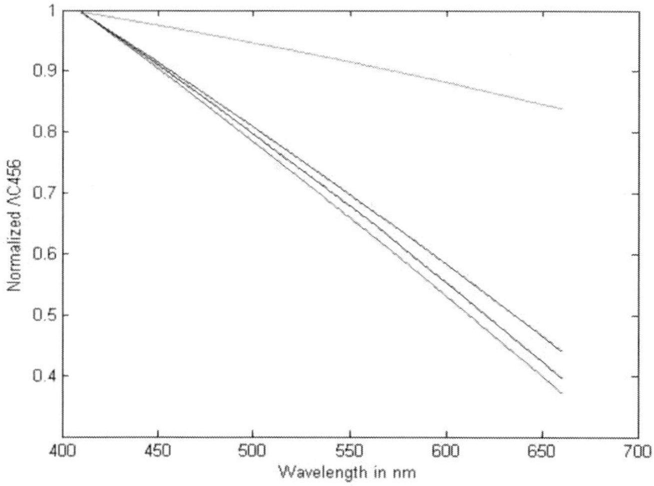

Figure 10.41. Plot of the normalized ΔC_{456}as a function of wavelength in Nano Wires of HD n-InSb (blue), HD n-InAs (red), HD $Hg_{1-x}Cd_xTe$ (green) and HD $In_{1-x}Ga_xAs_yP_{1-y}$ lattice matched to InP (black) where the unperturbed electrons obey the three band model of Kane. ($I = 1nWm^{-2}$, $d_z = 10$ nm, $d_y = 10$ nm and $n_o = 10^{11}$ m^{-1})

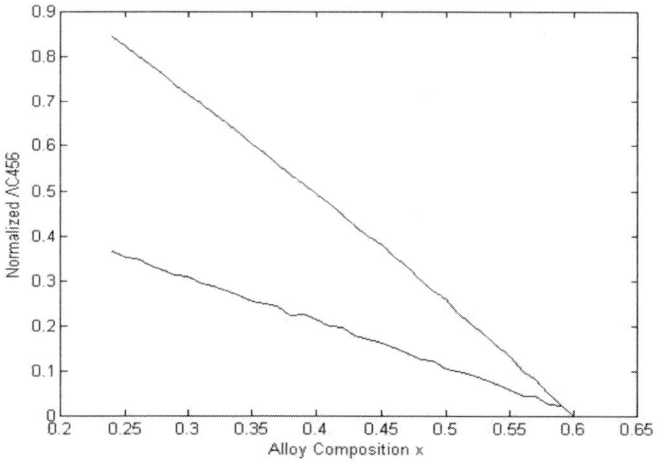

Figure 10.42. Plot of the normalized ΔC_{456}as a function of alloy composition in Nano Wires of HD $Hg_{1-x}Cd_xTe$ (green) and HD $In_{1-x}Ga_xAs_yP_{1-y}$ lattice matched to InP (black) where the unperturbed electrons obey the three band model of Kane. ($n_o = 10^{11}$ m^{-1}, $d_y = 20$ nm and $d_z = 15$ nm)

In Figure 10.34 we have plotted of the normalized ΔC_{456} as a function of electron concentration per unit volume in the presence of light waves for bulk specimens of HD n-InSb (blue),HD n-InAs (red),HD $Hg_{1-x}Cd_{x}Te$ (green) and HD $In_{1-x}Ga_{x}As_{y}P_{1-y}$ lattice matched to InP (black) respectively where the unperturbed electrons obey the three band model of Kane. ($I = 10^{-4} nWm^{-2}$ and $\lambda = 660$ nm). From the figure it appears that ΔC_{456} increases with increasing n_0 for all the four materials as concedered here. From (83) we can infer that the ΔC_{456} will increase with increase in n_0 if rate of decrease of G^2 is smaller than that of rate of increase of n_0 .In this context from Figure 10.1 we can infer that ΔC_{44} decreases with increasing electron Statistics This radical different behavior is due to the fact that from (82) one can infer that the ΔC_{44} will decrease with increase in n_0, if rate of decrease of G is greater than that of rate of increase of n_0. In Figure 17 we have plotted the normalized ΔC_{456} as a function of light intensity for bulk specimens of HD n-InSb (blue),HD n-InAs (red), HD $Hg_{1-x}Cd_{x}Te$ (green) and HD $In_{1-x}Ga_{x}As_{y}P_{1-y}$ lattice matched to InP (black) respectively where the unperturbed electrons obey the three band model of Kane. ($n_o = 10^{25}$-m^{-3} and $\lambda = 660$ nm).It appears from the Figure 10.35 that the ΔC_{456} increases with increasing light intensity and there are wide differences among the materials in the variation of ΔC_{456} as considered here reflecting the influence of the energy band constants of the different compounds. The plot 10.36 exhibits the variations of the normalized ΔC_{456} as a function of wave length for bulk specimens of HD n-InSb (blue),HD n-InAs (red), HD $Hg_{1-x}Cd_{x}Te$ (green) and HD $In_{1-x}Ga_{x}As_{y}P_{1-y}$ lattice matched to InP (black) respectively where the unperturbed electrons obey the three band model of Kane. ($n_o = 10^{25}$ m^{-3} and $I = 10^{-4}$ nWm^{-2}).

In Figure 10.38 we have plotted of the normalized ΔC_{456} as a function of $d_z$in the presence of light waves in Nano Wires of HD n-InSb (blue), HD n-InAs (red), HD $Hg_{1-x}Cd_{x}Te$ (green) and HD $In_{1-x}Ga_{x}As_{y}P_{1-y}$ lattice matched to InP (black) where the unperturbed electrons obey the three band model of Kane ($\lambda = 660$ nm, $I = 1 nWm^{-2}$, $n_o = 10^{11}$ m^{-1} and $d_y = 10$ nm). It appears from the figure that ΔC_{456} exhibits extreme spiky oscillation with respect to film thickness in Nano wires and influences of energy band constants are immaterial in this case. In Figure 10.39 we have plotted the normalized ΔC_{456}as a function of electron concentration in the presence of light waves in Nano Wires of HD n-InSb (blue), HD n-InAs (red), HD $Hg_{1-x}Cd_xTe$ (green) and HD $In_{1-x}Ga_xAs_yP_{1-y}$ lattice matched to InP (black) where the unperturbed electrons obey the three band model of Kane (λ = 660 nm, I = 1nWm^{-2}, d_z = 10 nm, d_y = 10 nm). It appears from the figure that ΔC_{456} exhibits extreme spiky oscillation with respect to electron statistics per unit length in Nano wires and influences of energy band constants are immaterial in this case. The Figure 10.40 exhibits the plot of the normalizedΔC_{456} as a function of light intensity in Nano Wires of HD n-InSb (blue), HD n-InAs (red), HD $Hg_{1-x}Cd_xTe$ (green) and HD $In_{1-x}Ga_xAs_yP_{1-y}$ lattice matched to InP (black) where the unperturbed electrons obey the three band model of Kane (λ = 660 nm, d_z = 10 nm,d_y = 10 nm and n_o = 10^8 m^{-1}). It appears from the figure that ΔC_{456} increases with decreasing light intensity for all the Nano wire materials as considered here. In Figure 10.41 we have plotted the normalized ΔC_{456}as a function of wavelength in Nano Wires of HD n-InSb (blue), HD n-InAs (red), HD $Hg_{1-x}Cd_xTe$ (green) and HD $In_{1-x}Ga_xAs_yP_{1-y}$ lattice matched to InP (black) where the unperturbed electrons obey the three band model of Kane (I = 1nWm^{-2}, d_z = 10 nm, d_y = 10 nm and n_o = 10^{11} m^{-1}). It appears from the figure that ΔC_{456} decreases with increasing light intensity for all the Nano wire materials as considered here. In Figure 10.42 we have plotted the normalized ΔC_{456}as a function of alloy composition in Nano Wires of HD $Hg_{1-x}Cd_xTe$ (green) and HD $In_{1-x}Ga_xAs_yP_{1-y}$ lattice matched to InP (black) where the unperturbed electrons obey the three band model of Kane (n_o = 10^{11} m^{-1}, d_y = 20 nm and d_z = 15 nm).We observe that ΔC_{456} decreases with increasing alloy composition for both the opto electronic materials.

The numerical results presented in this chapter would be different for other materials but the nature of variation would be unaltered. The theoretical results as given here would be useful in analyzing various other experimental data related to the transport phenomenon. We must note that the study of transport phenomena and the formulation of the electronic properties of low dimensional field aided HD compounds are based on the dispersion relations in such materials. It is worth remarking that this simplified formulation exhibits the basic qualitative features of the CECs' for field assisted low dimensional materials. The basic objective of this chapter is not solely to demonstrate the influence of quantum confinement on the CECs' for different quantum confined HD non-parabolic materials in the presence of external fields but also to formulate the appropriate DOS and effective electron mass in the most generalized form, since the transport and other phenomena in such nano structured materials having different band structures and the derivation of the expressions of many important electronic properties are based on the DOS in such compounds.

It is imperative to state that the present investigation excludes the many-body, hot electron, broadening and the allied effects in the simplified theoretical formalism due to the absence of proper analytical techniques for including them for generalized systems as considered here. Our simplified approach will be appropriate for the purpose of comparisons when the methods of tackling the formidable problems after inclusion of the said effects for the generalized systems emerge. The results of this simplified approach get transformed to the well- known formulation of the CECs' for wide gap materials having parabolic energy bands. This indirect test not only exhibits the mathematical compatibility of the formulation but also shows the fact that this simple analysis is more generalized one, since one can obtain the corresponding results for materials having parabolic energy bands under certain limiting conditions from the present derivation. Our experimental suggestion for the determination of the said elastic constants is valid for materials having arbitrary dispersion relations. Since the experimental curve of n_0 versus G is not available in the literature to the best of our knowledge for the present generalized systems, we cannot compare our theoretical formulation with the proposed experiment although the generalized analysis as presented in this context can be checked when the experimental investigation of G for the present system would appear in the literature.

References

[1] Ghatak, K. P., B. Mitra, *Physica Scripta*, 46, 182(1992).

[2] Nag, B., K. P. Ghatak, *Journal of Physics and Chemistry of Solids,* 58, 427(1997).

[3] Ghatak, K. P., B. Nag, *physica status solidi (b)* 205, 519 (1998).

[4] Ghatak, K. P., J. Y. Siddiqui, B Nag, *Physics Letters A* 282, 428 (2001).

[5] Ghatak, K. P., S. N. Biswas, *MRS Proceedings*, 308, 445, (1993).

[6] Ghatak, K. P., J. P. Banerjee, P. K. Chakrabarty, B. Nag, *Journal of Wave Material Interaction* 11, 166 (1996).

[7] Singh, L. J., S. Choudhary, A. Mallik, K. P. Ghatak, *Journal of Computational and Theoretical Nanoscience*, 2, 287 (2005).

[8] Mallik, A., K. P. Ghatak, S. Choudhary, L. J. Singh,.*Journal of Computational and Theoretical Nanoscience*, 2, 287 (2005).

[9] Ghatak, K. P. *International journal of electronics*, 71, 239(1991).

[10] Ghatak, K. P. *Acta Physica Hungarica*, 68, 253 (1990).

[11] Mitra, B., K. P. Ghatak, *physica status solidi (b),* 154, K35 (1989).

[12] Ghatak, K. P., P. K. Bose, *Journal of Wave Material Interaction,* 12, 53 (1997).

[13] Adhikari, S. M., KP Ghatak, *Journal of Nanoengineering and Nanomanufacturing,* 3, 48(2013).

[14] Chatterjee, B., S. Chakrabarti, S.K. Sen, M. Mitra, K.P. Ghatak, *Quantum Matter*, 5, 85 (2016).

[15] Ghatak, K. P., S. Bhattacharya, S. S. Roy, L. J. Singh, *Nonlinear Optics,* 32, 307(2010).

[16] Baruah, D., S. Choudhury, K. M. Singh, K. P. Ghatak, *Journal of Physics: Conference Series,* 61, 80(2007).

[17] Ghatak, K. P., S. N. Biswas, *Acta physica Slovaca,* 43, 425, (1993).

[18] Ghatak, K. P., M. Mondal, *physica status solidi (b),* 170, 57 (1992).

[19] Ghatak, K. P., D. K. Basu, B. Nag, Journ*al of Physics and Chemistry of Solids*, 58, 133(1997).

[20] Ghatak, K. P. *Physica status solidi (b)*, 154, K29 (1989).

[21] Ghatak, K. P., J. P. Banerjee, B. Goswami, B. Nag, *Nonlinear Optics-Reading*, 16, 241 (1996).

[22] Nag, B., K. P. Ghatak, *Nonlinear Optics-Reading*, 19, 1 (1998).

[23] Mitra, M., M. Chakraborty, S. Debbarma, S. Chakraborty, S. K. Sen, *Quantum Matter*, 5, 58(2016).

[24] Ghatak, K. P., B. Nag, *Nanostructured materials*, 10, 923(1998).

[25] Mitra, M., M Chakraborty, S Debbarma, S Chakraborty, SK Sen, *Quantum Matter*, 5, 58 (2016).

[26] Nag, B., K. P. Ghatak, *Physica Scripta*, 54, 657 (1996).

[27] Ghatak, K. P., S. Dutta, D. K. Basu, B Nag, *Il Nuovo Cimento D*, 20, 227 (1998).

[28] Bhattacharya, S., S. Chowdhury, S. Ghoshal, S. K. Biswas, D. De, K. P. Ghatak, *Journal of Computational and Theoretical Nanoscience*, 3, 423(2006); K. P. Ghatak, S. Bhattacharya, S. Pahari, S. N. Mitra, P. K. Bose, D. De, *Journal of Physics and Chemistry of Solids*, 70, 122(2009); K. P. Ghatak, S. Karmakar, D. De, S. Pahari, S. K. Charaborty, S. K. Biswas, *Journal of Computational and Theoretical Nanoscience*, 3, 153(2006); K. P. Ghatak, J. P. Banerjee, B. Nag, *Journal of applied physics*, 83, 1420 (1998); K. P. Ghatak, S. K. Biswas, D. De, S. Ghosal, S. Chatterjee, *Physica B*, 353, 127(2004); K. P. Ghatak, J. P. Banerjee, D. Bhattacharyya, B. Nag, *Nanotechnology*, 7, 110(1996); K. P. Ghatak, B. Nag, G. Mazumder, *MRS Proceedings,* 379, 109 (1995).

[29] Keyes, R. W. *IBM j. RES Dev.* 5,266 (1961); A. K. Sreedhar & S. C. Gupta, *phys. Rev,* 5B, 1360 (1972).

[30] Ghatak, K. P. and M. Mitra, Volume 1: *Electronic Properties, Series on Nanomaterials*, De Gruyter, Germany, (2018) pp 1-364.

[31] Nag, B. R. "*Electron Transport in Compound Semiconductors*" Springer, Heidelberg. (1980).

[32] Willardson, R. K. ed. by A. C. Beer, 1, *Semiconductors and Semimetals* (Academic Press, New York, 1966) p.102.

[33] E. O. Kane, *Phys. Rev.* 131,79 (1963), *Phys. Rev. B.*, 139, 343 (1965).

[34] V. L. Bonch Bruevich, *Sov. Phys. Sol. Stat.*, 4,53 (1963).

[35] Logan, R. A., A. G. Chynoweth, *Phys. Rev.,* 131, 89 (1963).
[36] C. J. Hwang, *J. Appl. Phys.* 40, 3731 (1969).
[37] J. I. Pankove, *Phys. Rev. A,* 130, 2059 (1965).
[38] Halperin, B. I., M. Lax, *Phys. Rev.,* 148,722 (1966).
[39] Abram, R. A., G. J. Rees, B. L. H. Wilson, *Adv. Phys,* 27,799 (1978).
[40] Shklovskii, B. I., A. L. Efros, *Electronic Properties of Doped Semiconductors*, Springer series in Solid State Sciences, vol.45 (Springer, Berlin, 1984).
[41] Kane, E. O. *Solid State Electron*, 28, 3 (1985).
[42] Chakraborty, P. K., J. C. Biswas, *J. Appl. Phys.* 82,3328 (1997).
[43] Bhattacharya, S. and Ghatak. K. P., *Fowler - Nordheim Field Emission Effects in Semiconductor Nanostructures*, Springer Series in Soild-State Science-170, Springer, Germany, (2012) pp. 1-338.
[44] Schmid, P. E. *Phys. Rev. B*, 23, 5531 (1981).
[45] Jellison, Jr. G. E., F. A. Modine, C. W. White, R. F. Wood, R. T. Young, *Phys. Rev. Lett.*, 46,1414 (1981).
[46] Fistul, V. I. *Heavily Doped Semiconductors* (Plenum, New York, 1969) ch 7.
[47] Hwang, C. J. *J. Appl. Phys,*41,2668 (1970); W. Sritrakool, H. R. Glyde, V. Sa Yakanit Can. *J. Phys,*60, 373 (1982); H. Ikoma, *J. Phys. Soc. Japan*, 27,514 (1969).
[48] Chen, M. H., C.C. Wu, C.J. Lin, *J. Low Temp. Phys.* 55, 127 (1984)
[49] Tsidilkovskii, I. M. *Band Structure of Semiconductors* (Pergamon Press, Oxford, 1982).
[50] Askerov, B. M., N. F. Gashimzcede, M. M. Panakhov, *Sov. Phys. Sol. State,* 29, 465 (1987).
[51] Ghatak, K. P., M. Mondal, *Phys. Stat. Sol. (b)*185, K5 (1994).
[52] Ghatak, K. P., B. Mitra, *IlNuovoCimento D*15, 97 (1993).
[53] Ghatak, K. P., S. N. Biswas, *Phys. Stat. Solidi (b)*140, K107 (1987).
[54] Ghatak, K. P. *IlNuovoCimento D* 13, 1321 (1991).
[55] Ghatak, K. P., A. Ghoshal, *Phys. Stat. Sol. (b)*170, K27 (1992).
[56] Ghatak, K. P., B. De, B. Nag, P. K. Chakraborty, *Molecular Crystals and Liquid Crystals Science and Technology Section B: Nonlinear Optics* 16, 221 (1996).

[57] GhatakK. P., M. Mitra, B. Goswami, B. Nag, *Molecular Crystals and Liquid Crystals Science and Technology Section B: Nonlinear Optics* 16, 167 (1996).

[58] Ghatak, K. P., D. K. Basu, D. Basu, B. Nag, *Nuovo Cimentodella Societa Italiana di Fisica D - Condensed Matter, Atomic, Molecular and Chemical Physics, Biophysics*18, 947 (1996).

[59] Mitra, B., K. P. Ghatak, *Phys. Letts. A*141, 81 (1989);S. K. Biswas, A. R. Ghatak, A. Neogi, A. Sharma, S. Bhattacharya, K. P. Ghatak, *Physica E:Low-Dimensional Systems and Nanostructures,* 36, 163 (2007).

[60] Mondal, M., A. Ghoshal, K. P. Ghatak, *Il Nuovo Cimento D*14, 63 (1992); K. P. Ghatak, M. Mondal*Phys. Stat. Sol. (b)*, 135, 819 (1986); L. J. Singh, S. Choudhury, D. Baruah, S. K. Biswas, S. Pahari, K. P. Ghatak, *Phys. B,*368, 188 (2005).

[61] Ghatak, K. P., B. De, M. Mondal, S. N. Biswas, *Materials Research Society Symposium – Proceedings* 198, 327(1990); K. P. Ghatak, *Proceedings of SPIE - The International Society for Optical Engineering*1584, 435 (1992).

[62] Ghatak, K. P., B. De, *Materials Research Society Symposium – Proceedings* 234, 55 and 59 (1991).

[63] Ghatak, K. P., B. De, M. Mondal, S. N. Biswas, *Materials Research Society Symposium – Proceedings* 184, 261 (1990).

[64] Sreedhar A. K. and S. C. Gupta, *Phys. Rev.* B 5, 3160 (1972); R. W. Keyes, *IBM. J. Res. Devlopm.* 5, 266 (1961); *Solid State Phys.* 20, 37 (1967).

[65] Bhattacharya, S., S. Chowdhury and K. P. Ghatak, *J. Compu. Theor. Nanosc.*, 3, 423 (2006); K. P. Ghatak, J. Y. Siddiqui and B. Nag, *Phys. Letts.* A, 282, 428 (2001); K. P. Ghatak, J.P. Banerjee and B. Nag, *J. Appl. Phys.,* 83, 1420 (1998); K. P. Ghatak and B. Nag, *Nanostruc. Material.,* 10, 923 (1998); B. Nag and K. P. Ghatak, *J. Phys. Chem. Sol.,* 58, 427 (1997).

[66] Ghatak, K. P., J. P. Banerjee, B. Goswami and B. Nag, in Nonlin. Opt. Quant. Opt., 16, 241 (1996); K. P. Ghatak, J.P. Banerjee, D. Bhattacharyya and B. Nag, N*anotechnology,* 7, 110 (1996); K. P.

Ghatak, J. P. Banerjee, M. Mitra and B. Nag, Nonlin. Opt. 17, 193 (1996); K. P. Ghatak, *Inter. J. Electron.,* 71, 239 (1991).

[67] Ghatak, K. P., B. De, S. N. Biswas and M. Mondal, Mechanical behavior of materials and structures in microelectronics, *MRS Symposium Proceedings,* Spring Meeting, 2216,191, (1991); K. P. Ghatak and B. De, *MRS Symposium Proceedings,* 226, 191 (1991); K. P. Ghatak, B. Nag and G. Majumdar, *MRS,* 379, 109, (1995); D. Baruah, S. Choudhury, K.M. Singh and K.P. Ghatak, J. *Phys., Conf. Series,* 61, 80, (2007).

[68] Ghatak, K. P. *Dispersion Relations in Heavily–Doped Nanostructures*, Springer Tracts in Modern Physics-265, Springer, Germany, (2016) pp- 1-625.

[69] Ghatak K. P. and S. Bhattacharya, *Heavily Doped 2D-Quantized Structures and the Einstein Relation*, Springer Tracts in Modern Physics-260, Springer, Germany, (2015) pp- 1-347.

[70] Ghatak, K. P. *Einstein's Photoemission Emission from Heavily-Doped Quantized Structures*, Springer Tracts in Modern Physics-262, Springer, Germany, (2015) pp- 1-495.

[71] Ghatak K. P. and S. Bhattacharya, *Debye Screening Length Effects of Nano Structured Materials*, Springer Tracts in Modern Physics-255, Springer, Germany, (2014) pp- 1-385.

About the Editor

Kamakhya Prasad Ghatak, D. Engg.
Senior Professor and Research Director
Department of Computer Science and Engineering
University of Engineering and Management
Institute of Engineering and Management
Calcutta University, Kolkata, India
Email: kamakhyaghatakcu@gmail.com

Professor Kamakhya Prasad Ghatak obtained his PhD (Tech) degree from the Institute of Radio Physics and Electronics of the University of Calcutta in 1988 on the basis of 27 published research papers in eminent SCI journals which is still a record of the said Institute. He was the first recipient of the Degree of Doctor of Engineering of Jadavpur University, Kolkata in 1991 since the University inception in 1955 and in the same year he received the INSA visiting fellowship to IIT-Kharagpur.

Prof. Ghatak is the principal co-author of more than 400 research papers in SCI journals and 60 research papers in the proceedings of the conferences of SPIE and MRS held in USA. He is the Principal co-author of 7 research monographs on various topics of Nano Technology from Springer-Verlag Germany (Vol.170 in Solid State Science, Vols. 116, 137 and 167 in Materials Science, Vols. 255 and 260 in Tracts in Modern Physics (TMP), 1 in Nano structured Science and Technology series) and vol. 1 in Nanomaterials series of De Gruyter, Germany. He is the solo Author of Vols. 262 and 265 in TMP of the Springer-Verlag, Germany and two research monographs (vols. 7 and 8) in the Series on the Foundations of Natural Science and Technology of World Scientific. He is the Principal Editor of the two edited monographs in series in Nanotechnology of NOVA, USA. This first time achievement of Prof. Ghatak is still a record in West Bengal.

The score in Vidwan portal unit, of central government of India, of Prof. K. P. Ghatak is 8.9 out of 10, the highest score among the nano-technologists of private engineering colleges and Universities in West Bengal.

Prof. Ghatak has successfully supervised more than 50 PhD students in different topics of Nanotechnology and the list includes Director, Vice-Chancellor, Principal, Professor and CEO of different reputed Institute, University, College and private organization respectively.

The AICTE has selected the first R & D project in his life for the best project award in electronics and second best research project award considering all the branches of engineering for the year 2006. In 2011, the UGC awarded a research project to him and placed him at the top in the list of awardees. He is the first recipient of the academic excellence award of the University of Engineering & Management, Kolkata from the Governor of West Bengal in its first convocation in 2017.

He was Assistant Professor in the Department of Radio physics and Electronics of the University of Calcutta from 1983-1987, Associate Professor in the Department of ETCE of Jadavpur University from 1987-1994, Professor and HOD of Electronic Science of the University of Calcutta from 1994-2012, HOD and Dean of the National Institute of Technology, Agartala from 2012 -2014 and from January 2015 he has joined in the IEM/UEM group, Kolkata as Research Director and Senior Professor. He was at the top of the Merit lists in all the above cases.

Prof. Ghatak is the referee, Editor and Guest Editor of different reputed international Journals. His present research interests are Nano Materials and Genetic Electronics respectively. His brief CV has been enlisted in many biographical references of USA and UK.

His Vision, Mission and Passion are to stay deeply in solitude. He is the firm believer of the principle of theoretical minimum of Landau. Rabindranath and Ramanajun are life time heroes of Prof. Ghatak.

INDEX

A

B

C

O

P

Q

R

S